A Panorama of Mathematics: Pure and Applied

CONTEMPORARY MATHEMATICS

658

A Panorama of Mathematics: Pure and Applied

Conference
Mathematics and its Applications
November 14–17, 2014
Kuwait University, Safat, Kuwait

Carlos M. da Fonseca
Dinh Van Huynh
Steve Kirkland
Vu Kim Tuan
Editors

American Mathematical Society
Providence, Rhode Island

EDITORIAL COMMITTEE

Dennis DeTurck, Managing Editor

Michael Loss   Kailash Misra   Catherine Yan

2000 *Mathematics Subject Classification.* Primary 03B15, 03F50, 15A23, 35A22, 35B30, 35Q35, 35R30, 49M25, 65F15, 83E15.

Library of Congress Cataloging-in-Publication Data

Names: Fonseca, Carlos M. da, 1968- editor.
Title: A panorama of mathematics : pure and applied : Conference on Mathematics and Its Applications, November 14-17, 2014, Kuwait University, Safat, Kuwait / Carlos M. da Fonseca [and three others], editors.
Description: Providence, Rhode Island : American Mathematical Society, [2016] | Series: Contemporary mathematics ; volume 658 | Includes bibliographical references and index.
Identifiers: LCCN 2015037686 | ISBN 9781470416683 (alk. paper)
Subjects: LCSH: Mathematics–Congresses. | AMS: Mathematical logic and foundations–General logic–Higher-order logic and type theory. msc | Mathematical logic and foundations–Proof theory and constructive mathematics–Metamathematics of constructive systems. msc | Linear and multilinear algebra; matrix theory–Basic linear algebra–Factorization of matrices. msc | Partial differential equations–General topics–Transform methods (e.g. integral transforms). msc | Partial differential equations–Qualitative properties of solutions–Dependence of solutions on initial and boundary data, parameters. msc | Partial differential equations–Equations of mathematical physics and other areas of application–PDEs in connection with fluid mechanics. msc | Partial differential equations–Miscellaneous topics–Inverse problems. msc | Calculus of variations and optimal control; optimization–Numerical methods–Discrete approximations. msc | Numerical analysis–Numerical linear algebra–Eigenvalues, eigenvectors. msc | Relativity and gravitational theory–Unified, higher-dimensional and super field theories–Kaluza-Klein and other higher-dimensional theories. msc
Classification: LCC QA1 .P164 2016 | DDC 510–dc23
LC record available at http://lccn.loc.gov/2015037686

Contemporary Mathematics ISSN: 0271-4132 (print); ISSN: 1098-3627 (online)

DOI: http://dx.doi.org/10.1090/conm/658

Copying and reprinting. Individual readers of this publication, and nonprofit libraries acting for them, are permitted to make fair use of the material, such as to copy select pages for use in teaching or research. Permission is granted to quote brief passages from this publication in reviews, provided the customary acknowledgment of the source is given.

Republication, systematic copying, or multiple reproduction of any material in this publication is permitted only under license from the American Mathematical Society. Permissions to reuse portions of AMS publication content are handled by Copyright Clearance Center's RightsLink® service. For more information, please visit: http://www.ams.org/rightslink.

Send requests for translation rights and licensed reprints to reprint-permission@ams.org.

Excluded from these provisions is material for which the author holds copyright. In such cases, requests for permission to reuse or reprint material should be addressed directly to the author(s). Copyright ownership is indicated on the copyright page, or on the lower right-hand corner of the first page of each article within proceedings volumes.

© 2016 by the American Mathematical Society. All rights reserved.
The American Mathematical Society retains all rights
except those granted to the United States Government.
Printed in the United States of America.

∞ The paper used in this book is acid-free and falls within the guidelines
established to ensure permanence and durability.
Visit the AMS home page at http://www.ams.org/

10 9 8 7 6 5 4 3 2 1   21 20 19 18 17 16

Contents

Preface	ix
An inverse problem in magnetohydrodynamics AMIN BOUMENIR	1
Combinatorial properties of circulant Hadamard matrices REINHARDT EULER, LUIS H. GALLARDO, and OLIVIER RAHAVANDRAINY	9
A new point of view on higher dimensional Kaluza-Klein theories AUREL BEJANCU	21
Location and size estimation of small rigid bodies using elastic far-fields FADHEL AL-MUSALLAM, DURGA PRASAD CHALLA, and MOURAD SINI	33
Zerofinding of analytic functions by structured matrix methods LUCA GEMIGNANI	47
Multi-frequency acousto-electromagnetic tomography GIOVANNI S. ALBERTI, HABIB AMMARI, and KAIXI RUAN	67
On the space-time fractional Schrödinger equation with time independent potentials SALEH BAQER and LYUBOMIR BOYADJIEV	81
A note on companion pencils J. L. AURENTZ, T. MACH, RAF VANDEBRIL, and D. S. WATKINS	91
Tame systems of linear and semilinear mappings and representation-tame biquivers T. KLIMCHUK, D. KOVALENKO, T. RYBALKINA, and V.V. SERGEICHUK	103
Invariance of total positivity of a matrix under entry-wise perturbation and completion problems MOHAMMAD ADM and JÜRGEN GARLOFF	115
Subsystems and regular quotients of C-systems VLADIMIR VOEVODSKY	127
Landweber-type operator and its properties ANDRZEJ CEGIELSKI	139

Recovering the conductances on grids: A theoretical justification
  C. Araúz, Á. Carmona, A. M. Encinas, and M. Mitjana 149

The unified transform in two dimensions
  Athanassios S. Fokas 167

Semi-homogeneous maps
  Wen-Fong Ke, Hubert Kiechle, Günter Pilz,
  and Gerhard Wendt 187

Adaptive numerical solution of eigenvalue problems arising from finite element models. AMLS vs. AFEM
  C. Conrads, V. Mehrmann, and A. Międlar 197

Interpreting the von Neumann entropy of graph Laplacians, and coentropic graphs
  Niel de Beaudrap, Vittorio Giovannetti, Simone Severini,
  and Richard Wilson 227

About the relative entropy method for hyperbolic systems of conservation laws
  Denis Serre and Alexis F. Vasseur 237

Smoothing techniques for exact penalty methods
  Christian Grossmann 249

Numerical aspects of sonic-boom minimization
  Navid Allahverdi, Alejandro Pozo, and Enrique Zuazua 267

Preface

The Conference on Mathematics and its Applications - 2014 was held at Kuwait University, Kuwait, November 14-17, 2014.

The main aim of the conference was to bring together excellence and expertise in mathematics in a bid to bridge different mathematical disciplines and to frame new forefronts of research in both pure and applied mathematics. The conference was therefore consciously chosen to be broadly-based and multi-themed in order to reflect the diversity in mathematics and stimulate interdisciplinary exchange. The main themes of the conference were Algebra, Analysis, Discrete Mathematics, and Inverse Problems and Imaging.

Nearly 100 researchers from 23 different countries attended the conference. The plenary speakers were: Thanasis Fokas, University of Cambridge, UK; Volker Mehrmann, Technische Universität Berlin, Germany; Carsten Thomassen, Technical University of Denmark, Lyngby, Denmark; Vladimir Voevodsky, Institute for Advanced Study, Princeton, New Jersey, USA, and Enrique Zuazua, Basque Center for Applied Mathematics, Bilbao, Basque Country, Spain.

The conference also featured an invited address by Habib Ammari, from the École Normale Supérieure, Paris, France, who was the winner of the 2013 Kuwait Prize in Basic Sciences - Mathematics awarded by Kuwait Foundation for the Advancement of Sciences (KFAS).

The plenary lectures detailed recent mathematical advances, emphasized the interrelations between algebraic geometry and algebraic topology; stressed the link between partial differential equations and control theory; exposed the growing trend between biomedical imaging and inverse problems; accentuated research and development in both linear and abstract algebra and their applications; and focused on modern aspects of graph theory and combinatorics. The parallel sessions covered a wide spectrum of mathematical areas including combinatorial matrix theory, numerical linear algebra, operator theory, discrete mathematics, the space-time fractional Schrödinger equation, hyperbolic systems, ill-posed problems and boundary value problems, and PDEs, just to name a few.

During the conference some recent advances in mathematics were highlighted with some interesting applications to science, engineering and industry, providing an extraordinary opportunity for young researchers to see much of the vast potential of mathematics.

In spite of the intensive scientific program of the conference, the participants had the opportunity to enjoy a social program that included an itinerary through Kuwait City, as well as a banquet dinner held at the Marine Sciences Center of Kuwait University.

The organizing committee of the conference did an excellent job. The committee consisted of Fadhel Al-Musallam (Chairman), Mansour Al-Zanaidi, Carlos M. da Fonseca, and Michael Johnson. All of them are members of the Department of Mathematics at Kuwait University.

Special thanks are extended to Habib Ammari, Gang Bao, Aurel Bejancu, Amin Boumenir, Marian Deaconescu, Wen-Fong Ke, Steve Kirkland, Denis Serre, and Holger Wendland who made up the Scientific Committee of the conference.

We acknowledge Kuwait Foundation for the Advancement of Sciences (KFAS) and Kuwait University for their financial support to the conference.

This volume of Contemporary Mathematics contains the main research contributions to the conference and presents a wide range of recent advances on pure and applied mathematics. We are grateful to the American Mathematical Society for publishing this volume.

The Editors
Carlos M. da Fonseca
Dinh Van Huynh
Steve Kirkland
Vu Kim Tuan

An inverse problem in magnetohydrodynamics

Amin Boumenir

ABSTRACT. We give a simple solution to an inverse problem related to the Grad-Sofranov equilibrium equation of plasma in a tokamak. We show that the solution is among the first eigenvalues of a family of Laplacians with a mixed boundary condition, and then show how it can be selected and computed.

Introduction. A tokamak is a toroidal magnetic reactor that holds plasma together to enable fusion reactions to occur and produce clean energy. The simplified equation of an equilibrium of plasma in a tokamak, as treated by Demidov and Moussaoui [6], is given by

$$\begin{cases} \Delta u = au + b \geq 0 & \text{on } \Omega \subset \mathbb{R}^n, \ n \geq 2 \\ u = 0 \quad \text{and} \quad (\partial_n u = \Phi) & \text{on } \partial\Omega \in C^{1,1} \end{cases} \tag{1}$$

where Ω is an open bounded connected domain in $\mathbb{R}^n$. The inverse problem is to find the values of the constants a and b from a single reading of the outer normal derivative given on the boundary, i.e. $\partial_n u = \Phi \in L^1(\partial\Omega)$. We shall refer to Φ as an observation, as admissible data should come from a solution generated by actual constants a and b.

In [4, 6], the idea is to rescale the solution $u = b\widetilde{u}$ and look at $\Delta\widetilde{u} = a\widetilde{u} + 1$. As shown in [3, 4], the new problem is then closely related to whether Ω has the Pompeiu property. This in turns is also related to Shiffer's conjecture, [8], which says that Ω must be a ball in $\mathbb{R}^n$ if there are infinitely many solutions. In his proof, [4, Theorem 1], Delmasso makes use of fact that a directional derivative $\partial_\kappa u$, where κ is a fixed unit vector, is solution of a Dirichlet eigenvalue problem, however no boundary conditions are used. Demidov an Moussaoui, [6, Proposition 3.2], using conformal mapping for planar domains, showed first that the map $a \to b$ is increasing and in case $\partial\Omega$ had a corner, then a partial knowledge of Φ close to the singularity is enough to help compute the value of b, which then, in principle, would give the value of a, since the map $a \to b$ is invertible. This is achieved by using lower order terms asymptotics of solutions of (1) close to the corner, and this yields a uniqueness result. All other known uniqueness results assume a smooth boundary, $\partial\Omega \in C^{2+k}$ and it is an open question whether one can actually compute these values (a, b) through a formula, at least b, and whether it would also hold in higher dimensions, see open questions in [6].

2010 *Mathematics Subject Classification.* Primary 35R30.

In this note, instead of rescaling the solution, we apply the Laplacian again to remove b from (1). The key idea here is to see that the positivity of $v = \Delta u \geq 0$ in (1), plays a major role. It implies at once that since $\Delta v = av$ holds, v must be the principal eigenfunction of Δ. It remains to find the appropriate boundary conditions, that guarantees its existence. This, as we shall see, requires the knowledge of Φ over all the boundary $\partial \Omega$, and leads to a standard minimization problem of a functional.

Our method helps us locate and compute the values a, and b when the boundary $\partial \Omega$ is $C^{1,1}$ smooth. We prove that the solutions (a, b) are points found on a certain curve parametrized by a parameter $\mu = -a/b$, and show how to select the points (a, b) on each line whose slope is $-1/\mu$. As we shall see, working with μ simplifies uniqueness results and also computations. Recall that uniqueness holds in some special cases only, for example when Ω has a corner, [**2, 6**], and is not expected to hold in the general case [**4, 14**]. The case when Ω has corners is also covered by our method, since they can be obtained as a limit of smooth domains. We point out to interesting shape optimization problems for a, and b as was done for the principal eigenvalue λ_1^D of Dirichlet Laplacian, i.e. $-\Delta \varphi_1 = \lambda_1^D \varphi_1$, [**11**], and also in [**1**], where not only shape but also topological optimizations for eigenvalues of the Laplacian are investigated.

1. Preliminaries

Observe that if $b = 0$, then $-a$ is a Dirichlet eigenvalue in (1), i.e. $a = -\lambda_1^D$ which is then a trivial case. On the other hand if $a = 0$, then (1) reduces to $\Delta u = b \geq 0$, which after integration yields

$$(2) \qquad b|\Omega| = \int_{\partial \Omega} \Phi(x) ds.$$

Therefore, in all that follows, we assume from the boundary condition in (1), that $b > 0$ and also $a \neq 0$, which means working inside the first and second quadrant of the ab-plane. Observe that in (1), Δu has the same smoothness as u and so we can apply the Laplacian again, which yields, after setting $v = \Delta u$

$$(3) \qquad \Delta v = av \quad \text{on} \quad \Omega.$$

From (1), we also have $v = au + b$ for any $x \in \Omega$, and taking the gradient yields $\nabla v = a \nabla u$. Thus moving to the boundary we deduce

$$(4) \qquad v(x) = b \quad \text{and also} \quad \partial_n v(x) = a \partial_n u(x) = a \Phi(x) \text{ for any } x \in \partial \Omega$$

which implies

$$(5) \qquad b \partial_n v(x) = a \Phi(x) v(x) \quad \text{on} \quad \partial \Omega$$

Combining equations (3) and (5) leads to a standard eigenvalue problem [**12**, lemma 2.2]

$$(6) \qquad \begin{cases} -\Delta v = -av \quad \text{and} \quad v \geq 0 \quad \text{on} \quad \Omega \\ \partial_n v + \mu \Phi v = 0 \quad \text{a.e. on} \quad \partial \Omega \end{cases}$$

where $\Phi \in L^1(\partial \Omega)$ and μ is a fixed parameter

$$(7) \qquad \mu = -a/b.$$

In other words, if u is a solution of (1) with given parameters (a, b), then $v = \Delta u$ is necessarily also a solution of (6) with the same values (a, b). The converse is also true provided $v(x) = b$ on $\partial\Omega$ since a, b and μ are related by (7). In fact if $v(x) = \text{constant} \neq 0$ on the boundary, then we can still rescale it back to $v(x) = b$ for $x \in \partial\Omega$. Then v is uniquely determined by (6) and solving

$$\Delta u = v \text{ in } \Omega \quad \text{while } u = 0 \text{ on } \partial\Omega$$

for u, yields from (6)

$$\Delta v = a\Delta u$$

which means that $v - au$ is harmonic in Ω and so constant there since $v - au = b$ on $\partial\Omega$. In other words,

$$\Delta u = v = au + b \quad \text{in } \Omega,$$

which is equation (1). It remains to see that from (6), we have $\partial_n v = -a\Phi$ on $\partial\Omega$ together with $v = au + b$ in Ω implies that $\partial_n u = \Phi$. We have just shown that problems (1) and (6) are equivalent whenever $v(x) = b$ on $\partial\Omega$.

PROPOSITION 1. *If u is a solution of (1) with $b > 0$, then $v = \Delta u$ solves (6) and furthermore $v(x) = b$ on $\partial\Omega$. Conversely if v solves (6) while $v(x) = \text{constant}\neq 0$ on $\partial\Omega$, then the solution u of $\Delta u = v$ while $u(x) = 0$ on $\partial\Omega$ solves (1) with b from (7).* □

It is readily seen now that an easier way to find (a, b) is to work with the eigenvalue problem in (6). To this end given an observation Φ, from (1), we need to solve eigenvalue problems (6), with all possible $\mu \in \mathbb{R}$ and filter out those eigenfunctions whose traces on the boundary satisfy $v(x) = \text{constant} \neq 0$. Thus from the eigenvalue problem we know the family (a_μ, μ) where a_μ denotes the dependence of a on the parameter μ. Next select the pair for which the eigenfunction $v(x)$ is a nonzero constant on $\partial\Omega$, which we rescale to $v(x) = b := -a/\mu$; this yields the solution u together with the sought values (a_μ, b) of the inverse problem in (1). The issue of global uniqueness is now simply given by the number of principal eigenfunctions that are constant, but nonzero on the boundary, as the parameter μ varies. Recall that the only positive (principal) eigenfunction of a Laplacian corresponds to the first eigenvalue of the Laplacian. For simplicity we start with the case $\mu > 0$, which means that the point (a, b) is in the second quadrant.

Remark: The existence of nonnegative solutions v in (6) is related to positivity preserving semigroups associated with the underlying Robin problem, as shown in [9].

Quadrant II. We now prove

PROPOSITION 2. *Assume u satisfies (1), with $b > 0$ and $a < 0$ then it necessary follows*

i) $0 < \Phi \in C(\partial\Omega)$

ii) $-a$ is the (unique) first positive eigenvalue of the Laplacian in (6) with $\mu = -a/b$

iii) $-a, b$ are computable using the knowledge of Ω, Φ, and μ

iv) $-\lambda_1^D < a < 0$

Proof. i) From (1), we have $\Delta u \geq 0$ in Ω and $u(x) = 0$ on $\partial\Omega$. It follows, from the maximum principle that 0 is an upper solution, i.e. that $u(x) < 0$ for $x \in \Omega$ and so $\Phi = \partial_n u(x) > 0$ for $x \in \partial\Omega$. As for the regularity of the solution we have, as for eigenfunctions, $u \in C^2(\Omega) \cap C^1(\overline{\Omega})$, since $\partial\Omega \in C^{1,1}$.

ii) Since v is an eigenfunction, and has a fixed sign, it must be the first eigenfunction. All other eigenfunctions must change sign to be orthogonal to the first eigenfunction.

iii) It follows from the Min-max principle of eigenvalues of (6) as $-a$, is uniquely defined, as the minimum of the functional, [**12**],

$$(8) \qquad -a_\mu = \inf_{v \in H^1(\Omega)} \frac{1}{\|v\|^2} \left(\int_\Omega |\nabla v(x)|^2 \, dx + \mu \int_{\partial\Omega} \Phi(x) |v(x)|^2 \, ds \right) > 0$$

where a_μ denotes the dependence of a on μ. Once a_μ and μ are known, then b is simply given by (7).

iv) Together $\mu \Phi(x) > 0$ on $\partial\Omega$ and $H_0^1(\Omega) \subset H^1(\Omega)$ imply that

$$(9) \qquad 0 < -a_\mu \leq \inf_{\substack{v \in H_0^1(\Omega) \\ 1 = \|v\|_{L^2(\Omega)}}} \int_\Omega |\nabla v(x)|^2 \, dx = \lambda_1^D. \qquad \Box$$

The algorithm to find a solution in the second quadrant is now simple. Given any observation $\Phi(x) > 0$, on $\partial\Omega$ consider the line with slope $-1/\mu < 0$. Then $(-a_\mu, b_\mu)$ is a well defined point on that line by (ii). We need to select the solution, as in proposition 1, that the corresponding eigenfunction is constant, but nonzero, on the boundary, which we then rescaleback to $v(x) = b_\mu$ on $\partial\Omega$. To this end denote these sought eigenvalues of (6) by

$$(10) \qquad a_\mu^* = a_\mu \quad \text{when } v(x) = \text{constant} \neq 0 \text{ on } \partial\Omega.$$

Thus the solution of the inverse problem are the points (a_μ^*, b_μ) which can be described with the help of an eigenfunction, which are special minimizers of (8),

$$-a_\mu^* \|v\|^2 = \int_\Omega |\nabla v(x)|^2 \, dx - a_\mu^* b_\mu \int_{\partial\Omega} \Phi(x) \, ds$$

which leads to

$$(11) \qquad a_\mu^* \left(b_\mu \int_{\partial\Omega} \Phi(x) \, ds - \|v\|^2 \right) = \int_\Omega |\nabla v(x)|^2 \, dx.$$

Thus from (11), in the second quadrant we have $b_\mu \int_{\partial\Omega} \Phi(x) ds < \|v\|^2$ and $v(x) > b_\mu$ in Ω, while in the first quadrant, i.e. $a_\mu^* > 0$, we must have $b_\mu \int_{\partial\Omega} \Phi(x) ds > \|v\|^2$ and $0 < v(x) < b_\mu$. Note also that if $a_\mu^* = 0$, then $\nabla v = 0$ in Ω and so $v(x) = b_\mu = \int_{\partial\Omega} \Phi(x) ds / |\Omega|$ in Ω. Thus we proved the following

PROPOSITION 3. *If $a_\mu^* < 0$ then $0 < b_\mu \int_{\partial\Omega} \Phi(x) ds < \|v\|^2$ and $v(x) > b_\mu$ in Ω while if $a_\mu^* > 0$ then $b_\mu \int_{\partial\Omega} \Phi(x) ds > \|v\|^2$ and $0 < v(x) < b_\mu$ in Ω. Finally if $a = 0$ then $b_0 = \int_{\partial\Omega} \Phi ds / |\Omega|$ and $v(x) = b_0$ in Ω.* $\Box$

We now use the maximum principle, [**13**, Theorem 17], to get a lower bound on eigenvalues $-a_\mu^* > 0$. Denote by $\rho > 0$ the radius of the smallest ball that contains Ω, let p be its center and assume that $p \in \Omega$. Let $r(x) = |x - p| \leq \rho$, be the euclidean distance of from $x \in \Omega$ to p. Also let us agree to write $\Phi_{\min} = \min_{x \in \partial\Omega} (\Phi(x)) > 0$, which exists since Φ is continuous, see (i) in Proposition 2.

PROPOSITION 4. *In the second quadrant we have*

$$\frac{2n\mu\Phi_{\min}}{\rho^2 \mu \Phi_{\min} + 2\rho} \leq -a_\mu^* \leq \lambda_1^D \quad \text{and} \quad 2b\rho \geq 2n\Phi_{\min} + a\rho^2.$$

Proof. The upper bound of the first inequality has been proved in Proposition 2(iv). For the lower bound, we need to verify that

$$w(x) = \tau + 1 - r^2(x)/\rho^2 \geq 0 \quad \text{where} \quad \tau = \frac{2}{\rho\mu\Phi_{\min}}$$

is an upper solution for (6). Clearly we have $w(x) \geq \tau > 0$ and it is also readily seen that

$$-\Delta w = 2n\rho^{-2} > 0.$$

On the boundary we have for $x \in \partial\Omega$, and $n(x)$ a unit outward normal vector at x

$$|(x-p).n| \leq r(x)$$

which implies that

$$\begin{aligned}
\partial_n w + \mu\Phi w &= -\rho^{-2}\nabla r^2.n + \mu\Phi\left(\tau + 1 - r^2/\omega^2\right) \\
&\geq -\frac{2}{\rho^2}(x-p).n + \mu\tau\Phi \\
&\geq -\frac{2r}{\rho^2} + \frac{2}{\rho}\frac{\Phi}{\Phi_{\min}} \\
&= \frac{2r}{\rho^2\Phi_{\min}}\left(\frac{\rho}{r}\Phi(x) - \Phi_{\min}\right) \geq 0.
\end{aligned}$$

It follows from the maximum principle that, since $w(x) \leq w(p) = \tau + 1$,

$$-a_\mu \geq \inf_{x \in \Omega} \frac{-\Delta w(x)}{w(x)} \geq \frac{2n}{(\tau+1)\rho^2} = \frac{2n\mu\Phi_{\min}}{(2\rho + \rho^2\mu\Phi_{\min})}.$$

Therefore for any given $\mu > 0$, by (10) we have a simple lower bound on a_μ^* in the second quadrant. The second inequality, is obtained by replacing $\mu = -a/b$ and says that the solution curve must be above a certain line. $\square$

We now have a clear picture in the second quadrant if we add the fact that the map $a \to b$ is increasing. Recall that when $b = 0$ then $a_\mu = -\lambda_1^D$, and as a increases and reaches $a = 0$, then $b = \frac{1}{|\Omega|}\int_{\partial\Omega} \Phi(x)ds$. Thus the curve (a_μ, b_μ) joins the point $\left(-\lambda_1^D, 0\right)$ to $\left(0, \frac{1}{|\Omega|}\int_{\partial\Omega} \Phi(x)ds\right)$ and is above the line $2b_\mu\rho \geq 2n\Phi_{\min} + a_\mu\rho^2$. Thus the sought solution (a_μ^*, b_μ) that belongs to the second quadrant is trapped in the region

$$0 \leq b_\mu \leq \frac{1}{|\Omega|}\int_{\partial\Omega}\Phi(x)ds \quad \text{and} \quad -\lambda_1^D \leq a_\mu \leq \frac{2b_\mu}{\rho\Phi_{\min}} - \frac{2n}{\rho^2}.$$

We now deal with the more difficult case when the point (a,b) is in the first quadrant and try to find an enclosure for the region.

Quadrant I. The point (a,b) being in the first quadrant and also on the line $b = -a/\mu$ means that $\mu < 0$. The existence of eigenvalues a_μ of (6), i.e. a minimizer of the functional in (8), is not a trivial matter anymore. To proceed further use the trace theorem

$$\alpha\int_{\partial\Omega}|v(x)|^2\,ds \leq \|v\|_{H^1(\Omega)}$$

where $\alpha > 0$ depends on Ω only. For any $v \in H^1(\Omega)$ we have

$$\begin{aligned}
J(v) &= \int_\Omega |\nabla v(x)|^2\, dx + \mu \int_{\partial\Omega} \Phi(x) |v(x)|^2\, ds \\
&= \int_\Omega |v(x)|^2\, dx + \int_\Omega |\nabla v(x)|^2\, dx + \mu \int_{\partial\Omega} \Phi(x) |v(x)|^2\, ds - \int_\Omega |v(x)|^2\, dx \\
&\geq \alpha \int_{\partial\Omega} |v(x)|^2\, ds + \mu \int_{\partial\Omega} \Phi(x) |v(x)|^2\, ds - \int_\Omega |v(x)|^2\, dx.
\end{aligned}$$

Thus the functional J is still bounded below, for $\mu < 0$ such that,

(12) $$\int_{\partial\Omega} (\alpha + \mu \Phi(x))\, ds = \alpha |\partial\Omega| + \mu \int_{\partial\Omega} \Phi(x)\, ds \geq 0$$

where $|\partial\Omega|$ denotes the length of $\partial\Omega$. However its minimum over $H^1(\Omega)$ maybe negative,

$$-a_\mu = \inf_{\substack{v \in H^1(\Omega) \\ 1 = \|v\|_{L^2(\Omega)}}} J(v) \geq -1 + \inf_{\substack{v \in H^1(\Omega) \\ 1 = \|v\|_{L^2(\Omega)}}} \int_{\partial\Omega} (\alpha + \mu \Phi(x)) |v(x)|^2\, ds$$

i.e. $0 < a_\mu < 1$ is possible which means we are in the first quadrant, or we have a lower level $a_\mu \leq 0$ which means that $b < 0$ and so is not applicable, as $b > 0$. For these possible positive a_μ^*, the analog of (11) is

$$a_\mu^* = \inf_{v \in H^1_{b_\mu}(\Omega)} \frac{1}{\|v\|^2} \left(a_\mu^* b_\mu \int_{\partial\Omega} \Phi(x)\, ds - \int_\Omega |\nabla v|^2\, dx \right) > 0.$$

Thus we are guaranteed the existence of a solution (a_μ, b_μ), i.e. eigenvalues to (6) if we restrict the values for μ to (12)

(13) $$-\alpha |\partial\Omega| \leq \mu \int_{\partial\Omega} \Phi(x)\, ds.$$

From (11) it follows from the trace theorem that in the first quadrant, $a_\mu^* > 0$, that we have

(14) $$\begin{aligned}
a_\mu^* b_\mu \int_{\partial\Omega} \Phi(x)\, ds &= \int_\Omega |\nabla v|^2\, dx + a_\mu^* \|v\|^2 \\
&\geq \alpha \min(1, a_\mu^*) \int_{\partial\Omega} |v(x)|^2\, ds \\
&\geq \alpha \min(1, a_\mu^*) b_\mu^2 |\partial\Omega|.
\end{aligned}$$

since $v(x) = b_\mu$ on the boundary $\partial\Omega$. Thus in the first quadrant, by dividing by b_μ^2 and using (7) we get

(15) $$\mu \int_{\partial\Omega} \Phi(x)\, ds \leq -\alpha \min(1, a_\mu^*) |\partial\Omega|$$

Thus the solution curve in quadrant I, is inside the rectangle $[0, 1] \times \left[0, \int_{\partial\Omega} \Phi(x)\, ds / \alpha |\partial\Omega|\right]$.

PROPOSITION 5. *When $0 < a_\mu^* \leq 1$, the solutions (a_μ^*, b_μ) in the first quadrant, if any, satisfy $\int_{\partial\Omega} \Phi(x)\, ds \geq \alpha b_\mu |\partial\Omega|$.* □

Conclusion. By using the parameter $\mu = -a/b$, and differentiating twice equation (1), the inverse problem is transformed into computing the principal eigenvalue of a Laplacian with mixed boundary condition.

The reconstruction algorithm is a follows: Given any observation $\Phi > 0$ on $\partial\Omega$ and μ as in (13), solve the eigenvalue problem, (6), and select the eigenvalues a_μ whose eigenfunctions have nonzero constant boundary traces. These would reveal all the sought all values $\left(a_\mu^*,\, b_\mu\right)$, with $-\lambda_1^D \leq a_\mu^* \leq 1$ in (1) that are generated by the observation Φ.

PROPOSITION 6. *To find the solution $\left(a_\mu^*, b_\mu\right)$ of the inverse problem (1), with $-\lambda_1^D \leq a_\mu^* \leq 1$, it is enough to find the first eigenfunction of (6) with μ as in (13) and whose boundary traces are nonzero constant.*

References

[1] Habib Ammari, Hyeonbae Kang, and Hyundae Lee, *Layer potential techniques in spectral analysis*, Mathematical Surveys and Monographs, vol. 153, American Mathematical Society, Providence, RI, 2009. MR2488135 (2010j:47062)

[2] S. I. Bezrodnykh and A. S. Demidov, *On the uniqueness of solution of Cauchy's inverse problem for the equation $\Delta u = au + b$*, Asymptot. Anal. **74** (2011), no. 1-2, 95–121. MR2850361 (2012j:35443)

[3] Carlos Alberto Berenstein, *An inverse spectral theorem and its relation to the Pompeiu problem*, J. Analyse Math. **37** (1980), 128–144, DOI 10.1007/BF02797683. MR583635 (82b:35031)

[4] R. Dalmasso, *A new result on the Pompeiu problem*, Trans. Amer. Math. Soc. **352** (2000), no. 6, 2723–2736, DOI 10.1090/S0002-9947-99-02533-7. MR1694284 (2000j:35060)

[5] Robert Dalmasso, *An inverse problem for an elliptic equation with an affine term*, Math. Ann. **316** (2000), no. 4, 771–792, DOI 10.1007/s002080050354. MR1758453 (2001d:35194)

[6] A. S. Demidov and M. Moussaoui, *An inverse problem originating from magnetohydrodynamics*, Inverse Problems **20** (2004), no. 1, 137–154, DOI 10.1088/0266-5611/20/1/008. MR2044610 (2005h:35357)

[7] Jin Cheng and Masahiro Yamamoto, *Determination of two convection coefficients from Dirichlet to Neumann map in the two-dimensional case*, SIAM J. Math. Anal. **35** (2004), no. 6, 1371–1393 (electronic), DOI 10.1137/S0036141003422497. MR2083783 (2005g:35296)

[8] Jian Deng, *Some results on the Schiffer conjecture in R^2*, J. Differential Equations **253** (2012), no. 8, 2515–2526, DOI 10.1016/j.jde.2012.06.002. MR2950461

[9] Fritz Gesztesy, Marius Mitrea, and Roger Nichols, *Heat kernel bounds for elliptic partial differential operators in divergence form with Robin-type boundary conditions*, J. Anal. Math. **122** (2014), 229–287, DOI 10.1007/s11854-014-0008-7. MR3183528

[10] David Gilbarg and Neil S. Trudinger, *Elliptic partial differential equations of second order*, Classics in Mathematics, Springer-Verlag, Berlin, 2001. Reprint of the 1998 edition. MR1814364 (2001k:35004)

[11] Antoine Henrot, *Extremum problems for eigenvalues of elliptic operators*, Frontiers in Mathematics, Birkhäuser Verlag, Basel, 2006. MR2251558 (2007h:35242)

[12] Hynek Kovařík, *On the lowest eigenvalue of Laplace operators with mixed boundary conditions*, J. Geom. Anal. **24** (2014), no. 3, 1509–1525, DOI 10.1007/s12220-012-9383-4. MR3223564

[13] Murray H. Protter and Hans F. Weinberger, *Maximum principles in differential equations*, Springer-Verlag, New York, 1984. Corrected reprint of the 1967 original. MR762825 (86f:35034)

[14] Michael Vogelius, *An inverse problem for the equation $\Delta u = -cu - d$* (English, with English and French summaries), Ann. Inst. Fourier (Grenoble) **44** (1994), no. 4, 1181–1209. MR1306552 (95h:35246)

DEPARTMENT OF MATHEMATICS, UWG, CARROLLTON GEORGIA 30118
E-mail address: boumenir@westga.edu

Combinatorial properties of circulant Hadamard matrices

Reinhardt Euler, Luis H. Gallardo, and Olivier Rahavandrainy

ABSTRACT. We describe combinatorial properties of the defining row of a circulant Hadamard matrix by exploiting its orthogonality to subsequent rows, and show how to exclude several particular forms of these matrices.

1. Introduction and basic definitions

Throughout this paper, the i-row of a matrix H of size $n \times n$ is denoted by H_i, $1 \leq i \leq n$. A (± 1)-matrix H of size $n \times n$ is called *Hadamard matrix* if its rows are mutually orthogonal. It is a *circulant matrix* if for $H_1 = (h_1, \ldots, h_n)$, its i-row is given by $H_i = (h_{1-i+1}, \ldots, h_{n-i+1})$, the subscripts being taken modulo n. In this case, the first row H_1 will be called the *defining row* of H.

An example of a 4×4 circulant Hadamard matrix is the following:

$$H = \begin{pmatrix} -1 & 1 & 1 & 1 \\ 1 & -1 & 1 & 1 \\ 1 & 1 & -1 & 1 \\ 1 & 1 & 1 & -1 \end{pmatrix}$$

A conjecture of Ryser (cf. [2, pp.97]) states that there is no circulant Hadamard matrix unless $n = 1$ or $n = 4$. This conjecture is still open although many partial results have been obtained. For recent results and further details, we refer to [1], [3] and [4].

The objective of this paper is to exhibit some combinatorial properties of the defining row H_1 of such a matrix H that we obtained by exploiting the orthogonality of H_1 to some of its subsequent rows. It turns out that these properties allow to exclude several particular forms for circulant Hadamard matrices. We think that a further development of this "combinatorial approach" may contribute to exclude other cases of such matrices and help to determine the maximum number k for which the first k rows of a circulant (± 1)-matrix are mutually orthogonal.

The order n (for $n > 1$) of a circulant Hadamard matrix H is well known to be of the form $n = 4m = 4h^2$, where h is an odd integer. Also, by orthogonality, the entries of any two different rows have the same sign in exactly $2m$ columns, and a

---

2010 *Mathematics Subject Classification.* Primary 15B34.
Key words and phrases. Hadamard matrix, circulant, combinatorial properties.

©2016 American Mathematical Society

different sign in the other $2m$ ones.

Finally, the property of being circulant allows any row of H to be represented as a circular sequence of "+" and "−". Figure 1 illustrates the first two rows of such a matrix H for $n = 12$.

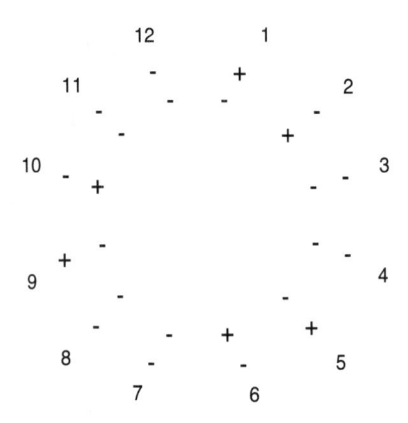

FIGURE 1. The first two rows of a circulant Hadamard matrix of order 12

Some further definitions are required. Let us consider the first row H_1 of a circulant Hadamard matrix H. A maximal sequence of consecutive +1's (respectively −1's) will be called a *positive block* (respectively *negative block*), denoted by B. If the size of such a block equals k, for some positive integer k, we will speak of a *positive* (respectively *negative*) *k-block*. The family of all blocks is denoted by $\mathcal{B}$, that of all k-blocks by $\mathcal{B}_k$ and that of all k-blocks with $k \geq i$, for a given positive integer i, by $\mathcal{B}_{\geq i}$. Thus, H_1 decomposes into a sequence of alternating, positive or negative blocks. For convenience, we suppose the first and last block of H_1 to be of opposite sign. Finally, a maximal alternating sequence of blocks in $\mathcal{B}_k$ (in $\mathcal{B}_{\geq k}$) will be called a k-*alternating sequence* ($\geq k$-*alternating sequence*, respectively), the *length* of which is given by the number of blocks forming it.

2. Our results

LEMMA 2.1. *The total number of blocks in the first row H_1 of a circulant Hadamard matrix H equals $2m$.*

Proof: We consider H_1 and H_2 of H and observe that the number of columns with entries of different sign equals the total number of blocks. Hence, $|\mathcal{B}| = 2m$. □

Alternatively, if we count the number of columns whose entries have equal sign, we obtain $\sum_{B \in \mathcal{B}}(|B| - 1)$, which equals $4m - |\mathcal{B}|$ and which by orthogonality, gives $2m$. Again, $|\mathcal{B}| = 2m$. □

LEMMA 2.2. *The total number of 1-blocks in H_1 equals m.*

Proof: We consider H_1 and H_3 of H and count the number of columns with entries of equal sign. We observe that:

- in a 1-alternating sequence, this number equals the length of the sequence minus 1, plus 1 arising from the ≥ 2-block preceding the sequence,
- a block $B \in \mathcal{B}_{\geq 3}$ contributes with $|B| - 2$,
- no block $B \in \mathcal{B}_2$ contributes to this number (except those preceding a 1-alternating path, whose contribution is already taken into account).

By orthogonality of H_1 and H_3, we obtain:

$$\begin{aligned}
2m &= |\mathcal{B}_1| + \sum_{B \in \mathcal{B}_{\geq 3}} (|B| - 2) \\
&= |\mathcal{B}_1| + \sum_{B \in \mathcal{B}_{\geq 2}} (|B| - 2) \\
&= |\mathcal{B}_1| + \left(\sum_{B \in \mathcal{B}_{\geq 2}} |B| \right) - 2|\mathcal{B}_{\geq 2}| \\
&= |\mathcal{B}_1| + (4m - |\mathcal{B}_1|) - 2(|\mathcal{B}| - |\mathcal{B}_1|) \\
&= 4m - 4m + 2|\mathcal{B}_1|.
\end{aligned}$$

Hence, $|\mathcal{B}_1| = m$.

Alternatively, we may count the number of columns of H_1 and H_3, with entries of different signs. We denote by α_1 the number of 1-alternating sequences, by $\alpha_{\geq 2}$ the number of ≥ 2-alternating sequences, by Λ the set of all ≥ 2-alternating sequences, and by $|\ell|$ the length of an element ℓ of Λ. We observe that:

- any 1-alternating sequence contributes with 2,
- any ≥ 2-alternating sequence ℓ contributes with $2(|\ell| - 1)$.

Thus, we obtain:

$$2m = 2\alpha_1 + 2\left(\sum_{\ell \in \Lambda} (|\ell| - 1) \right). \tag{1}$$

Since 1-alternating sequences and ≥ 2-alternating sequences are themselves alternating within H_1, $\alpha_1 = \alpha_{\geq 2}$, and we get from (1):

$$m = \alpha_1 + |\mathcal{B}_{\geq 2}| - \alpha_{\geq 2} = |\mathcal{B}_{\geq 2}| = |\mathcal{B}| - |\mathcal{B}_1| = 2m - |\mathcal{B}_1| \text{ by Lemma 2.1.}$$

Hence, $|\mathcal{B}_1| = m$. □

LEMMA 2.3. *The number of blocks of size 2 and the number of 1-alternating sequences add up to m:*

$$|\mathcal{B}_2| + \alpha_1 = m.$$

Proof: We consider H_1 and H_4 of H and count the number of columns with entries of equal sign. We observe that:

- any 1-alternating sequence is preceded and followed by a block of size ≥ 2,
- a block $B \in \mathcal{B}_{\geq 3}$ contributes with $|B| - 3$,

- any 2-block $B \in \mathcal{B}_2$ is preceded and followed by an element of opposite sign.

By orthogonality of H_1 and H_4 and by Lemmata 2.1 and 2.2, we obtain:

$$\begin{aligned} 2m &= \sum_{B \in \mathcal{B}_{\geq 3}} (|B| - 3) + |\mathcal{B}_2| + 2\alpha_1 \\ &= 4m - |\mathcal{B}_1| - 2|\mathcal{B}_2| - 3|\mathcal{B}_{\geq 3}| + |\mathcal{B}_2| + 2\alpha_1 \\ &= 3m - |\mathcal{B}_2| - 3(|\mathcal{B}| - m - |\mathcal{B}_2|) + 2\alpha_1 \\ &= 2|\mathcal{B}_2| + 2\alpha_1. \end{aligned}$$

Hence, $|\mathcal{B}_2| + \alpha_1 = m$. □

REMARK 2.4. *It would be interesting to know whether counting the number of columns with entries of different sign provides the same condition on $|\mathcal{B}_2|$, similar to that obtained for $|\mathcal{B}_1|$. For the sake of completeness and to make the reader more familiar with our approach, we have included such a condition at this place.*

For given n, let α_1, α_2 and $\alpha_{\geq 3}$ be the number of 1-, 2- and $\geq$ 3-alternating sequences, and let $\alpha_{2, \geq 3}$ denote the number of pairs of blocks B, B' with $B \in \mathcal{B}_2$, $B' \in \mathcal{B}_{\geq 3}$ or vice versa. We obtain the following:

LEMMA 2.5.
$$\alpha_1 + \alpha_{2, \geq 3} = \alpha_2 + \alpha_{\geq 3}.$$

Proof: Any 1-, 2- or $\geq$ 3-alternating sequence is preceded (followed) by a 2- or 3-, 1- or $\geq$ 3-, 1- or 2-alternating sequence, respectively.
Let us examine the first case, i.e. that of a 1-alternating sequence. Counting the number of columns with entries of different signs gives us a contribution of $(\ell - 1)$ for such a sequence, if ℓ denotes the length of it (see Figure 2).

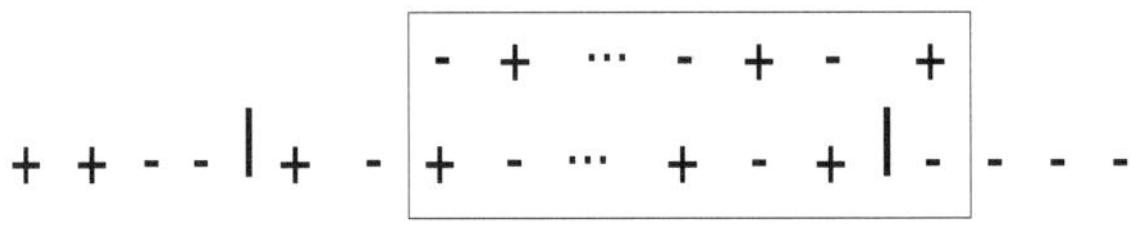

FIGURE 2. A 1-alternating sequence preceded (followed) by a 2 − ($\geq 3-$) one, and the contribution of $\ell - 1$.

For a 2-alternating sequence preceded and followed by a 1-alternating one, the total contribution will be $(\ell - 1)$ again. If, however, such a sequence is preceded or followed by a $\geq$ 3-alternating sequence (see Figure 3), one or two consecutive pairs of blocks B, B' with $B \in \mathcal{B}_2$, $B' \in \mathcal{B}_{\geq 3}$ or vice versa, will appear and contribute by an additional value of 1.

In that case, there is another contribution of 1 (encircled by a dashed line), that will be included in the contribution of the $\geq$ 3-alternating sequence.

Finally, for $\geq$ 3-alternating sequences, we obtain the situation depicted in Figure 4. The overall contribution of such a $\geq$ 3-alternating sequence is $3\ell - 1$. Also, observe that the pair of blocks B, B' with $B \in \mathcal{B}_{\geq 3}$, the last block of this sequence, and $B' \in \mathcal{B}_2$, the first of the following 2-alternating one, again contribute by 1 to

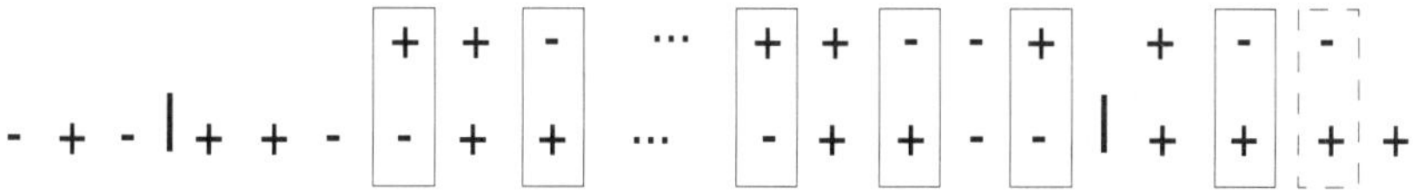

FIGURE 3. A 2-alternating sequence preceded (followed) by a $1-$ ($\geq 3-$) one.

the value $\alpha_{2,\geq 3}$.

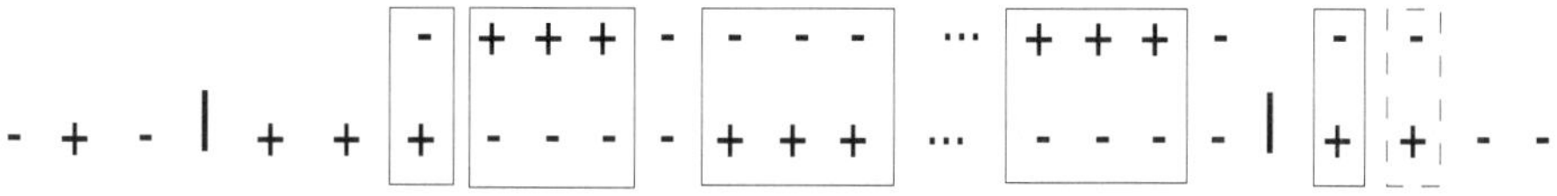

FIGURE 4. A ≥ 3-alternating sequence preceded (followed) by a $1-$ ($2-$) one.

Altogether, counting the number of columns with entries of different signs in H_1 and H_4 gives the condition:

$$\sum_{i\in I}(\ell_i-1)+\sum_{j\in J}(\ell_j-1)+\sum_{k\in K}(3\ell_k-1)+\alpha_{2,\geq 3}=2m,$$

where I, J, K, denote the sets of 1-, 2- and $\geq 3-$alternating sequences, respectively, and ℓ their length.

We obtain:

$$|\mathcal{B}_1|-\alpha_1+|\mathcal{B}_2|-\alpha_2+3|\mathcal{B}_{\geq 3}|-\alpha_{\geq 3}+\alpha_{2,\geq 3}=2m.$$

Since, by Lemma 2.1,

$$|\mathcal{B}_1|+|\mathcal{B}_2|+3|\mathcal{B}_{\geq 3}|=4m-2|\mathcal{B}_1|-2|\mathcal{B}_2|,$$

$$4m-\alpha_1-\alpha_2-\alpha_{\geq 3}+\alpha_{2,\geq 3}=2|\mathcal{B}_1|+2|\mathcal{B}_2|.$$

By Lemma 2.2, we obtain:

$$2m-\alpha_1-\alpha_2-\alpha_{\geq 3}+\alpha_{2,\geq 3}=2|\mathcal{B}_2|.$$

Since, by Lemma 2.3,

$$2m-2|\mathcal{B}_2|=2\alpha_1,$$

we obtain as stated

$$\alpha_1+\alpha_{2,\geq 3}=\alpha_2+\alpha_{\geq 3}.$$

□

We are now ready to show the nonexistence of a circulant Hadamard matrix for the following four particular situations:

i) $\alpha_1=1$, ii) $\alpha_1=m$, iii) $\alpha_1=m-1$, iv) $\alpha_1=2$

THEOREM 2.6. *If in H_1, there is only one 1-alternating sequence (of length m), or if there are m such sequences (all of length 1), or if the number of 1-alternating sequences is $m-1$, then there is no circulant Hadamard matrix H with H_1 as first row.*

Proof: We consider the three cases.

<u>Case $\alpha_1 = 1$</u>:

We know already, from Lemmata 2.1, 2.2 and 2.3, that:
- the total number of blocks is $2m$,
- that of 1-blocks is m,
- $|\mathcal{B}_2| + \alpha_1 = m$.

Therefore, since $n = 4m$, there is exactly one more block of size $m+2$, and if we suppose that this particular block precedes the 1-alternating sequence (clockwise), we obtain the configuration depicted in Figure 5:

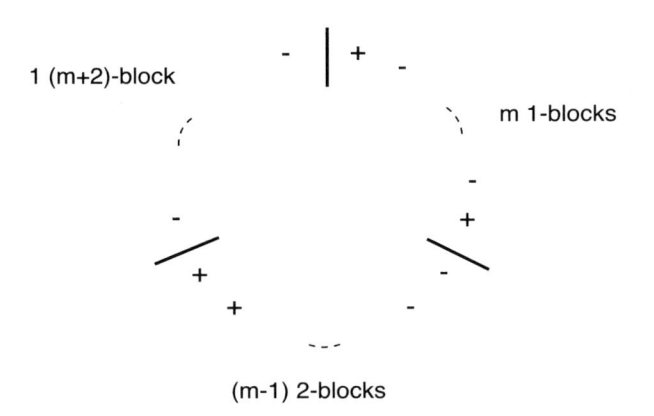

FIGURE 5. A first representation of H_1 for $\alpha_1 = 1$

Clockwise evaluation of the scalar product of H_1 and H_5 gives:

$$(m-4) + 0 + 2(m-3) + 0 + (m-2) + 0 = 4m - 12,$$

which allows for orthogonality only if $m = 3$, contradicting the condition on m to be an odd square integer.

By symmetry, we obtain the same result if the particular $(m+2)$-block follows the 1-alternating sequence.

If, now, the $(m+2)$-block is situated "in between" the sequence of 2-blocks, as depicted in Figure 6,

we may suppose that

$$m - 1 = k_1 + k_2, \text{with } k_1, k_2 \text{ both even or both odd.}$$

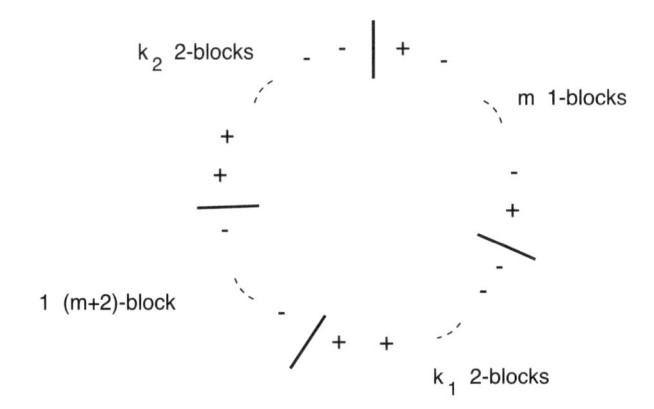

FIGURE 6. A second representation of H_1 for $\alpha_1 = 1$ and $(m-1) = k_1 + k_2$

Evaluating the scalar product of H_1 and H_5 gives:
$$(m-4) + 0 + 2(k_1 - 2) + 0 + (m-2) + 2(k_2 - 2) + 0 = 4m - 16.$$
Orthogonality implies $m = 4$, contradicting the condition on m to be odd.
We obtain the same result for the case $k_1 = 1$ or $k_2 = 1$, so that altogether, H_1 cannot be the defining row of a circulant Hadamard matrix.

Case $\alpha_1 = m$:

By Lemma 2.3, there is no 2-block in H_1 and m blocks remain to cover $3m$ elements, i.e. besides the m 1-blocks, there are exactly m 3-blocks, and both types alternate in H_1. But then the scalar product of H_1 and H_5 gives $4m > 0$ and H_1 cannot define a circulant Hadamard matrix either.

Case $\alpha_1 = m - 1$:

In this case there is exactly one consecutive pair of 1-blocks, and by Lemma 2.3, there is exactly one 2-block, so that $3m - 2$ elements have to be covered by $m - 1$ blocks of size ≥ 3. This is only possible with $(m - 2)$ 3-blocks and one 4-block. Moreover, the unique 2-block, say B, cannot be preceded *and* followed by a $\geq$ 3-block. Therefore, we come up with two possibilities:

a) B is preceded and followed by a 1-alternating sequence.

b) B is preceded or followed (but not both) by a $\geq$ 3-block.

Situation a) is depicted in Figure 7:

Just observe, that in this situation exactly two $\geq$ 3-blocks follow each other. We count the number of columns with different signs in H_1 and H_5 and observe that a contribution (of 2) to this number is given only by
- the two consecutive 1-blocks;

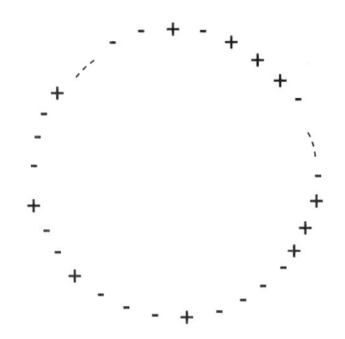

FIGURE 7. Illustration of possibility a)

- the unique 2-block;
- the unique 4-block;
- exactly one of the 3-blocks.

By orthogonality,
$$2m = 8,$$
a contradiction to the condition on m to be odd.

Situation b) is depicted in Figure 8:

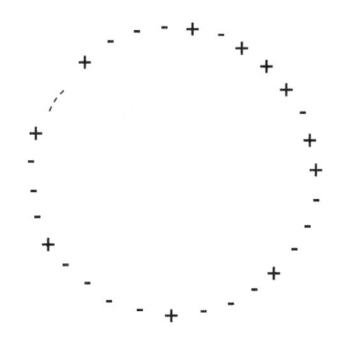

FIGURE 8. Illustration of possibility b)

In this situation, no 3-block contributes any more, and orthogonality of H_1 and H_5 gives
$$2m = 6,$$
a contradiction to the condition that m has to be an odd square integer. □

In the following, we will show how the orthogonality of H_1 and H_5 can be exploited to obtain a further condition involving $|\mathcal{B}_2|$ and $|\mathcal{B}_3|$, which we will use afterwards to exclude the situation where $\alpha_1 = 2$.

For this, we need to describe the possible types of 1-alternating and 2-alternating sequences which (up to symmetry) is done in Figures 9 and 10:

COMBINATORIAL PROPERTIES OF CIRCULANT HADAMARD MATRICES 17

```
- - -|+|- - -           + - -|+|- - -           + - -|+|- - +
    type 1                  type 2                  type 3
```

```
- - -|+ - + - ··· + -|+ + +    + - -|+ - + - ··· + -|- - -    + - -|+ - + - ··· + -|+ + -
         type 4                       type 5                       type 6
```

FIGURE 9. 1-alternating sequences of length 1 and ℓ, respectively

It is not difficult to verify that if we count the columns of H_1 and H_5 with equal sign, the contribution of these 6 types will be $3, 2, 1$ and ℓ, $\ell - 1$, $\ell - 2$, respectively.

For 2-alternating sequences, we obtain 3 similar types, depicted in Figure 10, whose contributions will be 2ℓ, $2\ell - 1$, $2\ell - 2$, respectively:

```
- - -|+ + - - ··· + +|- - -    + -|+ + - - ··· + +|- - -    + -|+ + - - ··· + +|- +
       type 1                       type 2                       type 3
```

FIGURE 10. 2-alternating sequences of length ℓ

with the properties that
- type 1 is preceded and followed by a $\geq$ 3-block;
- type 2 is preceded by a 1-alternating sequence and followed by a $\geq$ 3-block (or vice versa);
- type 3 is preceded and followed by a 1-alternating sequence.

If now α_1^i denotes the number of 1-alternating sequences of type i, for $i = 1, \ldots, 6$, and α_2^j the number of 2-alternating sequences of type j, for $j = 1, \ldots, 3$, we get

LEMMA 2.7. *If H is a circulant Hadamard matrix, then one has*
$$2|\mathcal{B}_2| + 2|\mathcal{B}_3| = (\alpha_1^2 + 2\alpha_1^3) + (2\alpha_1^4 + 3\alpha_1^5 + 4\alpha_1^6) + (\alpha_2^2 + 2\alpha_2^3).$$

Proof:
We can describe the total number of columns of H_1 and H_5, whose entries have equal sign, by the following expression
$$L := 3\alpha_1^1 + 2\alpha_1^2 + \alpha_1^3 + C + D,$$
where
$$C := \sum_{i \in I} \ell_i + \sum_{j \in J}(\ell_j - 1) + \sum_{k \in K}(\ell_k - 2),$$
and
$$D := \sum_{p \in P} 2\ell_p + \sum_{q \in Q}(2\ell_q - 1) + \sum_{r \in R}(2\ell_r - 2) + \sum_{B \in \mathcal{B}_{\geq 4}}(|B| - 4) + |\mathcal{B}_3|.$$

expression in which $\sum_{j \in J}(\ell_j - 1)$, for instance, relates to all 1-alternating sequences S_j, $j \in J$, whose contribution equals $\ell_j - 1$, i.e., which are of type 5.

We know the following:

i) $\alpha_1^1 + \alpha_1^2 + \alpha_1^3 + \sum_{i \in I} \ell_i + \sum_{j \in J} \ell_j + \sum_{k \in K} \ell_k = |\mathcal{B}_1| = m$, by Lemma 2.2,

ii) $\sum_{p \in P} \ell_p + \sum_{q \in Q} \ell_q + \sum_{r \in R} \ell_r = |\mathcal{B}_2|$, and

iii) $\alpha_1^1 + \alpha_1^2 + \alpha_1^3 + \alpha_1^4 + \alpha_1^5 + \alpha_1^6 = \alpha_1$.

Therefore,

$$\begin{aligned} 3\alpha_1^1 + 2\alpha_1^2 + \alpha_1^3 + C &= 2\alpha_1^1 + \alpha_1^2 + |\mathcal{B}_1| - \alpha_1^5 - 2\alpha_1^6 \\ &= \alpha_1 + m + \alpha_1^1 - \alpha_1^3 - \alpha_1^4 - 2\alpha_1^5 - 3\alpha_1^6. \end{aligned}$$

We also have

$$\sum_{p \in P} 2\ell_p + \sum_{q \in Q}(2\ell_q - 1) + \sum_{r \in R}(2\ell_r - 2) = 2|\mathcal{B}_2| - \alpha_2^2 - 2\alpha_2^3.$$

Thus, if we put

$$\delta := \alpha_1 + m + \alpha_1^1 - \alpha_1^3 - \alpha_1^4 - 2\alpha_1^5 - 3\alpha_1^6 + 2|\mathcal{B}_2| - \alpha_2^2 - 2\alpha_2^3,$$

then

$$\begin{aligned} L &= \delta + \sum_{B \in \mathcal{B}_{\geq 4}} |B| - 4|\mathcal{B}_{\geq 4}| + |\mathcal{B}_3| \\ &= \delta + 4m - |\mathcal{B}_1| - 2|\mathcal{B}_2| - 3|\mathcal{B}_3| - 4(|\mathcal{B}| - |\mathcal{B}_1| - |\mathcal{B}_2| - |\mathcal{B}_3|) + |\mathcal{B}_3| \\ &= \delta - m + 2|\mathcal{B}_2| + 2|\mathcal{B}_3|, \text{ by Lemmata 2.1 and 2.2} \\ &= \delta - \alpha_1 + |\mathcal{B}_2| + 2|\mathcal{B}_3|, \text{ by Lemma 2.3} \\ &= m + (\alpha_1^1 - \alpha_1^3 - \alpha_1^4 - 2\alpha_1^5 - 3\alpha_1^6) + 3|\mathcal{B}_2| + 2|\mathcal{B}_3| - \alpha_2^2 - 2\alpha_2^3. \end{aligned}$$

This last expression, by orthogonality, has to be equal to $2m$.

So, by Lemma 2.3 and by iii) above, we get

$$(\alpha_1^1 - \alpha_1^3 - \alpha_1^4 - 2\alpha_1^5 - 3\alpha_1^6) = (\alpha_1 - \alpha_1^2 - 2\alpha_1^3 - 2\alpha_1^4 - 3\alpha_1^5 - 4\alpha_1^6)$$

and

$$|\mathcal{B}_2| + \alpha_1 = m = \alpha_1 - \alpha_1^2 - 2\alpha_1^3 - 2\alpha_1^4 - 3\alpha_1^5 - 4\alpha_1^6 + 3|\mathcal{B}_2| + 2|\mathcal{B}_3| - \alpha_2^2 - 2\alpha_2^3,$$

which gives our result:

$$2|\mathcal{B}_2| + 2|\mathcal{B}_3| = (\alpha_1^2 + 2\alpha_1^3) + (2\alpha_1^4 + 3\alpha_1^5 + 4\alpha_1^6) + (\alpha_2^2 + 2\alpha_2^3).$$

$\square$

Lemma 2.7 can now be used to exclude another situation:

THEOREM 2.8. *If H_1 contains exactly two 1-alternating sequences, i.e. if $\alpha_1 = 2$, then H is not a circulant Hadamard matrix.*

Proof:
Lemma 2.3 implies that $|\mathcal{B}_2| = m - \alpha_1 = m - 2$. There are two blocks left, by Lemma 2.1, to cover the remaining $m + 4$ elements. Hence $|\mathcal{B}_3| \in \{0,1\}$ for $m \geq 3$. Also, $\alpha_2^1 + \alpha_2^2 + \alpha_2^3 \leq 4$, with equality if the two 1-alternating sequences and the two remaining blocks from $\mathcal{B}_{\geq 3}$ are all alternating with a 2-alternating sequence.

Since
$$(\alpha_1^2 + 2\alpha_1^3) + (2\alpha_1^4 + 3\alpha_1^5 + 4\alpha_1^6) + (\alpha_2^2 + 2\alpha_2^3) \leq 4 + 4 + 8 = 16,$$
we get by Lemma 2.7
$$m - 2 + |\mathcal{B}_3| = |\mathcal{B}_2| + |\mathcal{B}_3| \leq 8, \text{ with } |\mathcal{B}_3| \in \{0,1\}.$$
Hence $m \leq 10$ or $m \leq 9$, both cases for which no circulant Hadamard matrix exists. $\square$

3. Conclusion and further work

Future work should include an analysis of further rows of H along the same line. This is what we are currently doing.

References

[1] R. Craigen, G. Faucher, R. Low, and T. Wares, *Circulant partial Hadamard matrices*, Linear Algebra Appl. **439** (2013), no. 11, 3307–3317, DOI 10.1016/j.laa.2013.09.004. MR3119854
[2] Philip J. Davis, *Circulant matrices*, John Wiley & Sons, New York-Chichester-Brisbane, 1979. A Wiley-Interscience Publication; Pure and Applied Mathematics. MR543191 (81a:15003)
[3] Reinhardt Euler, Luis H. Gallardo, and Olivier Rahavandrainy, *Sufficient conditions for a conjecture of Ryser about Hadamard circulant matrices*, Linear Algebra Appl. **437** (2012), no. 12, 2877–2886, DOI 10.1016/j.laa.2012.07.022. MR2966605
[4] Ka Hin Leung and Bernhard Schmidt, *New restrictions on possible orders of circulant Hadamard matrices*, Des. Codes Cryptogr. **64** (2012), no. 1-2, 143–151, DOI 10.1007/s10623-011-9493-1. MR2914407

LAB-STICC UMR CNRS 6285, UNIVERSITY OF BREST, 20, AV. LE GORGEU, C.S. 93837, 29238 BREST CEDEX 3, FRANCE
E-mail address: `Reinhardt.Euler@univ-brest.fr`

LABORATOIRE DE MATHÉMATIQUES UMR CNRS 6205, UNIVERSITY OF BREST, 6, AVENUE LE GORGEU, C.S. 93837, 29238 BREST CEDEX 3, FRANCE
E-mail address: `Luis.Gallardo@univ-brest.fr`

LABORATOIRE DE MATHÉMATIQUES UMR CNRS 6205, UNIVERSITY OF BREST, 6, AVENUE LE GORGEU, C.S. 93837, 29238 BREST CEDEX 3, FRANCE
E-mail address: `Olivier.Rahavandrainy@univ-brest.fr`

A new point of view on higher dimensional Kaluza-Klein theories

Aurel Bejancu

ABSTRACT. The new approach we present in this paper is based on the Riemannian adapted connection and on the theory of adapted tensor fields on the total space of a principal bundle. We obtain the fully general $4D$ equations of motion and the $4D$ Einstein equations in a general gauge Kaluza Klein space. Note that all the previous studies on higher dimensional Kaluza-Klein theories, are particular cases of the general case considered in our paper.

1. Introduction

The classical Kaluza-Klein (KK) theory is known as a theory which unified the gravity with electromagnetism. In a modern setting, the theory is developed on a trivial principal bundle over the 4D spacetime, with $U(1)$ as fibre type. Two strong conditions have been imposed in the development of this theory:
(i) The "cylinder condition" which assumes that the metric on the total space of the principal bundle is independent on the fifth dimension.
(ii) The "compactification condition" which requires that the fibre must be a compact manifold.

Generalizations of KK theory have been done in the following directions:
(a) Remove the conditions (i) and (ii)
(b) Replace $U(1)$ by a non-abelian gauge group.

The first generalization of type (a) was presented by Einstein and Bergmann (cf.[10]). According to it, the local components of the $4D$ metric are periodic functions of the fifth coordinate. The brane-world theory (cf.[12]) and the space-time-matter (STM) theory (cf.[13], [16]) are also two generalizations of KK theory with respect to (a). Recently, we presented a new point of view on KK theory in a $5D$ space (cf. [1], [2], [3]). We removed both conditions (i) and (ii) and developed a new method of study based on what we called the Riemannian horizontal connection. Generalizations of type (b) have been considered by several people ([6], [7], [8], [9], [11], [14], [15]), but the studies have been developed under the conditions (i) and (ii).

In our paper we present a generalization of KK theory in both directions (a) and (b). A part of the results we emphasize here can be found with details in [4], which is the first paper in a series we will devote to this topic. Our new approach is developed on a principal bundle $\bar{M}$ over the $4D$ spacetime M, with an

2010 *Mathematics Subject Classification.* Primary 83E15, 53C80.

n-dimensional Lie group G as fibre type. In this way, the theory we develop here contains as particular cases all the other generalizations of KK theory.

The new point of view we develop here on higher dimensional KK theories is based on the Riemannian adapted connection that we constructed in [4], and on a $4D$ tensor calculus that we introduce via a natural splitting of the tangent bundle of the ambient space. We obtain, in a covariant form and in their full generality, the $4D$ equations of motion as part of the equations of motion in a $(4+n)D$ space. These equations show the existence of an extra force, which in a particular case, is perpendicular to the $4D$ velocity. Also, we obtain the $4D$ Einstein equations on a $(4+n)D$ relativistic gauge KK space. The high level of generality of the study, enables us to recover several results from earlier papers on this matter.

2. General Gauge Kaluza-Klein Space

The KK theory we present here is developed on a principal bundle $\bar{M}$ whose base is a $4D$ manifold M, and whose fibre is an n-dimensional Lie group G. Choose the local coordinates (x^α, y^i) on $\bar{M}$, where x^α and (y^i) are local coordinates on M and G respectively.

Throughout the paper we use the ranges of indices: $\alpha, \beta, \gamma, \cdots \in \{0,1,2,3\}$, $i,j,k,\cdots \in \{4,\cdots,n+3\}$. Also, for any vector bundle E over $\bar{M}$, we denote by $\Gamma(E)$ the $\mathcal{F}(\bar{M})$- module of smooth sections of E, where $\mathcal{F}(\bar{M})$ is the algebra of smooth functions on $\bar{M}$.

Denote by $V\bar{M}$ the *vertical distribution* on $\bar{M}$ of rank n, which is locally spanned by $\{\partial/\partial y^i\}$. Then, suppose that there exists on $\bar{M}$ a pseudo-Riemannian metric $\bar{g}$ whose restriction to $V\bar{M}$ is a Riemannian metric $g^\star$. Consider the complementary orthogonal distribution $H\bar{M}$ to $V\bar{M}$ in $T\bar{M}$, and call it the *horizontal distribution* on $\bar{M}$. Moreover, suppose that $H\bar{M}$ is invariant with respect to the action of G on $\bar{M}$, that is we have

$$(R_a)_\star (H\bar{M}) = H\bar{M}, \quad \forall a \in G,$$

where $(R_a)_\star$ is the differential of the right translation R_a of G. We call the triplet $(\bar{M}, \bar{g}, H\bar{M})$ a *general gauge Kaluza-Klein space*. Next, we consider the natural frame field $\{\partial/\partial x^\alpha, \partial/\partial y^i\}$ and denote by $\delta/\delta x^\alpha$ the projection of $\partial/\partial x^\alpha$ on $H\bar{M}$. Then we have

$$(2.1) \qquad \frac{\delta}{\delta x^\alpha} = \frac{\partial}{\partial x^\alpha} - L^i_\alpha \frac{\partial}{\partial y^i},$$

where L^i_α are local functions on $\bar{M}$. Also, $H\bar{M}$ is locally represented by the kernel of the 1-forms

$$(2.2) \qquad \delta y^i = dy^i + L^i_\alpha dx^\alpha.$$

We call $\{\delta/\delta x^\alpha, \partial/\partial y^i\}$ and $\{dx^\alpha, \delta y^i\}$ the *adapted frame* and the *adapted coframe* on $\bar{M}$, respectively. By direct calculations using (2.1) we obtain

$$(2.3) \qquad (a) \ \left[\tfrac{\delta}{\delta x^\beta}, \tfrac{\delta}{\delta x^\alpha}\right] = F^k_{\alpha\beta}\tfrac{\partial}{\partial y^k}, \quad (b) \ \left[\tfrac{\delta}{\delta x^\alpha}, \tfrac{\partial}{\partial y^i}\right] = L_i{}^k{}_\alpha \tfrac{\partial}{\partial y^k},$$

where we put

$$(2.4) \qquad (a) \ F^k_{\alpha\beta} = \frac{\delta L^k_\beta}{\delta x^\alpha} - \frac{\delta L^k_\alpha}{\delta x^\beta}, \quad (b) \ L_i{}^k{}_\alpha = \frac{\partial L^k_\alpha}{\partial y^i}.$$

Locally, the line element that represents the metric $\bar{g}$ is given by

$$(2.5) \qquad d\bar{s}^2 = g_{\alpha\beta}(x,y)dx^\alpha dx^\beta + g_{ij}(x,y)\delta y^i \delta y^j,$$

where we put:

(2.6) (a) $g_{\alpha\beta}(x,y) = \bar{g}\left(\dfrac{\delta}{\delta x^\alpha}, \dfrac{\delta}{\delta x^\beta}\right)$, (b) $g_{ij}(x,y) = \bar{g}\left(\dfrac{\partial}{\partial y^i}, \dfrac{\partial}{\partial y^j}\right)$.

Finally, we note that $g_{\alpha\beta}$ and g_{ij} satisfy the transformations:

(2.7) (a) $g_{\alpha\beta} = \tilde{g}_{\mu\nu} \dfrac{\partial \tilde{x}^\mu}{\partial x^\alpha} \dfrac{\partial \tilde{x}^\nu}{\partial x^\beta}$, (b) $g_{ij} = \tilde{g}_{hk} \dfrac{\partial \tilde{y}^h}{\partial y^i} \dfrac{\partial \tilde{y}^k}{\partial y^j}$,

with respect to the local changes of coordinates on $\bar{M}$ given by

(2.8) (a) $\tilde{x}^\alpha = \tilde{x}^\alpha(x^\mu)$, (b) $\tilde{y}^i = \tilde{y}^i(x^\mu, y^j)$.

3. Adapted Tensor Fields and Adapted Riemannian Connection

Let $(\bar{M}, \bar{g}, H\bar{M})$ be a general gauge KK space. In the present section we introduce the main geometric objects we need to develop a new point of view on higher dimensional KK theories.

First, we consider the dual vector bundles $H\bar{M}^*$ and $V\bar{M}^*$ of $H\bar{M}$ and $V\bar{M}$, respectively. Then, an $\mathcal{F}(\bar{M}) - (p+q+r+s)$-linear mapping

$$T : \Gamma(H\bar{M}^*)^p \times \Gamma(H\bar{M})^q \times \Gamma(V\bar{M}^*)^r \times \Gamma(V\bar{M})^s \longrightarrow \mathcal{F}(\bar{M}),$$

is an *adapted tensor field* of type (p, q; r, s). Note that any $\mathcal{F}(\bar{M}) - (q+r+s)$-linear mapping

$$T : \Gamma(H\bar{M})^q \times \Gamma(V\bar{M}^*)^r \times \Gamma(V\bar{M})^s \longrightarrow \Gamma(H\bar{M}),$$

defines an adapted tensor field of type $(1, q; r, s)$. Similarly, any $\mathcal{F}(\bar{M}) - (p+q+s)$-linear mapping

$$T : \Gamma(H\bar{M}^*)^p \times \Gamma(H\bar{M})^q \times \Gamma(V\bar{M})^s \longrightarrow \Gamma(V\bar{M}),$$

defines an adapted tensor field of type $(p, q; 1, s)$. More about adapted tensor fields can be found in [5].

Next, denote by h and v the projection morphisms of $T\bar{M}$ on $H\bar{M}$ and $V\bar{M}$, with respect to the Whitney sum

(3.1) $$T\bar{M} = H\bar{M} \oplus V\bar{M}.$$

Then, we consider the $\mathcal{F}(\bar{M})$-bilinear mapping

(3.2) $\begin{aligned} &F : \Gamma\left(H\bar{M}\right)^2 \to \Gamma\left(V\bar{M}\right); \\ &F(hX, hY) = -v[hX, hY], \quad \forall X, Y \in \Gamma(T\bar{M}). \end{aligned}$

Thus F is an adapted tensor field of type $(0, 2; 1, 0)$. Also, we define the $\mathcal{F}(\bar{M}) - 3$-linear mappings:

$$H : \Gamma\left(H\bar{M}\right)^2 \times \Gamma\left(V\bar{M}\right) \to \mathcal{F}(\bar{M}) \quad \text{and} \quad V : \Gamma\left(V\bar{M}\right)^2 \times \Gamma\left(H\bar{M}\right) \to \mathcal{F}(\bar{M}),$$

given by

(3.3) $\begin{aligned} H(hX, hY, vZ) &= \tfrac{1}{2}\{vZ\left(g(hX, hY)\right) - g(h[vZ, hX], hY) \\ &\quad - g(h[vZ, hY], hX)\}, \end{aligned}$

and

(3.4) $\begin{aligned} V(vX, vY, hZ) &= \tfrac{1}{2}\{hZ\left(g^\star(vX, vY)\right) - g^\star(v[hZ, vX], vY) \\ &\quad - g^\star(v[hZ, vY], vX)\}, \end{aligned}$

for all $X, Y, Z \in \Gamma(T\bar{M})$. Thus H and V given by (3.3) and (3.4) are adapted tensor fields of type $(0, 2; 0, 1)$ and $(0, 1; 0, 2)$, respectively. The Lorentz metric g on $H\bar{M}$ and the Riemannian metric g^* on $V\bar{M}$, enable us to define other two adapted tensor fields denoted by the same symbols H and V and given by

(3.5) $$g(hX, H(hY, vZ)) = H(hX, hY, vZ),$$

and

(3.6) $$g^*(vX, V(vY, hZ)) = V(vX, vY, hZ).$$

Now, let $\overset{h}{\nabla}$ and $\overset{v}{\nabla}$ be two linear connections on $H\bar{M}$ and $V\bar{M}$ respectively. Then $\nabla = (\overset{h}{\nabla}, \overset{v}{\nabla})$ given by

(3.7) $$\nabla_X Y = \overset{h}{\nabla}_Y hY + \overset{v}{\nabla}_X vY,$$

is called an *adapted linear connection* on $\bar{M}$. If moreover, ∇ satisfies

(3.8) $$(\nabla_X \bar{g})(Y, Z) = 0, \quad \forall X, Y, Z \in \Gamma(T\bar{M}),$$

is called a metric connection.

THEOREM 3.1 (cf.[**4**]). *Let $(\bar{M}, \bar{g}, H\bar{M})$ be a general gauge KK space. Then there exists a unique metric adapted linear connection $\nabla = (\overset{h}{\nabla}, \overset{v}{\nabla})$ whose torsion tensor field T is given by*

(3.9) $$\begin{aligned}(a) \quad & T(hX, hY) = F(hX, hY), \quad (b) \quad T(vX, vY) = 0, \\ (c) \quad & T(hX, vY) = V(vY, hX) - H(hX, vY),\end{aligned}$$

for all $X, Y \in \Gamma(T\bar{M})$.

Note that ∇ is given by

(3.10) $$\begin{aligned}(a) \quad & \overset{h}{\nabla}_{hX} hY = h\bar{\nabla}_{hX} hY, \\ (b) \quad & \overset{h}{\nabla}_{vX} hY = h[vX, hY] + H(hY, vX), \\ (c) \quad & \overset{v}{\nabla}_{vX} vY = v\bar{\nabla}_{vX} vY, \\ (d) \quad & \overset{v}{\nabla}_{hX} vY = v[hX, vY] + V(vY, hX),\end{aligned}$$

where $\bar{\nabla}$ is the Levi-Civita connection on $(\bar{M}, \bar{g}, H\bar{M})$. We call ∇ given by (3.10) the *Riemannian adapted connection*.

As we will apply these geometric objects to physics, we close this section with their local presentation. First, by using (2.3a) and (3.2), we obtain

(3.11) $$F(\frac{\delta}{\delta x^\alpha}, \frac{\delta}{\delta x^\beta}) = F^k_{\alpha\beta} \frac{\partial}{\partial y^k}.$$

Next, we put

(3.12) $$(a) \quad H(\frac{\delta}{\delta x^\beta}, \frac{\delta}{\delta x^\alpha}, \frac{\partial}{\partial y^i}) = H_{i\alpha\beta}, \quad (b) \quad V(\frac{\partial}{\partial y^j}, \frac{\partial}{\partial y^i}, \frac{\delta}{\delta x^\alpha}) = V_{\alpha ij},$$

and by using (3.3), (3.4) and (3.12) we obtain

(3.13) $$(a) \quad H_{i\alpha\beta} = \frac{1}{2} \frac{\partial g_{\alpha\beta}}{\partial y^i}, \quad (b) \quad V_{\alpha ij} = \frac{1}{2} \left\{ \frac{\delta g_{ij}}{\delta x^\alpha} - g_{kj} L_i{}^k{}_\alpha, -g_{ik} L_j{}^k{}_\alpha \right\}.$$

Finally, the Riemannian adapted connection is given by

(3.14)
(a) $\overset{h}{\nabla}_{\frac{\delta}{\delta x^\beta}} \frac{\delta}{\delta x^\alpha} = \Gamma_\alpha{}^\gamma{}_\beta \frac{\delta}{\delta x^\gamma}$, (b) $\overset{h}{\nabla}_{\frac{\partial}{\partial y^i}} \frac{\delta}{\delta x^\alpha} = \Gamma_\alpha{}^\gamma{}_i \frac{\delta}{\delta x^\gamma}$,

(c) $\overset{v}{\nabla}_{\frac{\partial}{\partial y^j}} \frac{\partial}{\partial y^i} = \Gamma_i{}^k{}_j \frac{\partial}{\partial y^k}$, (d) $\overset{v}{\nabla}_{\frac{\delta}{\delta x^\alpha}} \frac{\partial}{\partial y^i} = \Gamma_i{}^k{}_\alpha \frac{\partial}{\partial y^k}$,

where we put

(3.15)
(a) $\Gamma_\alpha{}^\gamma{}_\beta = \frac{1}{2} g^{\gamma\mu} \left\{ \frac{\delta g_{\mu\alpha}}{\delta x^\beta} + \frac{\delta g_{\mu\beta}}{\delta x^\alpha} - \frac{\delta g_{\alpha\beta}}{\delta x^\mu} \right\}$,

(b) $\Gamma_i{}^k{}_j = \frac{1}{2} g^{kh} \left\{ \frac{\partial g_{hi}}{\partial y^j} + \frac{\partial g_{hj}}{\partial y^i} - \frac{\partial g_{ij}}{\partial y^h} \right\}$,

(c) $\Gamma_\alpha{}^\gamma{}_i = H_{i\alpha}{}^\gamma$, (d) $\Gamma_i{}^k{}_\alpha = L_i{}^k{}_\alpha + V_{\alpha i}{}^k$.

4. 4D Equations of Motion in $(\bar{M}, \bar{g}, H\bar{M})$

The motions in $(\bar{M}, \bar{g}, H\bar{M})$ are given by the geodesics in $\bar{M}$ with respect to the Levi-Civita connection $\bar{\nabla}$. Locally, $\bar{\nabla}$ is expressed in terms of the geometric objects from the previous section as follows:

(4.1)
(a) $\bar{\nabla}_{\frac{\delta}{\delta x^\beta}} \frac{\delta}{\delta x^\alpha} = \Gamma_\alpha{}^\gamma{}_\beta \frac{\delta}{\delta x^\gamma} + \left(\frac{1}{2} F^k{}_{\alpha\beta} - H^k{}_{\alpha\beta} \right) \frac{\partial}{\partial y^k}$,

(b) $\bar{\nabla}_{\frac{\partial}{\partial y^i}} \frac{\delta}{\delta x^\alpha} = \left(H_{i\alpha}{}^\gamma + \frac{1}{2} F_{i\alpha}{}^\gamma \right) \frac{\delta}{\delta x^\gamma} + V_{\alpha i}{}^k \frac{\partial}{\partial y^k}$,

(c) $\bar{\nabla}_{\frac{\delta}{\delta x^\alpha}} \frac{\partial}{\partial y^i} = \left(H_{i\alpha}{}^\gamma + \frac{1}{2} F_{i\alpha}{}^\gamma \right) \frac{\delta}{\delta x^\gamma} + \Gamma_i{}^k{}_\alpha \frac{\partial}{\partial y^k}$,

(d) $\bar{\nabla}_{\frac{\partial}{\partial y^j}} \frac{\partial}{\partial y^i} = -V^\gamma{}_{ij} \frac{\delta}{\delta x^\gamma} + \Gamma_i{}^k{}_j \frac{\partial}{\partial y^k}$.

Next, consider a smooth curve $\bar{C}$ in $\bar{M}$ given locally by equations:

(4.2)
$$x^\alpha = x^\alpha(t), \quad y^i = y^i(t),$$
$$t \in [a, b], \quad \alpha \in \{0, 1, 2, 3\}, \quad i \in \{4, \cdots, 3+n\}.$$

The tangent vector field to $\bar{C}$ is expressed as follows with respect to the adapted frame field $\{\delta/\delta x^\alpha, \partial/\partial y^i\}$:

(4.3)
$$\frac{d}{dt} = \frac{dx^\alpha}{dt} \frac{\delta}{\delta x^\alpha} + \frac{\delta y^i}{\delta t} \frac{\partial}{\partial y^i},$$

where we put

(4.4)
$$\frac{\delta y^i}{\delta t} = \frac{dy^i}{dt} + L^i{}_\alpha \frac{dx^\alpha}{dt}.$$

Then by using (4.1) - (4.4) we deduce that the equations of motion in $(\bar{M}, \bar{g}, H\bar{M})$ are given by (cf. [**4**]):

(4.5)
(a) $\frac{d^2 x^\gamma}{dt^2} + \Gamma_\alpha{}^\gamma{}_\beta \frac{dx^\alpha}{dt} \frac{dx^\beta}{dt}$
$+ (2H_{i\alpha}{}^\gamma + F_{i\alpha}{}^\gamma) \frac{dx^\alpha}{dt} \frac{\delta y^i}{\delta t} - V^\gamma{}_{ij} \frac{\delta y^i}{\delta t} \frac{\delta y^j}{\delta t} = 0$,

(b) $\frac{d}{dt}\left(\frac{\delta y^k}{\delta t}\right) - H^k{}_{\alpha\beta} \frac{dx^\alpha}{dt} \frac{dx^\beta}{dt}$
$+ (\Gamma_i{}^k{}_\alpha + V_{\alpha i}{}^k) \frac{dx^\alpha}{dt} \frac{\delta y^i}{\delta t} + \Gamma_i{}^k{}_j \frac{\delta y^i}{\delta t} \frac{\delta y^j}{\delta t} = 0$.

We call (4.5a) the *4D equations of motion* on $(\bar{M}, \bar{g}, H\bar{M})$. To justify this name we consider two particular cases. First, suppose that the adapted tensor fields F, H and V vanish identically on $\bar{M}$. Then g becomes a metric on the base manifold and (4.5a) are just the equations of geodesics in the spacetime (M, g). Thus, in this case, *the projections of geodesics of $(\bar{M}, \bar{g})$ on M are geodesics of (M, g)*. Now, suppose that only H and V vanish identically on $\bar{M}$. Then (4.5a) becomes

$$(4.6) \qquad \frac{d^2 x^\gamma}{dt^2} + \Gamma_\alpha{}^\gamma{}_\beta(x^\mu) \frac{dx^\alpha}{dt} \frac{dx^\beta}{dt} + F_{i\alpha}{}^\gamma(x, y) \frac{dx^\alpha}{dt} \frac{\delta y^i}{\delta t} = 0,$$

which shows the existence of an extra force given by

$$(4.7) \qquad F^\gamma(x, y) = -F_{i\alpha}{}^\gamma(x, y) \frac{dx^\alpha}{dt} \frac{\delta y^i}{\delta t}.$$

Moreover, this force is perpendicular to the 4D velocity

$$U(t) = \frac{dx^\alpha}{dt} \frac{\delta}{\delta x^\alpha}.$$

This enables us to call (4.6) *the Lorentz force equations induced in the spacetime (M, g)*.

Among the geodesics of $(\bar{M}, \bar{g}, H\bar{M})$ there is a category which seems to have a great role in the dynamics developed in a higher dimensional KK theory. We say that a curve $\bar{C}$ in $\bar{M}$ is a *horizontal geodesic* if satisfies (4.5) and it is tangent to $H\bar{M}$ at any of its points. Then we deduce that $\bar{C}$ is a horizontal geodesic if and only if it satisfies the system

$$(4.8) \qquad \begin{array}{ll} (a) & \frac{d^2 x^\gamma}{dt^2} + \Gamma_\alpha{}^\gamma{}_\beta(x, y) \frac{dx^\alpha}{dt} \frac{dx^\beta}{dt} = 0, \\ (b) & H^k{}_{\alpha\beta} \frac{dx^\alpha}{dt} \frac{dx^\beta}{dt} = 0. \end{array}$$

Now, observe that horizontal geodesics in $\bar{M}$ are projectable on M. We call the projection C of $\bar{C}$ on M the *induced motion* on M by the motion $\bar{C}$ on $\bar{M}$. According to the two particular cases presented above, we may claim that *induced motions on M bring more information than both the motions from general relativity and the solutions of the Lorentz force equations*. This is due to the existence of extra dimensions and to the action of the Lie group G on $\bar{M}$.

5. 4D Einstein Gravitational Tensor field

Consider the Riemannian adapted connection ∇ and by using the Bianchi identities, we obtain

$$(5.1) \qquad \begin{array}{ll} (a) & \displaystyle\sum_{(hX, hY, hZ)} \{R(hX, hY, hZ) - H(hX, F(hY, hZ))\} = 0, \\ (b) & \displaystyle\sum_{(hX, hY, hZ)} \{(\nabla_{hX} R)(hY, hZ, hU) \\ & \qquad + R(F(hX, hY), hZ, hU)\} = 0, \end{array}$$

where R is the curvature tensor field. As the identities in (5.1) refer only to the horizontal vector fields, we call them the *4D Bianchi identities*. Now, consider

$$R(X, Y, Z, U) = \bar{g}(R(X, Y, U), Z),$$

and define the *4D Ricci tensor field*

$$\text{(5.2)} \quad Ric(hX, hY) = \sum_{\gamma=0}^{3} \epsilon_\gamma R(E_\gamma, hX, E_\gamma, hY),$$

and the *internal Ricci tensor*

$$\text{(5.3)} \quad Ric(vX, vY) = \sum_{k=4}^{n+3} R(E_k, vX, E_k, vY),$$

where $\{E_\gamma\}$ and $\{E_k\}$ are orthogonal bases in $\Gamma(H\bar{M})$ and $\Gamma(V\bar{M})$, respectively, and ϵ_γ is the signature of E_γ. This enables us to define the *4D scalar curvature* $\overset{h}{\mathbf{R}}$ and the *4D internal scalar curvature* $\overset{v}{\mathbf{R}}$ by

$$\text{(5.4)} \quad (a) \ \overset{h}{\mathbf{R}} = \sum_{\gamma=0}^{3} \epsilon_\gamma Ric(E_\gamma, E_\gamma), \quad (b) \ \overset{v}{\mathbf{R}} = \sum_{k=4}^{n+3} Ric(E_k, E_k).$$

Finally, define the horizontal and vertical tensor fields G and $G^\star$ by

$$\text{(5.5)} \quad \begin{aligned} (a) & \quad G(hX, hY) = Ric(hX, hY) - \tfrac{\overset{h}{\mathbf{R}}}{2} g(hX, hY), \\ (b) & \quad G^\star(vX, vY) = Ric(vX, vY) - \tfrac{\overset{v}{\mathbf{R}}}{2} g^\star(vX, vY). \end{aligned}$$

We call G the *4D Einstein gravitational tensor field*.

Next, we need all these objects expressed by their local components. First, we put

$$\text{(5.6)} \quad \begin{aligned} (a) & \quad R\left(\tfrac{\delta}{\delta x^\gamma}, \tfrac{\delta}{\delta x^\beta}, \tfrac{\delta}{\delta x^\alpha}\right) = R_\alpha{}^\mu{}_{\beta\gamma} \tfrac{\delta}{\delta x^\mu}, \\ (b) & \quad R\left(\tfrac{\partial}{\partial y^i}, \tfrac{\delta}{\delta x^\beta}, \tfrac{\delta}{\delta x^\alpha}\right) = R_\alpha{}^\mu{}_{\beta i} \tfrac{\delta}{\delta x^\mu}. \end{aligned}$$

Then by using (3.11), (3.12a) and (5.6) into (5.1) we obtain the local expressions of the *4D Bianchi identities*

$$\text{(5.7)} \quad \begin{aligned} (a) & \sum_{(\alpha,\beta,\gamma)} \left\{ R_\alpha{}^\mu{}_{\beta\gamma} + F^k{}_{\alpha\beta} H_k{}^\mu{}_\gamma \right\} = 0, \\ (b) & \sum_{(\alpha,\beta,\gamma)} \left\{ R_\nu{}^\mu{}_{\alpha\beta|\gamma} - R_\nu{}^\mu{}_{\alpha k} F^k{}_{\beta\gamma} \right\} = 0. \end{aligned}$$

Now, we say that $(\bar{M}, \bar{g}, H\bar{M})$ is a *relativistic gauge KK space* if the following conditions are satisfied

$$\text{(A)} \sum_{(\alpha,\beta,\gamma)} \left\{ F^k{}_{\alpha\beta} H_k{}^\mu{}_\gamma \right\} = 0, \quad \text{(B)} \sum_{(\alpha,\beta,\gamma)} \left\{ R_\nu{}^\mu{}_{\alpha k} F^k{}_{\beta\gamma} \right\} = 0.$$

Note that most of the higher dimensional KK spaces are relativistic. Suppose that the Lie group G admits n Killing vector fields as generators. Then $H = 0$ on $\bar{M}$, and both conditions (A) and (B) are satisfied. This is the case which was intensively studied in earlier literature (cf. [6], [7], [8], [9], [11], [14], [15]). Another important class of rgKK spaces is the one for which $H\bar{M}$ is integrable. Then $F = 0$ on $\bar{M}$ and thus (A) and (B) are also satisfied. Examples of such spaces can be found in the book of Wesson [16] and in the survey paper of Overduin and Wesson [13].

Now, we put

$$R_{\alpha\beta} = Ric\left(\frac{\delta}{\delta x^\beta}, \frac{\delta}{\delta x^\alpha}\right), \quad R_{ij} = Ric\left(\frac{\partial}{\partial y^j}, \frac{\partial}{\partial y^i}\right),$$

$$G_{\alpha\beta} = G\left(\frac{\delta}{\delta x^\beta}, \frac{\delta}{\delta x^\alpha}\right), \quad G_{ij} = G^\star\left(\frac{\partial}{\partial y^j}, \frac{\partial}{\partial y^i}\right),$$

and obtain

(5.8) $\quad$ (a) $\quad G_{\alpha\beta} = R_{\alpha\beta} - \dfrac{\overset{h}{\mathbf{R}}}{2} g_{\alpha\beta}, \quad$ (b) $\quad G_{ij} = R_{ij} - \dfrac{\overset{v}{\mathbf{R}}}{2} g_{ij}.$

Finally, by using (A) and (B) we deduce that *the 4D Einstein gravitational tensor field is symmetric and of horizontal divergence zero.*

6. 4D Einstein Equations in a rgKK space $(\bar{M}, \bar{g}, H\bar{M})$

Denote by $\bar{R}$ and R the curvature tensor fields of $\bar{\nabla}$ and ∇, respectively. Then the structure equations for the immersions of the distributions $H\bar{M}$ and $V\bar{M}$ in the tangent bundle of a rgKK space are given by

(6.1)

(a) $\quad \bar{R}_{\alpha\beta\gamma\mu} = R_{\alpha\beta\gamma\mu} + H^k{}_{\alpha\mu} H_{k\beta\gamma} - H^k{}_{\alpha\gamma} H_{k\beta\mu}$

$\quad\quad\quad + \frac{1}{4}\left\{F^k{}_{\alpha\gamma} F_{k\mu\beta} - F^k{}_{\alpha\mu} F_{k\gamma\beta} - 2F^k{}_{\gamma\mu} F_{k\alpha\beta}\right\},$

(b) $\quad \bar{R}_{\alpha i\gamma\mu} = \frac{1}{2}\left\{F_{i\alpha\gamma|\mu} - F_{i\alpha\mu|\gamma}\right\} + H_{i\alpha\mu|\gamma} - H_{i\alpha\gamma|\mu} - V_{\alpha i k} F^k{}_{\gamma\mu},$

(c) $\quad \bar{R}_{\alpha i\gamma j} = \frac{1}{2} F_{i\alpha\gamma|j} - H_{i\alpha\gamma|j} - V_{\alpha ij|\gamma} + \frac{1}{2}\left\{F_{i\alpha\mu} H_{j\gamma}{}^\mu\right.$

$\quad\quad\quad \left. + H_{i\gamma\mu} F_{j\alpha}{}^\mu\right\} - H_{i\alpha\mu} H_{j\gamma}{}^\mu - V_{\alpha i k} V_{\gamma j}{}^k + \frac{1}{4} F_{i\gamma\mu} F_{j\alpha}{}^\mu,$

(d) $\quad \bar{R}_{i\alpha k h} = V_{\alpha i h|k} - V_{\alpha i k|h} + \frac{1}{2}\left\{V^\mu{}_{ih} F_{k\mu\alpha} - V^\mu{}_{ik} F_{h\mu\alpha}\right\},$

(e) $\quad \bar{R}_{ijkh} = R_{ijkh} + V^\mu{}_{ih} V_{\mu kj} - V^\mu{}_{ik} V_{\mu hj}.$

Now, by using (6.1) and the 4D and internal Ricci tensors, we deduce that the local components of the Ricci tensor of $\bar{\nabla}$ with respect to the adapted frame field $\{\delta/\delta x^\alpha, \partial/\partial y^i\}$ are given by

(6.2)

(a) $\quad \bar{R}_{\alpha\beta} = R_{\alpha\beta} - \frac{1}{2} F^k{}_{\alpha\mu} F_{k\beta}{}^\mu - H^k{}_{\alpha\beta} TrH_k - V_{\alpha h k} V_\beta{}^{hk} - H^k{}_{\alpha\beta|k}$

$\quad\quad\quad + \frac{1}{2}\left\{F^k{}_{\alpha\mu} H_{k\beta}{}^\mu + F^k{}_{\beta\mu} H_{k\alpha}{}^\mu - (TrV_\alpha)_{|\beta} - (TrV_\beta)_{|\alpha}\right\},$

(b) $\quad \bar{R}_{\alpha i} = \frac{1}{2} F_{i\alpha}{}^\mu{}_{|\mu} + H_{i\alpha}{}^\mu{}_{|\mu} - (TrH_i)_{|\alpha} - (TrV_\alpha)_{|i}$

$\quad\quad\quad + V_{\alpha i}{}^k{}_{|k} + \frac{1}{2}\left\{F_{i\alpha}{}^\mu TrV_\mu + F^k{}_{\alpha\mu} V^\mu{}_{ki}\right\},$

(c) $\quad \bar{R}_{ij} = R_{ij} + \frac{1}{4} F_{i\mu\nu} F_j{}^{\mu\nu} - H_{i\mu\nu} H_j{}^{\mu\nu}$

$\quad\quad\quad - V^\mu{}_{ij} TrV_\mu - V^\mu{}_{ij|\mu} - (TrH_i)_{|j},$

where we put

$$TrH_i = H_{i\alpha\beta} g^{\alpha\beta} \quad \text{and} \quad TrV_\alpha = V_{\alpha ij} g^{ij}.$$

Due to (6.2), the scalar curvature $\bar{\mathbf{R}}$ of $\bar{\nabla}$ is expressed as follows

(6.3) $\quad\quad\quad \bar{\mathbf{R}} = \overset{h}{\mathbf{R}} + \overset{v}{\mathbf{R}} - D - \frac{1}{4} F_{k\mu\nu} F^{k\mu\nu},$

where we put

$$D = H_{k\mu\nu}H^{k\mu\nu} + V_{\mu hk}V^{\mu hk} + TrH^k TrH_k + TrV^\mu TrV_\mu$$
$$+ 2\left\{(TrH^k)_{|k} + (TrV^\mu)_{|\mu}\right\}.$$

Finally, the local components of the Einstein gravitational tensor field $\bar{G}$ of $\bar{\nabla}$ are given by

(6.4)
(a) $\bar{G}_{\alpha\beta} = G_{\alpha\beta} - \frac{1}{2}(\overset{v}{\mathbf{R}} - D)g_{\alpha\beta} + \frac{1}{2}E_{\alpha\beta} - H^k{}_{\alpha\beta}TrH_k$
$\quad - V_{\alpha hk}V_\beta{}^{hk} - H^k{}_{\alpha\beta|k} + \frac{1}{2}\left\{F^k{}_{\alpha\mu}H_{k\beta}{}^\mu + F^k{}_{\beta\mu}H_{k\alpha}{}^\mu\right.$
$\quad \left. - (TrV_\alpha)_{|\beta} - (TrV_\beta)_{|\alpha}\right\},$

(b) $\bar{G}_{\alpha i} = \frac{1}{2}F_{i\alpha}{}^\mu{}_{|\mu} + H_{i\alpha}{}^\mu{}_{|\mu} - (TrH_i)_{|\alpha} - (TrV_\alpha)_{|i}$
$\quad + V_{\alpha i}{}^k{}_{|k} + \frac{1}{2}\left\{F_{i\alpha}{}^\mu TrV_\mu + F^k{}_{\alpha\mu}V^\mu{}_{ki}\right\},$

(c) $\bar{G}_{ij} = G_{ij} - \frac{1}{2}(\overset{v}{\mathbf{R}} - D)g_{ij} + \frac{1}{4}E_{ij} - H_{i\mu\nu}H_j{}^{\mu\nu}$
$\quad - V^\mu{}_{ij}TrV_\mu - V^\mu{}_{ij|\mu} - (TrH_i)_{|j},$

where we put

$$E_{\alpha\beta} = \frac{1}{4}g_{\alpha\beta}F_{k\mu\nu}F^{k\mu\nu} - F^k{}_{\alpha\mu}F_{k\beta}{}^\mu,$$
$$E_{ij} = \frac{1}{2}g_{ij}F_{k\mu\nu}F^{k\mu\nu} + F_{i\mu\nu}F_j{}^{\mu\nu}.$$

It is well known that the STM theory of relativity developed by Wesson and his co-workers (cf. [**13**], [**16**]) is an approach for $5D$ relativity which asserts that the $4D$ matter is a consequence of the geometry of a $5D$ space. According to the theory we developed here, we may claim that the $4D$ matter is a consequence of the geometry of a $(4+n)$-dimensional rgKK space $(\bar{M}, \bar{g}, H\bar{M})$. As in STM theory, we consider the vacuum Einstein equations in $(\bar{M}, \bar{g}, H\bar{M})$ and by using (6.4) obtain

(6.5)
(a) $G_{\alpha\beta} = \overset{h}{\Lambda} g_{\alpha\beta} + \overset{h}{k} T_{\alpha\beta}$

(b) $\frac{1}{2}F_{i\alpha}{}^\mu{}_{|\mu} + H_{i\alpha}{}^\mu{}_{|\mu} - (TrH_i)_{|\alpha} - (TrV_\alpha)_{|i}$
$\quad + V_{\alpha i}{}^k{}_{|k} + \frac{1}{2}\left\{F_{i\alpha}{}^\mu TrV_\mu + F^k{}_{\alpha\mu}V^\mu{}_{ki}\right\} = 0,$

(c) $G_{ij} = \overset{v}{\lambda} g_{ij} + \overset{v}{k} T_{ij},$

where $\overset{h}{k}$ and $\overset{v}{k}$ are constants, and we put

(6.6)
(a) $\overset{h}{\Lambda} = \tfrac{1}{2}(\overset{v}{\mathbf{R}} - D)$, (b) $\overset{v}{\Lambda} = \tfrac{1}{2}(\overset{h}{\mathbf{R}} - D)$,

(c) $\overset{h}{k} T_{\alpha\beta} = -\tfrac{1}{2}E_{\alpha\beta} + H^k{}_{\alpha\beta}TrH_k + V_{\alpha hk}V_\beta{}^{hk}$
$\quad + H^k{}_{\alpha\beta|k} - \tfrac{1}{2}\left\{F^k{}_{\alpha\mu}H_{k\beta}{}^\mu + F^k{}_{\beta\mu}H_{k\alpha}{}^\mu\right.$
$\quad \left. - (TrV_\alpha)_{|\beta} - (TrV_\beta)_{|\alpha}\right\}$,

(d) $\overset{v}{k} T_{ij} = -\tfrac{1}{4}E_{ij} + H_{i\mu\nu}H_j{}^{\mu\nu} + V^\mu{}_{ij}TrV_\mu + V^\mu{}_{ij|\mu} + (TrH_i)_{|j}$.

We call (6.5a) the *4D Einstein equations* in $(\bar{M}, \bar{g}, H\bar{M})$. We also call $\overset{h}{\Lambda}$ and $T_{\alpha\beta}$ the *4D cosmological function* and the *4D stress-energy tensor field* in $(\bar{M}, \bar{g}, H\bar{M})$, respectively.

In particular, if the adapted tensor field H vanishes identically on $\bar{M}$, then from (3.13a) we deduce that the local components $g_{\alpha\beta}$ of the Lorentz metric g on $H\bar{M}$ are independent of (y^i). This enables us to consider the base manifold M endowed with the Lorentz metric $g = g_{\alpha\beta}(x^\mu)$. Also, $\overset{h}{\Lambda}$ and $\overset{h}{k}T_{\alpha\beta}$ are given by

(6.7) $\overset{h}{\Lambda} = -\dfrac{1}{2}\left\{\dfrac{1}{4}F_{k\mu\nu}F^{k\mu\nu} + V_{\mu hk}V^{\mu hk} + (TrV^\mu)_{|\mu}\right\}$,

and

(6.8) $\overset{h}{k} T_{\alpha\beta} = -\dfrac{1}{2}E_{\alpha\beta} + V_{\alpha hk}V_\beta{}^{hk} + \dfrac{1}{2}\left\{(TrV_\alpha)_{|\beta} + (TrV_\beta)_{|\alpha}\right\}$.

As $G_{\alpha\beta}$ are independent of (y^i) from (6.5a) we conclude that the 4D cosmological function and the 4D stress-energy tensor field must satisfy the constraints

(6.9) $\dfrac{\partial \overset{h}{\Lambda}}{\partial y^i}g_{\alpha\beta} + \overset{h}{k}\dfrac{\partial T_{\alpha\beta}}{\partial y^i} = 0$.

This proves once more that the "cylinder condition" from previous higher dimensional Kaluza-Klein theories (cf. [6], [7], [8], [9], [11], [14], [15]) should be removed. As a consequence, the constraints (6.9) do not appear into the study.

7. Conclusions

In the first part of the paper we presented the fully general equations of motion in a $(4+n)$-dimensional gauge Kaluza-Klein space (cf. (4.5)). The 4D equations of motion (4.5a) have some extra terms constructed by means of adapted tensor fields $(F_{i\alpha}{}^\gamma, H_{i\alpha}{}^\gamma, V^\gamma{}_{ij})$, which can be used to test the theory. This means that the dynamics in such a space is under the effect of an extra force whose existence is guaranteed by the extra dimensions. In a particular case (see Section 4) we have seen that such a force does not contradict the 4D physics. It remains as open question whether this result is still valid in the general case.

The 4D Einstein equations (6.5a) induced by the vacuum Einstein equations in a $(4+n)D$ rgKK space are presented in the second part of the paper. At the present stage, physical interpretations of these equations are difficult to point out.

However, they support the Einstein's vision that the "base-wood" of matter can be transformed in the "marble" of the geometry.

Now, we note that this is the first study of both the $4D$ equations of motion and the $4D$ Einstein equations in a higher dimensional KK theory, wherein both conditions (i) and (ii) from Introduction have been removed. Thus some results from earlier papers are particular cases of the corresponding ones stated in the present paper. For $n = 1$, the equations of motion from (4.5) are the equations in (5.6) from [**2**]. The formula (6.3) for the scalar curvature $\widetilde{\mathbf{R}}$ of $\widetilde{M}$ is a generalization of the following formulas:(2.6) of [**11**], (24) of [**6**], (4.6) of [**14**] and (3.5.7) of [**8**]. Also, the $4D$ Einstein equations from the $5D$ STM theory obtained in [**17**] and [**3**], are particular cases of the equations (6.5a) from the present paper.

Finally, we should mention that we concentrate our study mainly on the mathematical part of higher dimensional Kaluza-Klein theories. Further studies should be developed on both the $4D$ equations of motion and the $4D$ Einstein equations, in order to find out the interrelations between extra dimensions and real matter.

References

[1] A. Bejancu, *A new point of view on general Kaluza-Klein theories*, Progress of Theoretical Physics, vol.128, no.3,(2012), 541-585.

[2] Aurel Bejancu, *4D equations of motion and fifth force in a 5D general Kaluza-Klein space*, Gen. Relativity Gravitation **45** (2013), no. 11, 2273–2290, DOI 10.1007/s10714-013-1592-z. MR3115476

[3] Aurel Bejancu, *4D Einstein equations in a 5D general Kaluza-Klein space*, Gen. Relativity Gravitation **45** (2013), no. 11, 2291–2308, DOI 10.1007/s10714-013-1591-0. MR3115477

[4] Aurel Bejancu, *On higher dimensional Kaluza-Klein theories*, Adv. High Energy Phys. (2013), Art. ID 148417, 12. MR3141431

[5] Aurel Bejancu and Hani Reda Farran, *Foliations and geometric structures*, Mathematics and Its Applications (Springer), vol. 580, Springer, Dordrecht, 2006. MR2190039 (2006j:53034)

[6] Y. M. Cho, *Higher-dimensional unifications of gravitation and gauge theories*, J. Mathematical Phys. **16** (1975), no. 10, 2029–2035. MR0384043 (52 #4920)

[7] Y. M. Cho and P. G. O. Freund, *Average energy expended per ion pair in liquid xenon*, Phys. Rev. D, vol. 12, no.5 (1975), 1771-1775.

[8] R. Coquereaux and A. Jadczyk, *Geometry of multidimensional universes*, Comm. Math. Phys. **90** (1983), no. 1, 79–100. MR714613 (85e:83053)

[9] B. De Witt, *Dynamical Theories of Groups and Fields*, Gordon and Breach, New York, 1965.

[10] A. Einstein and P. Bergmann, *On a generalization of Kaluza's theory of electricity*, Ann. of Math. (2) **39** (1938), no. 3, 683–701, DOI 10.2307/1968642. MR1503432

[11] Ryszard Kerner, *Generalization of the Kaluza-Klein theory for an arbitrary non-abelian gauge group* (English, with French summary), Ann. Inst. H. Poincaré Sect. A (N.S.) **9** (1968), 143–152. MR0264961 (41 #9550)

[12] R. Maartens and K. Koyama, *Brane-world gravity*, Living Rev.Relativ., vol.13, no.5(2010), 1-124.

[13] J. M. Overduin and P. S. Wesson, *Kaluza-Klein gravity*, Phys. Rep. **283** (1997), no. 5-6, 303–378, DOI 10.1016/S0370-1573(96)00046-4. MR1450560 (98g:83096)

[14] Abdus Salam and J. Strathdee, *On Kaluza-Klein theory*, Ann. Physics **141** (1982), no. 2, 316–352, DOI 10.1016/0003-4916(82)90291-3. MR673985 (83m:83053)

[15] A. Trautman, *Fibre bundles associated with space-time*, Rep. Mathematical Phys. **1** (1970/1971), no. 1, 29–62. MR0281472 (43 #7189)

[16] Paul S. Wesson, *Space—time—matter: modern Kaluza-Klein theory*, World Scientific Publishing Co., Inc., River Edge, NJ, 1999. MR1704285 (2000k:83054)

[17] Paul S. Wesson and J. Ponce de León, *Kaluza-Klein equations, Einstein's equations, and an effective energy-momentum tensor*, J. Math. Phys. **33** (1992), no. 11, 3883–3887, DOI 10.1063/1.529834. MR1185866 (93i:83087)

Department of Mathematics, Kuwait University, Kuwait
E-mail address: aurel.bejancu@ku.edu.kw

Location and size estimation of small rigid bodies using elastic far-fields

Fadhel Al-Musallam, Durga Prasad Challa, and Mourad Sini

ABSTRACT. We are concerned with the linearized, isotropic and homogeneous elastic scattering problem by (possibly many) small rigid obstacles of arbitrary Lipschitz regular shapes in 3D. Based on the Foldy-Lax approximation, valid under a sufficient condition on the number of the obstacles, the size and the minimum distance between them, we show that any of the two body waves, namely the pressure waves P or the shear waves S, is enough for solving the inverse problem of detecting these scatterers and estimating their sizes. Further, it is also shown that the shear-horizontal part SH or the shear vertical part SV of the shear waves S are also enough for the location detection and the size estimation. Under some extra assumption on the scatterers, as the convexity assumption, we derive finer size estimates as the radius of the largest ball contained in each scatterer and the one of the smallest ball containing it. The two estimates measure, respectively, the thickness and length of each obstacle.

1. Introduction and statement of the results

Let $B_1, B_2, \ldots, B_M$ be M open, bounded and simply connected sets in $\mathbb{R}^3$ with Lipschitz boundaries, containing the origin. We assume that their sizes and Lipschitz constants are uniformly bounded. We set $D_m := \epsilon B_m + z_m$ to be the small bodies characterized by the parameter $\epsilon > 0$ and the locations $z_m \in \mathbb{R}^3$, $m = 1, \ldots, M$.

Assume that the Lamé coefficients λ and μ are constants satisfying $\mu > 0$ and $3\lambda + 2\mu > 0$. Let U^i be a solution of the Navier equation $(\Delta^e + \omega^2)U^i = 0$ in $\mathbb{R}^3$, $\Delta^e := (\mu\Delta + (\lambda + \mu)\nabla \operatorname{div})$. We denote by U^s the elastic field scattered by the M small bodies $D_m \subset \mathbb{R}^3$ due to the incident field U^i. We restrict ourselves to the scattering by rigid bodies. Hence the total field $U^t := U^i + U^s$ satisfies the following exterior Dirichlet problem of the elastic waves

$$(1.1) \qquad (\Delta^e + \omega^2)U^t = 0 \text{ in } \mathbb{R}^3 \setminus \left(\bigcup_{m=1}^M \bar{D}_m \right),$$

$$(1.2) \qquad U^t|_{\partial D_m} = 0, \ 1 \leq m \leq M$$

2010 *Mathematics Subject Classification.* Primary 35R30, 35P25, 35Q61.

Key words and phrases. Elastic wave scattering, small-scatterers, Foldy-Lax approximation, capacitance, MUSIC algorithm.

with the Kupradze radiation conditions (K.R.C)

(1.3) $\quad \lim_{|x|\to\infty} |x|^{\frac{d-1}{2}}(\frac{\partial U_p}{\partial |x|} - i\kappa_{p^\omega} U_p) = 0$, and $\lim_{|x|\to\infty} |x|^{\frac{d-1}{2}}(\frac{\partial U_s}{\partial |x|} - i\kappa_{s^\omega} U_s) = 0$,

where the two limits are uniform in all the directions $\hat{x} := \frac{x}{|x|} \in \mathbb{S}^2$ and $\mathbb{S}^2$ is the unit sphere. Also, we denote $U_p := -\kappa_{p^\omega}^{-2}\nabla(\nabla \cdot U^s)$ to be the longitudinal (or the pressure or P) part of the field u and $U_s := \kappa_{s^\omega}^{-2}\nabla \times (\nabla \times U^s)$ to be the transversal (or the shear or S) part of the field U^s corresponding to the Helmholtz decomposition $U^s = U_p + U_s$. The constants $\kappa_{p^\omega} := \frac{\omega}{c_p}$ and $\kappa_{s^\omega} := \frac{\omega}{c_s}$ are known as the longitudinal and transversal wavenumbers, $c_p := \sqrt{\lambda + 2\mu}$ and $c_s := \sqrt{\mu}$ are the corresponding phase velocities, respectively and ω is the frequency.

The scattering problem (1.1-1.3) is well posed in the Hölder or Sobolev spaces, see [3, 10, 13] for instance, and the scattered field U^s has the following asymptotic expansion:

(1.4) $\quad U^s(x) := \frac{e^{i\kappa_{p^\omega}|x|}}{|x|} U_p^\infty(\hat{x}) + \frac{e^{i\kappa_{s^\omega}|x|}}{|x|} U_s^\infty(\hat{x}) + O(\frac{1}{|x|^2}), \; |x| \to \infty$

uniformly in all directions $\hat{x} \in \mathbb{S}^{d-1}$. The longitudinal part of the far-field, i.e. $U_p^\infty(\hat{x})$ is normal to $\mathbb{S}^2$ while the transversal part $U_s^\infty(\hat{x})$ is tangential to $\mathbb{S}^2$. As usual in scattering problems we use plane incident waves in this work. For the Lamé system, the full plane incident wave is of the form $U^i(x, \theta) := \alpha\theta\, e^{i\kappa_{p^\omega}\theta \cdot x} + \beta\theta^\perp e^{i\kappa_{s^\omega}\theta \cdot x}$, where $\theta^\perp$ is any direction in $\mathbb{S}^2$ perpendicular to the incident direction $\theta \in \mathbb{S}^2$, α, β are arbitrary constants. In particular, the pressure and shear incident waves are given as follows;

(1.5) $\quad U^{i,p}(x,\theta) := \theta e^{i\kappa_{p^\omega}\theta \cdot x}$ and $U^{i,s}(x,\theta) := \theta^\perp e^{i\kappa_{s^\omega}\theta \cdot x}$.

Pressure incident waves propagate in the direction of θ, whereas shear incident waves propagate in the direction of $\theta^\perp$. In the two dimensional case, the shear waves have only one direction. But in the three dimensional case, they have two orthogonal components called vertical and horizontal shear directions denoted by $\theta^{\perp v}$ and $\theta^{\perp h}$ respectively. So, $\theta^\perp = \tilde{\theta}^\perp/|\tilde{\theta}^\perp|$ with $\tilde{\theta}^\perp := \alpha\theta^{\perp h} + \beta\theta^{\perp v}$ for arbitrary constants α and β. To give the explicit forms of $\theta^{\perp h}$ and $\theta^{\perp v}$, we recall the Euclidean basis $\{e_1, e_2, e_3\}$ where $e_1 := (1, 0, 0)^T, e_2 := (0, 1, 0)^T$ and $e_3 := (0, 0, 1)^T$, write $\theta := (\theta_x, \theta_y, \theta_z)^T$ and set $r^2 := \theta_x^2 + \theta_y^2$. Let $\mathcal{R}_3 = \mathcal{R}_3(\theta)$ be the rotation map transforming θ to e_3. Then in the basis $\{e_1, e_2, e_3\}$, $\mathcal{R}_3$ is given by the matrix

(1.6) $\quad \mathcal{R}_3 = \frac{1}{r^2}\begin{bmatrix} \theta_2^2 + \theta_1^2\theta_z & -\theta_x\theta_y(1-\theta_z) & -\theta_x r^2 \\ -\theta_x\theta_y(1-\theta_z) & \theta_1^2 + \theta_2^2\theta_z & -\theta_y r^2 \\ \theta_x r^2 & \theta_y r^2 & \theta_z r^2 \end{bmatrix}.$

It satisfies $\mathcal{R}_3^T \mathcal{R}_3 = I$ and $\mathcal{R}_3 \theta = e_3$. Correspondingly, we write $\theta^{\perp h} := \mathcal{R}_3^T e_1$ and $\theta^{\perp v} := \mathcal{R}_3^T e_2$. These two directions represent the horizontal and the vertical directions of the shear wave and they are given by

(1.7)
$\theta^{\perp h} = \frac{1}{r^2}(\theta_y^2 + \theta_x^2\theta_z, \theta_x\theta_y(\theta_z - 1), -r^2\theta_x)^\top, \; \theta^{\perp v} = \frac{1}{r^2}(\theta_x\theta_y(\theta_z - 1), \theta_x^2 + \theta_y^2\theta_z, -r^2\theta_y)^\top.$

The functions $U_p^\infty(\hat{x}, \theta) := U_p^\infty(\hat{x})$ and $U_s^\infty(\hat{x}, \theta) := U_s^\infty(\hat{x})$ for $(\hat{x}, \theta) \in \mathbb{S}^2 \times \mathbb{S}^2$ are called the P-part and the S-part of the far-field pattern respectively.

DEFINITION 1.1. *We define*

(1) $a := \max\limits_{1 \leq m \leq M} diam(D_m) \left[= \epsilon \max\limits_{1 \leq m \leq M} diam(B_m) \right]$,

(2) $d := \min\limits_{\substack{m \neq j \\ 1 \leq m,j \leq M}} d_{mj}$, where $d_{mj} := dist(D_m, D_j)$.

(3) $\omega_{\max}$ *as the upper bound of the used frequencies, i.e.* $\omega \in [0, \omega_{\max}]$.

(4) Ω *to be a bounded domain in* $\mathbb{R}^3$ *containing the small bodies* D_m, $m = 1, \ldots, M$.

Our goal in this work is to justify the following results.

THEOREM 1.2. *The matrix* $(U_p^\infty(\hat{x}_j, \theta_l))_{j,l=1}^N$ *(or* $(U_s^\infty(\hat{x}_j, \theta_l))_{j,l=1}^N$*), for N large enough, corresponding to one of the incident waves in* (1.5) *is enough to localize the centers z_j and estimate the sizes of the obstacles D_j, $j = 1, \ldots, M$.*

Further, the matrix $(U_{SV}^\infty(\hat{x}_j, \theta_l))_{j,l=1}^N$ *(or* $(U_{SH}^\infty(\hat{x}_j, \theta_l))_{j,l=1}^N$*) is also enough to localize the obstacles and estimate their sizes. Here $U_{SV}^\infty(\cdot, \cdot)$ and $U_{SH}^\infty(\cdot, \cdot)$ are respectively the Shear-Horizontal and the Shear-Vertical parts of the shear parts of the far-fields.*

The approach we use to justify these results is based on two steps.

(1) In the first step, we derive the asymptotic expansion of the far-fields in terms of the three parameters modeling the collection of scatterers, namely M, a and d. This is sometimes called the Foldy-Lax approximation.

(2) We use the dominant term of this approximation coupled with the so-called MUSIC algorithm to detect the locations of the obstacles.

This approach is known for a decade, see [5] for instance. Our contribution to this approach is twofold corresponding to the two steps mentioned above. Regarding the first step, we provide the asymptotic expansion in terms of the three parameters M, a and d, while in the previous literature the two parameters M and d are assumed to be fixed, [3, 4]. This expansion is justified for the Lamé model, under consideration here, in our previous work [7]. Regarding the second step, which is the object of this paper, we apply the MUSIC algorithm, see [5, 9], to the P-parts (respectively the S-parts) of the elastic far-fields to localize the centers of the scatterers. Further, we extract the elastic capacitances of the obstacles from these data. Finally, from these capacitances we derive lower and upper estimates of the scaled perimeter of the scatterers, see Theorem 3.5. If in addition the obstacles are convex, then we derive an upper estimate of the largest ball contained in each obstacle and a lower bound of the smallest ball containing it, see Theorem 3.6. The two estimates measure, respectively, the thickness and length of each obstacle. It seems to us that these two estimates are new in the literature.

Let us also emphasize that our results mean that any of the two body waves (pressure or shear waves) is enough to localize and estimate the sizes of the scatterers.

The rest of paper is organized as follows. In section 2, we recall, from [7], the Foldy-Lax approximation of the elastic fields. In section 3, we use these approximations to justify Theorem 1.2.

2. Forward Problem

2.1. The asymptotic expansion of the far-fields. The forward problem is to compute the P-part, $U_p^\infty(\hat{x}, \theta)$, and the S-part, $U_s^\infty(\hat{x}, \theta)$, of the far-field pattern associated with the Lamé system (1.1-1.3) for various incident and the observational directions. The main result is the following theorem, see [7, Theorem 1.2], which justifies the Foldy-Lax approximation, in order to represent the scattering by small scatterers taking into account the three parameters M, a and d.

THEOREM 2.1. *There exist two positive constants a_0 and c_0 depending only on the size of Ω, the Lipschitz character of $B_m, m = 1, \ldots, M$, $d_{\max}$ and $\omega_{\max}$ such that if*

$$(2.1) \qquad a \leq a_0 \text{ and } \sqrt{M-1}\frac{a}{d} \leq c_0$$

then the P-part, $U_p^\infty(\hat{x}, \theta)$, and the S-part, $U_s^\infty(\hat{x}, \theta)$, of the far-field pattern have the following asymptotic expressions

(2.2)
$$U_p^\infty(\hat{x}, \theta) = \frac{1}{4\pi c_p^2} (\hat{x} \otimes \hat{x}) \left[\sum_{m=1}^M e^{-i\frac{\omega}{c_p}\hat{x}\cdot z_m} Q_m \right.$$
$$\left. + O\left(Ma^2 + M(M-1)\frac{a^3}{d^2} + M(M-1)^2\frac{a^4}{d^3}\right) \right],$$

(2.3)
$$U_s^\infty(\hat{x}, \theta) = \frac{1}{4\pi c_s^2} (I - \hat{x} \otimes \hat{x}) \left[\sum_{m=1}^M e^{-i\frac{\omega}{c_s}\hat{x}\cdot z_m} Q_m \right.$$
$$\left. + O\left(Ma^2 + M(M-1)\frac{a^3}{d^2} + M(M-1)^2\frac{a^4}{d^3}\right) \right].$$

uniformly in $\hat{x}$ and θ in $\mathbb{S}^2$. The constant appearing in the estimate $O(.)$ depends only on the size of Ω, the Lipschitz character of the reference bodies, a_0, c_0 and ω_{max}. The vector coefficients Q_m, $m = 1, \ldots, M$, are the solutions of the following linear algebraic system

$$(2.4) \qquad C_m^{-1} Q_m + \sum_{\substack{j=1 \\ j \neq m}}^M \Gamma^\omega(z_m, z_j) Q_j = -U^i(z_m, \theta),$$

for $m = 1, \ldots, M$, with Γ^ω denoting the Kupradze matrix of the fundamental solution to the Navier equation with frequency ω, $C_m := \int_{\partial D_m} \sigma_m(s) ds$ and σ_m is the solution matrix of the integral equation of the first kind

$$(2.5) \qquad \int_{\partial D_m} \Gamma^0(s_m, s)\sigma_m(s) ds = I, \; s_m \in \partial D_m,$$

with I the identity matrix of order 3. The algebraic system (2.4) is invertible under the condition:

$$(2.6) \qquad \frac{a}{d} \leq c_1 t^{-1}$$

with

$$t := \left[\frac{1}{c_p^2} - 2diam(\Omega)\frac{\omega}{c_s^3}\left(\frac{1-\left(\frac{1}{2}\kappa_{s\omega}diam(\Omega)\right)^{N_\Omega}}{1-\left(\frac{1}{2}\kappa_{s\omega}diam(\Omega)\right)} + \frac{1}{2^{N_\Omega-1}}\right)\right.$$
$$\left. -diam(\Omega)\frac{\omega}{c_p^3}\left(\frac{1-\left(\frac{1}{2}\kappa_{p\omega}diam(\Omega)\right)^{N_\Omega}}{1-\left(\frac{1}{2}\kappa_{p\omega}diam(\Omega)\right)} + \frac{1}{2^{N_\Omega-1}}\right)\right],$$

which is assumed to be positive and $N_\Omega := [2diam(\Omega)\max\{\kappa_{s\omega},\kappa_{p\omega}\}e^2]$, where $[\cdot]$ denotes the integral part and $\ln e = 1$. The constant c_1 depends only on the Lipschitz character of the reference bodies B_m, $m = 1, \ldots, M$.

We call the system (2.4) the elastic Foldy-Lax algebraic system. The matrix $C_m := \int_{\partial D_m}\sigma_m(s)ds$, where σ_m solves (2.5), is called the elastic capacitance of the set D_m. One of the interests of the expansions (2.2) and (2.3) is that we can reduce the computation of the elastic fields due to small obstacles to solving an algebraic system (i.e. (2.4)) and inverting a first kind integral equation (i.e. (2.5)). Another goal in deriving the expansion in terms of the three parameters is the quantification of the equivalent effective medium, without homogeneization (i.e. with no periodicity assumption on the distribution of the scatterers), see [1] for the acoustic model.

2.2. The fundamental solution. The Kupradze matrix $\Gamma^\omega = (\Gamma_{ij}^\omega)_{i,j=1}^3$ of the fundamental solution to the Navier equation is given by

$$(2.7) \quad \Gamma^\omega(x,y) = \frac{1}{\mu}\Phi_{\kappa_s\omega}(x,y)\mathbf{I} + \frac{1}{\omega^2}\nabla_x\nabla_x^\top[\Phi_{\kappa_s\omega}(x,y) - \Phi_{\kappa_p\omega}(x,y)],$$

where $\Phi_\kappa(x,y) = \frac{\exp(i\kappa|x-y|)}{4\pi|x-y|}$ denotes the free space fundamental solution of the Helmholtz equation $(\Delta + \kappa^2)u = 0$ in $\mathbb{R}^3$. The asymptotic behavior of Kupradze tensor at infinity is given as follows
(2.8)
$$\Gamma^\omega(x,y) = \frac{1}{4\pi c_p^2}\hat{x}\otimes\hat{x}\frac{e^{i\kappa_{p\omega}|x|}}{|x|}e^{-i\kappa_{p\omega}\hat{x}\cdot y} + \frac{1}{4\pi c_s^2}(I-\hat{x}\otimes\hat{x})\frac{e^{i\kappa_{s\omega}|x|}}{|x|}e^{-i\kappa_{s\omega}\hat{x}\cdot y} + O(|x|^{-2})$$

with $\hat{x} = \frac{x}{|x|} \in \mathbb{S}^2$, see [10] for instance.

3. Inverse problem

3.1. Scalar far-field patterns. We define the scalar P-part, $U_p^\infty(\hat{x},\theta)$, and the scalar S-part, $U_s^\infty(\hat{x},\theta)$, of the far-field pattern of the problem (1.1-1.3) respectively as

$$U_p^\infty(\hat{x},\theta) := 4\pi c_p^2\left(\hat{x}\cdot U_p^\infty(\hat{x},\theta)\right)$$
$$(3.1) \quad = \sum_{m=1}^M \hat{x}e^{-i\frac{\omega}{c_p}\hat{x}\cdot z_m}Q_m + O\left(Ma^2 + M(M-1)\frac{a^3}{d^2} + M(M-1)^2\frac{a^4}{d^3}\right),$$

$$U_s^\infty(\hat{x},\theta) := 4\pi c_s^2\left(\hat{x}^\perp \cdot U_s^\infty(\hat{x},\theta)\right)$$
$$(3.2) \quad = \sum_{m=1}^M \hat{x}^\perp e^{-i\frac{\omega}{c_s}\hat{x}\cdot z_m}Q_m + O\left(Ma^2 + M(M-1)\frac{a^3}{d^2} + M(M-1)^2\frac{a^4}{d^3}\right).$$

From (3.1) and (3.2), we can write the scalar P and the scalar S parts of the far-field pattern as

$$\text{U}_p^\infty(\hat{x},\theta) = \sum_{m=1}^{M} \hat{x} e^{-i\frac{\omega}{c_p}\hat{x}\cdot z_m} Q_m \tag{3.3}$$

$$\text{U}_s^\infty(\hat{x},\theta) = \sum_{m=1}^{M} \hat{x}^\perp e^{-i\frac{\omega}{c_s}\hat{x}\cdot z_m} Q_m \tag{3.4}$$

with the error of order $O\left(Ma^2 + M(M-1)\frac{a^3}{d^2} + M(M-1)^2\frac{a^4}{d^3}\right)$ and Q_m can be obtained from the linear algebraic system (2.4).

3.2. The elastic Foldy-Lax algebraic system. We can rewrite the algebraic system (2.4),

$$C_m^{-1} Q_m = -U^i(z_m) - \sum_{\substack{j=1\\j\neq m}}^{M} \Gamma^\omega(z_m,z_j) C_j (C_j^{-1} Q_j), \tag{3.5}$$

for all $m = 1, 2, \ldots, M$. It can be written in a compact form as

$$\mathbf{B} Q = U^I, \tag{3.6}$$

where $Q, U^I \in \mathbb{C}^{3M\times 1}$ and $\mathbf{B} \in \mathbb{C}^{3M\times 3M}$ are defined as
(3.7)
$$\mathbf{B} := \begin{pmatrix} -C_1^{-1} & -\Gamma^\omega(z_1,z_2) & -\Gamma^\omega(z_1,z_3) & \cdots & -\Gamma^\omega(z_1,z_M) \\ -\Gamma^\omega(z_2,z_1) & -C_2^{-1} & -\Gamma^\omega(z_2,z_3) & \cdots & -\Gamma^\omega(z_2,z_M) \\ \cdots & \cdots & \cdots & \cdots & \cdots \\ -\Gamma^\omega(z_M,z_1) & -\Gamma^\omega(z_M,z_2) & \cdots & -\Gamma^\omega(z_M,z_{M-1}) & -C_M^{-1} \end{pmatrix},$$

$Q := \begin{pmatrix} Q_1^\top & Q_2^\top & \cdots & Q_M^\top \end{pmatrix}^\top$ and $U^I := \begin{pmatrix} U^i(z_1)^\top & U^i(z_2)^\top & \cdots & U^i(z_M)^\top \end{pmatrix}^\top$.

The above linear algebraic system is solvable for the 3D vectors Q_j, $1 \leq j \leq M$, when the matrix $\mathbf{B}$ is invertible. The invertibility of $\mathbf{B}$ is discussed in [7, Corollary 4.3].

Let us denote the inverse of $\mathbf{B}$ by $\mathcal{B}$ and the corresponding 3×3 blocks of $\mathcal{B}$ by $\mathcal{B}_{mj}$, $m, j = 1, \ldots, M$. Then we can rewrite (3.3) and (3.4), with the same error, as follows

$$\text{U}_p^\infty(\hat{x},\theta) = \sum_{m=1}^{M}\sum_{j=1}^{M} e^{-i\frac{\omega}{c_p}\hat{x}\cdot z_m} \hat{x}^\top \mathcal{B}_{mj} U^i(z_j,\theta) \tag{3.8}$$

$$\text{U}_s^\infty(\hat{x},\theta) = \sum_{m=1}^{M}\sum_{j=1}^{M} e^{-i\frac{\omega}{c_s}\hat{x}\cdot z_m} (\hat{x}^\perp)^\top \mathcal{B}_{mj} U^i(z_j,\theta) \tag{3.9}$$

for a given incident direction θ and observation direction $\hat{x}$. From (3.8) and (3.9), we can get the scalar P and the scalar S parts of the far-field patterns corresponding to plane incident P-wave $U^{i,p}(x,\theta)$ and S-wave $U^{i,s}(x,\theta)$, that we denote respectively by $\text{U}_p^{\infty,p}(\hat{x},\theta)$, $\text{U}_s^{\infty,p}(\hat{x},\theta)$, $\text{U}_p^{\infty,s}(\hat{x},\theta)$, $\text{U}_s^{\infty,s}(\hat{x},\theta)$ as below

$$\text{U}_p^{\infty,p}(\hat{x},\theta) = \sum_{m=1}^{M}\sum_{j=1}^{M} e^{-i\frac{\omega}{c_p}\hat{x}\cdot z_m} \hat{x}^\top \mathcal{B}_{mj} \theta\, e^{i\frac{\omega}{c_p}\theta\cdot z_j}, \tag{3.10}$$

$$(3.11) \quad U_s^{\infty,p}(\hat{x},\theta) = \sum_{m=1}^{M}\sum_{j=1}^{M} e^{-i\frac{\omega}{c_s}\hat{x}\cdot z_m} (\hat{x}^\perp)^\top \mathcal{B}_{mj}\theta\, e^{i\frac{\omega}{c_p}\theta\cdot z_j},$$

$$(3.12) \quad U_p^{\infty,s}(\hat{x},\theta) = \sum_{m=1}^{M}\sum_{j=1}^{M} e^{-i\frac{\omega}{c_p}\hat{x}\cdot z_m} \hat{x}^\top \mathcal{B}_{mj}\theta^\perp\, e^{i\frac{\omega}{c_s}\theta\cdot z_j},$$

$$(3.13) \quad U_s^{\infty,s}(\hat{x},\theta) = \sum_{m=1}^{M}\sum_{j=1}^{M} e^{-i\frac{\omega}{c_s}\hat{x}\cdot z_m} (\hat{x}^\perp)^\top \mathcal{B}_{mj}\theta^\perp\, e^{i\frac{\omega}{c_s}\theta\cdot z_j}.$$

All the far-field patterns (3.10-3.13) are valid with the same error which is equal to the error in (3.3-3.4). Now onwards, let $U^\infty(\hat{x},\theta)$ represents any one of the scattered fields mentioned above.

3.3. Localization of D_m's via the MUSIC algorithm. The MUSIC algorithm is a method to determine the locations $z_m, m = 1, 2, \ldots, M$, of the scatterers $D_m, m = 1, 2, \ldots, M$ from the measured far-field pattern $U^\infty(\hat{x},\theta)$ for a finite set of incidence and observation directions, i.e. $\hat{x}, \theta \in \{\theta_j, j = 1, \ldots, N\} \subset \mathbb{S}^2$. We refer the reader to the monograph [11] for more information about this algorithm. We follow the way in [9] which is based on the presentation in [11].

3.3.1. *The factorization of the response matrix.* We assume that the number of scatterers is not larger than the number of incident and observation directions, precisely $N \geq 3M$. We define the response matrix $F \in \mathbb{C}^{N\times N}$ by

$$(3.14) \quad F_{jl} := U^\infty(\theta_j, \theta_l).$$

The P-part and the S-part of the response matrix F by F_p and F_s respectively. From (3.8-3.9) and (3.14), we can write

$$(F_p)_{jl} := U_p^\infty(\theta_j,\theta_l) = \sum_{m=1}^{M}\sum_{j=1}^{M} e^{-i\frac{\omega}{c_p}\theta_j\cdot z_m} \theta_j^\top \mathcal{B}_{mj} U^i(z_j,\theta_l)$$
$$= \left[\theta_j^\top e^{-i\frac{\omega}{c_p}\theta_j\cdot z_1}, \theta_j^\top e^{-i\frac{\omega}{c_p}\theta_j\cdot z_2}, \cdots, \theta_j^\top e^{-i\frac{\omega}{c_p}\theta_j\cdot z_M} \right]$$
$$\times \mathcal{B}\left[(U^i(z_j,\theta_l))^\top, (U^i(z_2,\theta_l))^\top, \cdots, (U^i(z_m,\theta_l))^\top\right]^\top,$$

$$(F_s)_{jl} := U_s^\infty(\theta_j,\theta_l) = \sum_{m=1}^{M}\sum_{j=1}^{M} e^{-i\frac{\omega}{c_s}\theta_j\cdot z_m} (\theta_j^\perp)^\top \mathcal{B}_{mj} U^i(z_j,\theta_l)$$
$$= \left[(\theta_j^\perp)^\top e^{-i\frac{\omega}{c_s}\theta_j\cdot z_1}, (\theta_j^\perp)^\top e^{-i\frac{\omega}{c_s}\theta_j\cdot z_2}, \cdots, (\theta_j^\perp) e^{-i\frac{\omega}{c_s}\theta_j\cdot z_M} \right]$$
$$\times \mathcal{B}\left[(U^i(z_j,\theta_l))^\top, (U^i(z_2,\theta_l))^\top, \cdots, (U^i(z_m,\theta_l))^\top\right]^\top.$$

for all $j, l = 1, \ldots, N$. In PP, PS, SS and SP scatterings, denote the response matrix F by F_p^p, F_s^p, F_s^s and F_p^s respectively and these can be factorized as

$$(3.15) \quad F_p^p = H^{p*}\mathcal{B}H^p, \quad F_s^p = H^{s*}\mathcal{B}H^p, \quad F_s^s = H^{s*}\mathcal{B}H^s, \quad \text{and } F_p^s = H^{p*}\mathcal{B}H^s.$$

Here, the matrices $H^p \in \mathbb{C}^{3M \times N}$ and $H^s \in \mathbb{C}^{3M \times N}$ are defined as,

$$H^p := \begin{pmatrix} \theta_1 e^{i\frac{\omega}{c_p}\theta_1 \cdot z_1} & \theta_2 e^{i\frac{\omega}{c_p}\theta_2 \cdot z_1} & \cdots & \theta_N e^{i\frac{\omega}{c_p}\theta_N \cdot z_1} \\ \theta_1 e^{i\frac{\omega}{c_p}\theta_1 \cdot z_2} & \theta_2 e^{i\frac{\omega}{c_p}\theta_2 \cdot z_2} & \cdots & \theta_N e^{i\frac{\omega}{c_p}\theta_N \cdot z_2} \\ \cdots & \cdots & \cdots & \cdots \\ \theta_1 e^{i\frac{\omega}{c_p}\theta_1 \cdot z_M} & \theta_2 e^{i\frac{\omega}{c_p}\theta_2 \cdot z_M} & \cdots & \theta_N e^{i\frac{\omega}{c_p}\theta_N \cdot z_M} \end{pmatrix},$$

and

$$H^s := \begin{pmatrix} \theta_1^\perp e^{i\frac{\omega}{c_s}\theta_1 \cdot z_1} & \theta_2^\perp e^{i\frac{\omega}{c_s}\theta_2 \cdot z_1} & \cdots & \theta_N^\perp e^{i\frac{\omega}{c_s}\theta_N \cdot z_1} \\ \theta_1^\perp e^{i\frac{\omega}{c_s}\theta_1 \cdot z_2} & \theta_2^\perp e^{i\frac{\omega}{c_s}\theta_2 \cdot z_2} & \cdots & \theta_N^\perp e^{i\frac{\omega}{c_s}\theta_N \cdot z_2} \\ \cdots & \cdots & \cdots & \cdots \\ \theta_1^\perp e^{i\frac{\omega}{c_s}\theta_1 \cdot z_M} & \theta_2^\perp e^{i\frac{\omega}{c_s}\theta_2 \cdot z_M} & \cdots & \theta_N^\perp e^{i\frac{\omega}{c_s}\theta_N \cdot z_M} \end{pmatrix}.$$

In order to determine the locations z_m, we consider a 3D-grid of sampling points $z \in \mathbb{R}^3$ in a region containing the scatterers $D_1, D_2, \ldots, D_M$. For each point z, we define the vectors $\phi_{z,p}^j$ and $\phi_{z,s}^j$ in $\mathbb{C}^N$ by

(3.16)
$$\phi_{z,p}^j := \left((\theta_1 \cdot e_j) e^{-i\frac{\omega}{c_p}\theta_1 \cdot z}, (\theta_2 \cdot e_j) e^{-i\frac{\omega}{c_p}\theta_2 \cdot z}, \ldots, (\theta_N \cdot e_j) e^{-i\frac{\omega}{c_p}\theta_N \cdot z} \right)^T,$$

(3.17)
$$\phi_{z,s}^j := \left((\theta_1^\perp \cdot e_j) e^{-i\frac{\omega}{c_s}\theta_1 \cdot z}, (\theta_2^\perp \cdot e_j) e^{-i\frac{\omega}{c_s}\theta_2 \cdot z}, \ldots, (\theta_N^\perp \cdot e_j) e^{-i\frac{\omega}{c_s}\theta_N \cdot z} \right)^T, \forall j = 1,2,3.$$

3.3.2. *MUSIC characterization of the response matrix.* Recall that MUSIC is essentially based on characterizing the range of the response matrix (signal space), forming projections onto its null (noise) spaces, and computing its singular value decomposition. In other words, the MUSIC algorithm is based on the property that the test vector $\phi_{z,r}^j$ is in the range $\mathcal{R}(F_r)$ of F_r if and only if z is at one of locations of the scatterers, see [9]. Here, $F_r := F_r^p$ or $F_p := F_r^s$ and $r \in \{p, s\}$.

It can be proved based on the non-singularity of the scattering matrix $\mathcal{B}$ in the factorizations (3.15) of $F_{r_2}^{r_1}$, $r_1, r_2 \in \{p, s\}$. Due to this, the standard linear algebraic argument yields that, if $N \geq 3M$ and the if the matrix H^r has maximal rank $3M$, then the ranges $\mathcal{R}(H^{r^*})$ and $\mathcal{R}(F_r)$ coincide.

For sufficiently large number N of incident and the observational directions by following the same lines as in [4, 9, 11], the maximal rank property of H can be justified. In this case MUSIC algorithm is applicable for our response matrices F_p^p, F_s^p, F_p^s and F_s^s.

From the above discussion, MUSIC characterization of the locations of the small scatterers in elastic exterior Drichlet problem can be written as the following

THEOREM 3.1. *For $N \geq 3M$ sufficiently large, we have*
(3.18)
$$z \in \{z_1, \ldots, z_M\} \iff \phi_{z,t}^j \in \mathcal{R}(H^{t^*}), \text{ for some } j = 1, 2, 3 \text{ and for all } t \in \{p, s\}.$$

Furthermore, the ranges of H^{t^*} and F_t^r coincide and thus

(3.19) $\quad z \in \{z_1, \ldots, z_M\} \iff \phi_{z,t}^j \in \mathcal{R}(F_t^r) \iff \mathcal{P}_t \phi_{z,t}^j = 0, \text{ for some } j = 1, 2, 3$
$$\text{and for all } r, t \in \{p, s\}$$

where $\mathcal{P}_t : \mathbb{C}^N \to \mathcal{R}(F_t^r)^\perp = \mathcal{N}(F_t^{r^*})$ is the orthogonal projection onto the null space $\mathcal{N}(F_t^{r^*})$ of $F_t^{r^*}$.

From Theorem 3.1, the MUSIC algorithm holds for the response matrices corresponding to the PP, PS, SS and SP scatterings. To make the best use of the singular value decomposition in SP and PS scatterings, we apply the MUSIC algorithm to $F_p^s F_p^{s*}$ (resp. $F_p^{s*} F_p^s$) and $F_s^{p*} F_s^p$ (resp. $F_s^p F_s^{p*}$) with the help of the test vectors $\phi_{z,p}^j$ (resp. $\phi_{z,s}^j$) respectively.

As we are dealing with the 3D case, while dealing with S incident wave or S-part of the far-field pattern, it is enough to use one of its horizontal (S^h) or vertical (S^v) parts. Hence, it is enough to study the far-field pattern of any of the PP, PS^h, PS^v, $S^h S^h$, $S^h S^v$, $S^v S^h$, $S^v S^v$, $S^h P$, $S^v P$ elastic scatterings to locate the scatterers. In other words, in three dimensional case, instead of using the full incident wave and the full far-field pattern, it is enough to study one combination of pressure (P), horizontal shear (S^h) or vertical shear (S^v) parts of the elastic incident wave and a corresponding part of the elastic far-field patterns, see [9].

Indeed, define the vectors $\phi_{z,s^h}^j, \phi_{z,s^v}^j \in \mathbb{C}^N$ and the matrices $H^{s^h}, H^{s^v} \in \mathbb{C}^{3M \times N}$ exactly as $\phi_{z,s}^j$ and H^s replacing $\theta_i^\perp$ for $i = 1, \ldots, N$ by $\theta_i^{\perp h}$ and $\theta_i^{\perp v}$ respectively, see (1.7). We denote the response matrices by $F_{s^h}^p$, $F_p^{s^h}$, $F_{s^v}^p$, $F_p^{s^v}$, $F_{s^h}^{s^h}$, $F_{s^h}^{s^v}$, $F_{s^v}^{s^h}$, and $F_{s^v}^{s^v}$ in the elastic PS^h, $S^h P$, PS^v, $S^v P$, $S^h S^h$, $S^h S^v$, $S^v S^h$, $S^v S^v$ scatterings respectively, then we can state the following theorem related to the MUSIC algorithm for sufficiently large number of incident and observation angles,

THEOREM 3.2. *For $N \geq 3M$ sufficiently large, we have*
$$(3.20) \quad z \in \{y_1, \ldots, y_M\} \iff \phi_{z,t}^j \in \mathcal{R}(H^{t*}), \text{ for some } j = 1, 2, 3 \text{ and for all } t \in \{p, s^h, s^v\}.$$

Furthermore, the ranges of H^{t*} and F_t^r coincide and thus
$$(3.21) \quad z \in \{y_1, \ldots, y_M\} \iff \phi_{z,t}^j \in \mathcal{R}(F_t^r) \iff \mathcal{P}_t \phi_{z,t}^j = 0, \text{ for some } j = 1,2,3$$
$$\text{and for all } r, t \in \{p, s^h, s^v\}$$

where $\mathcal{P}_t : \mathbb{C}^N \to \mathcal{R}(F_t^r)^\perp = \mathcal{N}(F_t^{r*})$ is the orthogonal projection onto the null space $\mathcal{N}(F_t^{r*})$ of F_t^{r*}.

The proof of the previous two theorems can be carried out in the same lines as in [9].

3.4. Estimating the sizes.

3.4.1. *Recovering the capacitances.* Once we locate the scatterers from the given far-field patterns using the MUSIC algorithm, we can recover the capacitances C_m of D_m from the factorization $F_t^r = H^{t*} \mathcal{B} H^r$ of $F_t^r \in \mathbb{C}^{N \times N}$, $r, t \in \{p, s^h, s^v\}$. Indeed, we know that the matrix H^t has maximal rank, see Theorem 3.1 and Theorem 3.2 of [9]. So, the matrix $H^t H^{t*} \in \mathbb{C}^{3M \times 3M}$ is invertible. Let us denote its inverse by I_{H^t}. Once we locate the scatterers through finding the locations $z_1, z_2, \ldots, z_M$ by using the MUSIC algorithm for the given far-field patterns, we can recover I_{H^t} and hence the matrix $\mathcal{B} \in \mathbb{C}^{3M \times 3M}$ given by $\mathcal{B} = I_{H^t} H^t F_t^r H^{r*} I_{H^r}$, where $I_{H^t} H^t$ (resp. $H^{r*} I_{H^r}$) is the pseudo inverse of H^{t*} (resp. H^r). As we know the structure of $\mathbf{B} \in \mathbb{C}^{3M \times 3M}$, the inverse of $\mathcal{B} \in \mathbb{C}^{3M \times 3M}$, we can recover the capacitance matrices $C_1, C_2, \ldots, C_M$ of the small scatterers $D_1, D_2, \ldots, D_M$ from the diagonal blocks of **B**, see (3.7). From these capacitances, we can estimate the size of the obstacles as follows.

3.4.2. *Estimating the sizes of the obstacles from the capacitances.* Let us first start with the following lemma which compares the elastic and the acoustic capacitances [1], see [**7, 15**].

LEMMA 3.3. *Let $\lambda_{eig_m}^{min}$ and $\lambda_{eig_m}^{max}$ be the minimal and maximal eigenvalues of the elastic capacitance matrices C_m, for $m = 1, 2, \ldots, M$. Denote by C_m^a the capacitance of each scatterer in the acoustic case, then we have the following estimate;*

$$(3.22) \quad \mu C_m^a \leq \lambda_{eig_m}^{min} \leq \lambda_{eig_m}^{max} \leq (\lambda + 2\mu) C_m^a, \quad \text{for} \quad m = 1, 2, \ldots, M.$$

Now, let us derive lower and upper bounds of the sizes of the obstacles in terms of the acoustic capacitances.

Assume that D_j's are balls of radius ρ_j, and center 0 for simplicity, then we know that $\int_{\{y:|y|=\rho_j\}} \frac{dS_y}{|x-y|} = 4\pi\rho_j$, for $|x| = \rho_j$, as observed in [**14**, formula (5.12)]. Hence $\sigma_j(s) = \rho_j^{-1}$ and then $C_j^a = \int_{\partial D_j} \rho_j^{-1} ds = 2\pi\rho_j$ from which we can estimate the radius ρ_j. Other geometries, as cylinders, for which one can estimate exactly the size, from the capacitance, are shown in chapter 4 of [**16**].

For general geometries, we proceed as follows. First, we recall the following result, see [**7**].

LEMMA 3.4. *For every $1 \leq j \leq M$, the capacitance C_j^a is of the form*

$$(3.23) \quad C_j^a = C_{B_j}^a \epsilon$$

where $C_{B_j}^a$ is the acoustic capacitance of B_j.

Now let us consider a single obstacle $D := \epsilon B + z$. Since $C_B^a := \int_{\partial B} \sigma(s) ds$ and $\int_{\partial B} \frac{\sigma(s)}{4\pi|t-s|} ds = 1$, from the invertibility of the single layer potential $S : L^2(\partial B) \to H^1(\partial B)$, defined as $Sf(t) = \int_{\partial B} \frac{f(s)}{4\pi|t-s|} ds = 1$, we deduce that

$$(3.24) \quad C_B^a \leq |\partial B|^{\frac{1}{2}} \|\sigma\|_{L^2(\partial B)} \leq |\partial B|^{\frac{1}{2}} \|S^{-1}\|_{\mathcal{L}(H^1(\partial B), L^2(\partial B))} |\partial B|^{\frac{1}{2}}$$
$$= \|S^{-1}\|_{\mathcal{L}(H^1(\partial B), L^2(\partial B))} |\partial B|.$$

On the other hand, we recall the following lower estimate, see Theorem 3.1 in [**16**] for instance,

$$(3.25) \quad C_B^a \geq \frac{4\pi|\partial B|^2}{J}$$

where $J := \int_{\partial B} \int_{\partial B} \frac{1}{|s-t|} ds dt$. Remark that $J = 4\pi \int_{\partial B} S(1)(s) ds$. Hence

$$J \leq 4\pi |\partial B|^{\frac{1}{2}} \|S\|_{\mathcal{L}(L^2(\partial B), H^1(\partial B))} \|1\|_{H^1(\partial B)} \leq 4\pi \|S\|_{\mathcal{L}(L^2(\partial B), H^1(\partial B))} |\partial B|$$

and using (3.25) we obtain the lower bound

$$(3.26) \quad C_B^a \geq \|S\|_{\mathcal{L}(L^2(\partial B), H^1(\partial B))}^{-1} |\partial B|.$$

Finally combining (3.24) and (3.26), we derive the estimate

$$(3.27) \quad \|S\|_{\mathcal{L}(L^2(\partial B), H^1(\partial B))}^{-1} |\partial B| \leq C_B^a \leq \|S^{-1}\|_{\mathcal{L}(H^1(\partial B), L^2(\partial B))} |\partial B|$$

---

[1] Recall that, for $m = 1, \ldots, M$, $C_m^a := \int_{\partial D_m} \sigma_m(s) ds$ and σ_m is the solution of the integral equation of the first kind $\int_{\partial D_m} \frac{\sigma_m(s)}{4\pi|t-s|} ds = 1$, $t \in \partial D_m$, see [**8, 15**].

Using Lemma 3.4 and the relation $|\partial D_\epsilon| = \epsilon^2 |\partial B|$ we obtain the following size estimation:

$$(3.28) \qquad \|S^{-1}\|^{-1}_{\mathcal{L}(H^1(\partial B), L^2(\partial B))} \epsilon C_\epsilon^a \leq |\partial D_\epsilon| \leq \|S\|_{\mathcal{L}(L^2(\partial B), H^1(\partial B))} \epsilon C_\epsilon^a.$$

Now, using Lemma 3.3, we derive the following lower and upper bounds of the sizes of the obstacles D_m

(3.29)
$$\|S^{-1}\|^{-1}_{\mathcal{L}(H^1(\partial B_m), L^2(\partial B_m))} (\lambda + 2\mu)^{-1} \lambda_{eig_m}^{max} \leq \frac{|\partial D_m|}{\epsilon}$$
$$\leq \|S\|_{\mathcal{L}(L^2(\partial B_m), H^1(\partial B_m))} \mu^{-1} \lambda_{eig_m}^{min}.$$

Observe that one can estimate the norms of the operators appearing in (3.29) in terms of (only) the Lipschitz character. We summarize this result in the following theorem

THEOREM 3.5. *There exist two constants $c(Lip)$ and $C(Lip)$ depending only on the Lipschitz character of $B_1, \ldots, B_M$, such that*

$$(3.30) \qquad c(Lip)(\lambda + 2\mu)^{-1} \lambda_{eig_m}^{max} \leq \frac{|\partial D_m|}{\epsilon} \leq C(Lip) \mu^{-1} \lambda_{eig_m}^{min}.$$

Precisely $C(Lip)$ and $c(Lip)$ are characterized respectively by $\|S\|_{\mathcal{L}(L^2(\partial B_m), H^1(\partial B_m))} \leq C(Lip)$ and $\|S^{-1}\|_{\mathcal{L}(H^1(\partial B_m), L^2(\partial B_m))} \leq c^{-1}(Lip)$. We can use the estimate (3.29) to provide the lower and the upper estimates of the scaled 'size' of scatterers $\frac{|\partial D_m|}{\epsilon}$. Under some conditions of the reference obstacles, we can derive more explicit size estimates. Let us first define

$$(3.31) \qquad \delta(x) := \min_{y \in \partial B} |x - y|, \ x \in \mathbb{R}^3 \text{ and } R_i(B) := \sup_{x \in B} \delta(x).$$

The quantity $R_i(B)$ is the radius of the largest ball contained in B. Now, we set

$$(3.32) \qquad R_e(B) := \frac{1}{2} \max_{x,y \in \bar{B}} |x - y|.$$

The quantity $R_e(B)$ is the radius of the smallest ball containing B.

By the Gauss theorem, we see that

$$(3.33) \qquad |B| = \frac{1}{3} \int_B div(x) \, dx = \frac{1}{3} \int_{\partial B} s \cdot n(s) ds$$

hence $|B| \leq \frac{1}{3} |\partial B| \max_{s \in \partial B} |s| \leq \frac{2}{3} |\partial B| R_e(B)$ since $\max_{s \in \partial B} |s| \leq 2R_e(B)$. Hence

$$(3.34) \qquad |\partial B| \geq \frac{3}{2} \frac{|B|}{R_e(B)}$$

To derive the upper bound for $|\partial B|$, we use the following argument borrowed from ([**12**], section 4). If we assume that B is convex, then $\delta(\cdot)$ is a concave function in B and then, see [**6**],

$$(3.35) \qquad \nabla \delta(x) \cdot (y - x) \geq \delta(y) - \delta(x), \ x, y \in \bar{B}.$$

But $\nabla \delta(x) = -\nu(x)$ for $x \in \partial B$, where ν is the external unit normal to ∂B. Let us now assume, in addition to the convexity property, that $B(0, R_i(B)) \subset B$ which roughly means that the origin is the 'center' of B. With this assumption, taking

$y = 0$ in (3.35), we obtain $s \cdot \nu(s) \geq \delta(0) \geq R_i(B)$, (since $B(0, R_i(B)) \subset B$). Replacing in (3.33), we obtain $|B| \geq \frac{1}{3}|\partial B|R_i(B)$ hence

$$|\partial B| \leq 3\frac{|B|}{R_i(B)}. \tag{3.36}$$

Replacing (3.34) and (3.36) in (3.27), we have

$$\frac{3}{2}\|S\|^{-1}_{\mathcal{L}(L^2(\partial B), H^1(\partial B))}\frac{|B|}{R_e(B)} \leq C_B^a \leq 3\|S^{-1}\|_{\mathcal{L}(H^1(\partial B), L^2(\partial B))}\frac{|B|}{R_i(B)}.$$

Using the double inequality $\frac{4}{3}\pi R_i^3(B) \leq |B| \leq \frac{4}{3}\pi R_e^3(B)$ we obtain

$$2\pi\|S\|^{-1}_{\mathcal{L}(L^2(\partial B), H^1(\partial B))}\frac{R_i^2(B)}{R_e(B)}R_i(B) \leq C_B^a \tag{3.37}$$

$$\leq 4\pi\|S^{-1}\|_{\mathcal{L}(H^1(\partial B), L^2(\partial B))}\frac{R_e^2(B)}{R_i(B)}R_e(B).$$

Now, we apply these double estimates to D instead of B, knowing that the two assumptions on B are inherited by D, then we obtain

$$2\pi\|S\|^{-1}_{\mathcal{L}(L^2(\partial D), H^1(\partial D))}\frac{R_i^2(D)}{R_e(D)}R_i(D) \leq C_D^a \tag{3.38}$$

$$\leq 4\pi\|S^{-1}\|_{\mathcal{L}(H^1(\partial D), L^2(\partial D))}\frac{R_e^2(D)}{R_i(D)}R_e(D).$$

Observe that $\frac{R_i^2(D)}{R_e(D)}$ scales as $\|S\|_{\mathcal{L}(L^2(\partial D), H^1(\partial D))}$ and $\frac{R_e^2(D)}{R_i(D)}$ scales as $\|S^{-1}\|^{-1}_{\mathcal{L}(H^1(\partial B), L^2(\partial B))}$ since obviously $R_i(D) = \epsilon R_i(B)$, $R_e(D) = \epsilon R_e(B)$ and we have $\|S\|_{\mathcal{L}(L^2(\partial D), H^1(\partial D))} \leq \epsilon\|S\|_{\mathcal{L}(L^2(\partial B), H^1(\partial B))}$ and $\|S^{-1}\|_{\mathcal{L}(H^1(\partial D), L^2(\partial D))} \leq \epsilon^{-1}\|S^{-1}\|_{\mathcal{L}(H^1(\partial B), L^2(\partial B))}$, see [8, Lemma 2.4 and Lemma 2.5] for the last two inequalities. Using these properties, we deduce that

$$2\pi\|S\|^{-1}_{\mathcal{L}(L^2(\partial B), H^1(\partial B))}\frac{R_i^2(B)}{R_e(B)}R_i(D) \leq C_D^a \tag{3.39}$$

$$\leq 4\pi\|S^{-1}\|_{\mathcal{L}(H^1(\partial B), L^2(\partial B))}\frac{R_e^2(B)}{R_i(B)}R_e(D).$$

We can estimate $2\pi\|S\|^{-1}_{\mathcal{L}(L^2(\partial B), H^1(\partial B))}\frac{R_i^2(B)}{R_e(B)}$ from below by a constant $c(Lip)$ depending only the Lipschitz character of B and $4\pi\|S^{-1}\|_{\mathcal{L}(H^1(\partial B), L^2(\partial B))}\frac{R_e^2(B)}{R_i(B)}$ by a constant $C(Lip)$ also depending only on the Lipschitz character of B. With these apriori bounds (3.39) becomes

$$c(Lip)R_i(D) \leq C_D^a \leq C(Lip)R_e(D). \tag{3.40}$$

Using Lemma 3.3, we deduce the following result.

THEOREM 3.6. *Assume that the reference obstacles $B_m, m = 1, \ldots, M$, are convex and satisfy the properties $B(0, R_i(B_m)) \subset B_m$. Then, there exist two constants $c(Lip)$ and $C(Lip)$ depending only on the Lipschitz character of the obstacles*

$B_m, m = 1, \ldots, M$, such that we have the estimates

(3.41) $$R_i(D_m) \leq c^{-1}(Lip)(\lambda + 2\mu)^{-1}\lambda_{eig_m}^{max},$$

(3.42) $$R_e(D_m) \geq C^{-1}(Lip)\mu^{-1}\lambda_{eig_m}^{min}.$$

The estimate (3.41) means that the largest ball contained in D_m has a radius not exceeding $c^{-1}(Lip)(\lambda + 2\mu)^{-1}\lambda_{eig_m}^{max}$. Hence (3.41) measures the thickness of D_m. The estimate (3.42) means that the radius of the smallest ball containing D_m is not lower than $C^{-1}(Lip)\mu^{-1}\lambda_{eig_m}^{min}$. Hence (3.42) measures the length of the obstacle D_m.

Conclusion

Based on the asymptotic expansion of the elastic waves by small rigid obstacles, derived in [7], we have shown that any of the two elastic waves, i.e. P-waves or S-waves (precisely SH-waves or SV-waves in 3D elasticity), is enough to localize the obstacles and estimate their respective sizes. Compared to the existing literature, see for instance [5], we allow the obstacles to be close and the cluster to be spread in any given region. In addition, the derived precise size estimates seem to be new compared to the related literature. We stated the MUSIC algorithm based on the mentioned measurements and we believe that performing this algorithm will provide us with accurate numerical tests, see our previous work [9] on point-like obstacles. Let us also mention that, using our techniques, we can write down a topological derivative based imaging approach from only the shear or compressional part of the elastic wave, compare to [2]. We think that these two points deserve to be studied.

References

[1] Bashir Ahmad, Durga Prasad Challa, Mokhtar Kirane, and Mourad Sini, *The equivalent refraction index for the acoustic scattering by many small obstacles: with error estimates*, J. Math. Anal. Appl. **424** (2015), no. 1, 563–583, DOI 10.1016/j.jmaa.2014.11.020. MR3286580

[2] Habib Ammari, Elie Bretin, Josselin Garnier, Wenjia Jing, Hyeonbae Kang, and Abdul Wahab, *Localization, stability, and resolution of topological derivative based imaging functionals in elasticity*, SIAM J. Imaging Sci. **6** (2013), no. 4, 2174–2212, DOI 10.1137/120899303. MR3123825

[3] H. Ammari, E. Bretin, J. Garnier, H. Kang, H. Lee, and A. Wahab *Mathematical Methods in Elasticity Imaging*, Princeton Series in Applied Mathematics. To appear.

[4] Habib Ammari, Pierre Calmon, and Ekaterina Iakovleva, *Direct elastic imaging of a small inclusion*, SIAM J. Imaging Sci. **1** (2008), no. 2, 169–187, DOI 10.1137/070696076. MR2486036 (2010d:35388)

[5] Habib Ammari and Hyeonbae Kang, *Polarization and moment tensors*, Applied Mathematical Sciences, vol. 162, Springer, New York, 2007. With applications to inverse problems and effective medium theory. MR2327884 (2009f:35339)

[6] Stephen Boyd and Lieven Vandenberghe, *Convex optimization*, Cambridge University Press, Cambridge, 2004. MR2061575 (2005d:90002)

[7] D. P. Challa and M. Sini. The Foldy-Lax approximation of the scattered waves by many small bodies for the Lamé system. *Preprint,arXiv: 1308. 3072*.

[8] Durga Prasad Challa and Mourad Sini, *On the justification of the Foldy-Lax approximation for the acoustic scattering by small rigid bodies of arbitrary shapes*, Multiscale Model. Simul. **12** (2014), no. 1, 55–108, DOI 10.1137/130919313. MR3158777

[9] Durga Prasad Challa and Mourad Sini, *Inverse scattering by point-like scatterers in the Foldy regime*, Inverse Problems **28** (2012), no. 12, 125006, 39, DOI 10.1088/0266-5611/28/12/125006. MR2997015

[10] George Dassios and Ralph Kleinman, *Low frequency scattering*, Oxford Mathematical Monographs, The Clarendon Press, Oxford University Press, New York, 2000. Oxford Science Publications. MR1858914 (2003a:35001)

[11] Andreas Kirsch and Natalia Grinberg, *The factorization method for inverse problems*, Oxford Lecture Series in Mathematics and its Applications, vol. 36, Oxford University Press, Oxford, 2008. MR2378253 (2009k:35322)

[12] Hynek Kovařík, *On the lowest eigenvalue of Laplace operators with mixed boundary conditions*, J. Geom. Anal. **24** (2014), no. 3, 1509–1525, DOI 10.1007/s12220-012-9383-4. MR3223564

[13] V. D. Kupradze, T. G. Gegelia, M. O. Basheleĭshvili, and T. V. Burchuladze. *Three-dimensional problems of the mathematical theory of elasticity and thermoelasticity*, volume 25 of *North-Holland Series in Applied Mathematics and Mechanics*. North-Holland Publishing Co., Amsterdam, 1979.

[14] V. Maz'ya and A. Movchan, *Asymptotic treatment of perforated domains without homogenization*, Math. Nachr. **283** (2010), no. 1, 104–125, DOI 10.1002/mana.200910045. MR2598596 (2011a:35038)

[15] Vladimir Maz'ya, Alexander Movchan, and Michael Nieves, *Green's kernels and meso-scale approximations in perforated domains*, Lecture Notes in Mathematics, vol. 2077, Springer, Heidelberg, 2013. MR3086675

[16] Alexander G. Ramm, *Wave scattering by small bodies of arbitrary shapes*, World Scientific Publishing Co. Pte. Ltd., Hackensack, NJ, 2005. MR2311425 (2008a:35002)

DEPARTMENT OF MATHEMATICS, KUWAIT UNIVERSITY, P.O. BOX 13060, SAFAT, KUWAIT
E-mail address: musallam@sci.kuniv.edu.kw

DEPARTMENT OF MATHEMATICS, INHA UNIVERSITY, INCHEON 402-751, S. KOREA
E-mail address: durga.challa@inha.ac.kr

RICAM, AUSTRIAN ACADEMY OF SCIENCES, ALTENBERGERSTRASSE 69, A-4040, LINZ, AUSTRIA
E-mail address: mourad.sini@oeaw.ac.at
Current address: Department of Mathematics, Kuwait University, Kuwait
E-mail address: msini@sci.kuniv.edu.kw

Zerofinding of analytic functions by structured matrix methods

Luca Gemignani

ABSTRACT. We propose a fast and numerically robust algorithm based on structured numerical linear algebra technology for the computation of the zeros of an analytic function inside the unit circle in the complex plane. At the core of our method there are two matrix algorithms: (a) a fast reduction of a certain linearization of the zerofinding problem to a matrix eigenvalue computation involving a perturbed CMV–like matrix and (b) a fast variant of the QR eigenvalue algorithm suited to exploit the structural properties of this latter matrix. We illustrate the reliability of the proposed method by several numerical examples.

1. Introduction

Processes as population dynamics that evolve in discrete time steps are usually modeled using discrete dynamical systems. In this case the stability analysis for an equilibrium state of the system leads to the problem of computing all the zeros of a certain analytic or meromorphic function $f\colon \Omega \subseteq \mathbb{C} \to \mathbb{C}$ within a bounded domain in the complex plane. Ideally, increasingly accurate approximations of the zeros can be found after linearization by means of some efficient eigenvalue solver.

This approach is pursued in [3] where two linearization techniques based on the evaluation of $f(z)$ at roots of unity are investigated. Given the values $f_k = f(e^{2\pi(k-1)/n})$, $1 \leq k \leq n$, attained by $f(z)$ at the n-th roots of unity, polynomial and rational interpolants of $f(z)$ are constructed whose zeros provide approximations of the zeros of $f(z)$ inside the unit circle. The resulting rootfinding computations can then be recast in a matrix setting as a (generalized) eigenproblem for companion–type matrices. If the coefficients of the interpolating polynomial are available then a customary companion matrix or pencil can be formed. Otherwise, working directly from values a generalized eigenvalue problem with arrowhead structure is defined [18, 28].

Although these two matrix eigenproblems are equivalent in the sense that they share the same finite spectrum, for the purpose of numerical computation we favor the second option based on the following reasons. First of all, the conditioning of the eigenvalues depends on the representation of the interpolating polynomial. From our numerical experience, the representation by values generally leads to a better conditioning for the approximations of the roots around the origin in the complex

2010 *Mathematics Subject Classification.* Primary 65H05, 30C15.

plane. This part of the spectrum is relevant to establish the stability features of the system. A theoretical analysis supporting this claim is presented in [11]. Secondly, for large n the balancing of the vector of sampled values can directly be related with the behavior of the absolute value of the function on the unit circle. A similar conclusion does not hold for the coefficient vector which can be highly unbalanced even for small varying functions. Balancing is usually recommended for achieving efficiency and accuracy in numerical computations.

Whatever linearization technique we apply, it results in a rank–structured (generalized) eigenproblem. The exploitation of the structural properties of the resulting matrix eigenvalue problem also leads to a significant improvement of the computational efficiency. In recent years based on the concept of rank structure many authors have provided fast adaptations of the QR/QZ iteration applied to small rank modifications of Hermitian or unitary matrices/pencils [8–10, 12, 17, 20, 32, 33]. However, despite the common framework, there are several differences between the Hermitian and the unitary case which makes the latter much more involved computationally. To circumvent this drawback in [7] a novel approach for solving the perturbed unitary eigenproblem has been proposed. The approach relies upon an initial transformation of the matrix by unitary congruence into a modified CMV–like form [15, 25] with staircase shape. The CMV-like form of a unitary matrix is particularly suited for the application of the QR eigenvalue algorithm [14]. The staircase shape also reveals invariance properties under the same algorithm [2]. Combining these facts together yields a very nice data–sparse parametrization of the matrix which is maintained under the iterative process.

In this paper we extend the CMV–based stuff to solving the generalized eigenproblem associated with the point–value representation of the interpolating polynomial of a given analytic function at the roots of unity. More specifically, we design a fast composite algorithm relying on the following two steps.

(1) The initial arrowhead pencil is converted into a modified CMV–like form. We propose a novel method for the reduction of a unitary diagonal matrix into a CMV–like form by unitary congruence having prescribed the first column of the associated transformation matrix. It turns out that this method applied to the initial matrix pair (A, B) performs the reduction into a modified pair $(\widehat A, B)$, where $\widehat A$ is a CMV–like matrix perturbed by a rank–one correction in its first row. Then we adjust the deflation procedure described in [27] to the unitary setting by transforming the generalized eigenproblem $\widehat A \boldsymbol{x} = \lambda B \boldsymbol{x}$ into a classical eigenproblem $\tilde A \boldsymbol{x} = \lambda \boldsymbol{x}$ having the same finite spectrum. In addition, the matrix $\tilde A$ still inherits the modified CMV–like structure of $\widehat A$. The complexity of the reduction process is quadratic in the size of the matrix.

(2) The fast adaptation of the QR scheme proposed in [7] is applied for the solution of the resulting matrix eigenproblem involving $\tilde A$. The modified CMV–like structure of $\tilde A$ induces the rank structure of the matrices generated under the QR process applied to $\tilde A$. By using a suitable entrywise parametrization of these matrices each iteration can be carried out at a linear time so that the overall complexity remains quadratic.

The composite algorithm is numerically reliable, since each computational step relies on unitary transformations. The computational efficiency is enhanced by the use of data–sparse representations of the involved matrices that are easy to manipulate and update. The overall complexity is quadratic using a linear memory storage.

A detailed outline of the paper is as follows. In Section 2 we formulate the computational problem and develop our algorithm for performing the reductions carried out at step 1. In Section 3 we briefly restate the fast QR iteration presented in [7]. In Section 4 we discuss several numerical examples indicating the robustness and the efficiency of the proposed numerical approach. Finally, the conclusion and further developments are drawn in Section 5.

2. Problem Statement and Basic Reductions

In this section we first introduce the matrix eigenproblem with arrowhead shape resulting from the polynomial interpolation at the roots of unity of an analytic function $f(z)$. Then, we describe the basic reductions of this problem into a modified CMV–like form.

The unique polynomial of degree less than n interpolating the function $f(z)$ at the n–th roots of unity $z_k = e^{2\pi(k-1)/n}$, $1 \leq k \leq n$, can be expressed as

$$p(z) = (z^n - 1) \sum_{j=1}^{n} \frac{w_j f_j}{z - z_j},$$

where

$$f_j = f(z_j), \quad w_j = \left(\prod_{k=1, k \neq j} (z_j - z_k) \right)^{-1} = z_j/n, \quad 1 \leq j \leq n.$$

In [18] it was shown that the roots of $p(z)$ are the finite eigenvalues of the matrix pencil given by

(2.1) $$T(z) = A - zB, \quad A, B \in \mathbb{C}^{(n+1)\times(n+1)},$$

where

(2.2) $$A = \begin{bmatrix} 0 & -\xi_1 f_1 & \cdots & -\xi_n f_n \\ w_1/\xi_1 & z_1 & & \\ \vdots & & \ddots & \\ w_n/\xi_n & & & z_n \end{bmatrix}, \quad B = \begin{bmatrix} 0 & & & \\ & 1 & & \\ & & \ddots & \\ & & & 1 \end{bmatrix},$$

where $\xi_1, \ldots, \xi_m$ are nonzero additional parameters introduced for balancing and symmetry purposes. Observe that since the size of the matrices A, B is $n+1$, then $T(z)$ in (2.1), (2.2) has at least two spurious infinite eigenvalues. In the sequel of this section we present a method to transform the generalized eigenproblem for the matrix pair (2.2) into a classical eigenvalue problem for a perturbed CMV–like matrix. This reduction also incorporates a reliable numerical technique for deflating some possibly infinite eigenvalues.

CMV matrices, introduced in [15] and [30] where the term was coined, are defined as product of Givens transformations arranged in suitable patterns. For a given pair $(\gamma, k) \in \mathbb{D} \times \mathbb{I}_n$, $\mathbb{D} = \{z \in \mathbb{C} : |z| < 1\}$, $\mathbb{I}_n = \{1, 2, \ldots, n-1\}$, we set

(2.3) $$\mathcal{G}_k(\gamma) = I_{k-1} \oplus \begin{bmatrix} \bar{\gamma} & \sigma \\ \sigma & -\gamma \end{bmatrix} \oplus I_{n-k-1} \in \mathbb{C}^{n \times n},$$

where $\sigma \in \mathbb{R}, \sigma > 0$ and $|\gamma|^2 + \sigma^2 = 1$. Similarly, if $\gamma \in \mathbb{S}^1 = \{z \in \mathbb{C} \colon |z| = 1\}$ then denote
$$\mathcal{G}_n(\gamma) = I_{n-1} \oplus \gamma \in \mathbb{C}^{n \times n}.$$

The following definition identifies an important class of structured unitary matrices.

DEFINITION 2.1. [15] For a given coefficient sequence $(\gamma_1, \ldots, \gamma_{n-1}, \gamma_n) \in \mathbb{D}^{n-1} \times \mathbb{S}^1$ we introduce the unitary block diagonal matrices
$$\mathcal{L} = \mathcal{G}_1(\gamma_1) \cdot \mathcal{G}_3(\gamma_3) \cdots \mathcal{G}_{2\lfloor \frac{n+1}{2} \rfloor - 1}(\gamma_{2\lfloor \frac{n+1}{2} \rfloor - 1}), \quad \mathcal{M} = \mathcal{G}_2(\gamma_2) \cdot \mathcal{G}_4(\gamma_4) \cdots \mathcal{G}_{2\lfloor \frac{n}{2} \rfloor}(\gamma_{2\lfloor \frac{n}{2} \rfloor}),$$
and define
$$(2.4) \qquad \mathcal{C} = \mathcal{L} \cdot \mathcal{M}$$
as the CMV matrix associated with the prescribed coefficient list.

The decomposition (2.4) of a unitary matrix was first investigated for eigenvalue computation in [1] and [14]. The shape of CMV matrices is analyzed in [25] where the next definition is given.

DEFINITION 2.2. [25] A matrix $F \in \mathbb{C}^{n \times n}$ has CMV shape if the possibly nonzero entries exhibit the following pattern where + denotes a positive entry:

$$F = \begin{bmatrix} \star & \star & + & & & & & \\ + & \star & \star & & & & & \\ & & \star & \star & \star & + & & \\ & & + & \star & \star & \star & & \\ & & & & \star & \star & \star & + \\ & & & & + & \star & \star & \star \\ & & & & & & \star & \star & \star \\ & & & & & & + & \star & \star \end{bmatrix}, \quad (n = 2k),$$

or

$$F = \begin{bmatrix} \star & \star & + & & & & \\ + & \star & \star & & & & \\ & & \star & \star & \star & + & \\ & & + & \star & \star & \star & \\ & & & & \star & \star & \star & + \\ & & & & + & \star & \star & \star \\ & & & & & & \star & \star \end{bmatrix}, \quad (n = 2k-1).$$

The definition is useful for computational purposes since shapes are easier to check than comparing the entries of the matrix. Obviously, CMV matrices have a CMV shape and, conversely, it is shown that a unitary matrix with CMV shape is CMV [16]. The positiveness of the complementary parameters σ_k in (2.3) as well as of the entries marked with + in Definition 2.2 is necessary to establish the connection of CMV matrices with corresponding sequences of orthogonal polynomials on the unit circle [25]. From the point of view of eigenvalue computation, however, this condition can be relaxed. In [6] we simplify the above definition by skipping the positiveness condition on the entries denoted as +. The fairly more general class of matrices considered in [6] is referred to as CMV–like matrices. There it is also shown that the block Lanczos method can be used to reduce a unitary matrix into the direct sum of CMV–like matrices. Here we pursue a different approach relying upon the fact that the reduction of the arrow matrix pencil (2.1), (2.2) into a perturbed CMV–like form follows from computing a unitary matrix $Q \in \mathbb{C}^{n \times n}$ such that

(2.5) $$Q^H D Q = F, \quad Q^H \boldsymbol{w} = \alpha \boldsymbol{e}_1,$$

where $D = \mathrm{diag}\,[z_1, \ldots, z_n]$, F is CMV–like and, moreover,

$$A = \begin{bmatrix} 0 & -\xi_1 f_1 & \cdots & -\xi_n f_n \\ w_1/\xi_1 & z_1 & & \\ \vdots & & \ddots & \\ w_n/\xi_n & & & z_n \end{bmatrix} = \left[\begin{array}{c|c} 0 & -\boldsymbol{f}^H \\ \hline \boldsymbol{w} & D \end{array}\right].$$

Observe that the bordered unitary matrix

$$\widehat{Q} = \left[\begin{array}{c|c} 1 & \boldsymbol{0}^T \\ \hline \boldsymbol{0} & Q \end{array}\right]$$

satisfies

$$\widehat{Q}^H A \widehat{Q} = \left[\begin{array}{c|c} 0 & \widehat{\boldsymbol{f}}^H \\ \hline \alpha \boldsymbol{e}_1 & F \end{array}\right], \quad \widehat{Q}^H B \widehat{Q} = B.$$

In addition from (2.5) there follows that

$$Q^H D \boldsymbol{w} = F Q^H \boldsymbol{w} = \alpha F \boldsymbol{e}_1,$$

which means that the entries of $Q^H D \boldsymbol{w}$ from 3 to n are zero.

In the next subsection we present an efficient solution of (2.5) based on the recognition of (2.5) as an inverse eigenvalue problem (IEP) associated with the reduction to CMV–like form.

2.1. Efficient Recursive Solution of the Associated IEP.
The solvability of (2.5) is addressed in [6] by using a constructive approach relying upon the block–Lanczos method. We just recall the main result here. For the sake of simplicity we consider the case where n is even.

THEOREM 2.3. *Let $\mathcal{D}_+ = \frac{D+D^H}{2}$ and $\mathcal{D}_- = \frac{D-D^H}{2}$ be such that $D = \mathcal{D}_+ + \mathcal{D}_-$. Then the block–Lanczos procedure applied to $\mathcal{D}_+$ with initial vectors $[\boldsymbol{w}|D\boldsymbol{w}]$ produces a unitary matrix $Q \in \mathbb{C}^{n \times n}$ satisfying:*

(1) $[\boldsymbol{w}|D\boldsymbol{w}] = Q(:,1:2) \cdot R$ for an upper triangular R;

(2) $Q^H \mathcal{D}_+ Q$, $Q^H \mathcal{D}_- Q$ and, a fortiori, $Q^H D Q$ are block tridiagonal matrices with 2×2 diagonal and off-diagonal blocks.

(3) By denoting with B_i, C_i, $1 \leq i \leq n/2 - 1$, the subdiagonal and superdiagonal blocks, respectively, in the block tridiagonal reduction of D it is found that $\mathrm{rank}\,(B_i) \leq 1$ and $\mathrm{rank}\,(C_i) \leq 1$ and, moreover,

$$B_i = \begin{bmatrix} 0 & \star \\ 0 & \star \end{bmatrix}, \quad C_i = \begin{bmatrix} \star & 0 \\ \star & 0 \end{bmatrix},$$

where $\star$ denotes a possibly nonzero entry.

The construction advocated in the previous theorem can suffer from numerical instabilities and premature breakdowns demanding for restarting techniques. An Householder–style implementation of the Lanczos reduction is described in [25]. An alternative approach based on the Householder (block) tridiagonalization algorithm applied to $\mathcal{D}_+$ is also presented in [6] in order to circumvent these difficulties. Hereafter we further elaborate on this modification by devising a novel reduction scheme using unitary transformations which has the advantage of working directly on the matrix D rather than $\mathcal{D}_+$. Our approach relies upon the recognition of (2.5) as an inverse eigenvalue problem (IEP). Recursive solution methods are customary in this field (see [13], [31] and the references given therein) and can be extended to deal with the present case.

The construction of Q can be carried out by a recursive algorithm which acts on the enlarged matrix

$$\mathbb{C}^{(n+2)\times(n+2)} \ni \widehat{A} = \begin{bmatrix} 0 & 0 & \mathbf{0}^T \\ 0 & 0 & -\boldsymbol{f}^H \\ \hline \boldsymbol{w} & D\boldsymbol{w} & D \end{bmatrix}$$

by annihilating entries of the first two columns and then by returning the trailing submatrix to CMV–like form. For the sake of simplicity we restrict ourselves to the case where n is even. The procedure is recursive. Suppose that $Q_{n-k} \in \mathbb{C}^{(n-k)\times(n-k)}$ solves (2.5) for the input data $D_{n-k} = D(k+1\colon n, k+1\colon n)$ and $\boldsymbol{w}^{(n-k)} = \boldsymbol{w}(k+1\colon n)$, that is,

$$Q_{n-k}^H D_{n-k} Q_{n-k} = F_{n-k}, \quad Q_{n-k}^H \boldsymbol{w}^{(n-k)} = \alpha_{n-k} \boldsymbol{e}_1.$$

It is found that the unitary matrix $\widehat{Q}_{n-k} = I_{k+2} \oplus Q_{n-k}$ satisfies $\widehat{Q}_{n-k}^H \widehat{A} \widehat{Q}_{n-k} = \widehat{A}_{n-k}$, where

$$\widehat{A}_{n-k}(k-1\colon n, 1\colon n+2) = \left[\begin{array}{c|cc|c} W_{n-k} & z_{k-1} & & \\ & & z_k & \\ \hline R_{n-k} & & & F_{n-k} \end{array}\right],$$

where $W_{n-k}, R_{n-k} \in \mathbb{C}^{2\times 2}$ and R_{n-k} is upper triangular. Then we can determine a unitary matrix $\mathcal{G}_{n-k}$ such that

$$\mathcal{G}_{n-k}^H \begin{bmatrix} W_{n-k} \\ R_{n-k} \end{bmatrix} = \widehat{R}_{n-k-2} = \begin{bmatrix} R_{n-k+2} \\ 0 & 0 \\ 0 & 0 \end{bmatrix}.$$

Let

$$(I_k \oplus \mathcal{G}_{n-k} \oplus I_{n-k-2})^H \widehat{A}_{n-k} (I_k \oplus \mathcal{G}_{n-k} \oplus I_{n-k-2}) = \tilde{A}_{n-k}$$

be partitioned as follows

$$\tilde{A}_{n-k}(k-1\colon n, 1\colon n+2) = \left[\begin{array}{c|c|c} R_{n-k+2} & O_{n-k+2,k-2} & \tilde{F}_{n-k+2} \end{array}\right],$$

where

$$\tilde{F}_{n-k+2} = (\mathcal{G}_{n-k} \oplus I_{n-k-2})^H \left[\begin{array}{cc|c} z_{k-1} & & \\ & z_k & \\ \hline & & F_{n-k} \end{array}\right] (\mathcal{G}_{n-k} \oplus I_{n-k-2}).$$

By denoting

$$\widehat{Q}_{n-k+2} = (I_2 \oplus Q_{n-k}) \cdot (\mathcal{G}_{n-k} \oplus I_{n-k-2}),$$

this implies that

(2.6) $$\widehat{Q}_{n-k+2}^H D_{n-k+2} \widehat{Q}_{n-k+2} = \tilde{F}_{n-k+2},$$

and, moreover,

(2.7) $$\widehat{Q}_{n-k+2}^H \boldsymbol{w}^{(n-k+2)} = (R_{n-k+2})_{1,1} \boldsymbol{e}_1,$$

$$\hat{Q}^H_{n-k+2} D_{n-k+2} \boldsymbol{w}^{(n-k+2)} = (R_{n-k+2})_{1,2} \boldsymbol{e}_1 + (R_{n-k+2})_{2,2} \boldsymbol{e}_2. \quad (2.8)$$

Now, the following properties can easily be deduced.

(1) As $\tilde{F}_{n-k+2}, \hat{Q}_{n-k+2}$ fulfill (2.6), (2.7) and (2.8) for a partial set of input data we obtain that the first column of $\tilde{F}_{n-k+2}$ is of the form $\rho \boldsymbol{e}_1 + \xi \boldsymbol{e}_2$ for suitable ρ, ξ.

(2) Since $\tilde{F}_{n-k+2}$ is unitary, from the *nullity theorem* [22] this also implies that $\tilde{F}_{n-k+2}(1\colon 2, 3\colon n-k+2)$ is a matrix of rank one at most.

(3) As $\tilde{F}_{n-k+2}(1\colon 2, 4\colon 5)$ is annihilated by a unitary transformation on the right then due to the *nullity theorem* a staircase pattern is identified in the lower triangular portion of the modified $\tilde{F}_{n-k+2}$ and, therefore, the rank properties are transmitted to the trailing submatrices.

By combining these three facts together we find that there exists a matrix $\tilde{Q}_{n-k+2}$ such that $\tilde{Q}^H_{n-k+2} \tilde{F}_{n-k+2} \tilde{Q}_{n-k+2} = F_{n-k+2}$ is a CMV–like matrix and, moreover, $\tilde{Q}_{n-k+2}$ only acts on the columns of $\tilde{F}_{n-k+2}$ in position 3 through $n-k+2$. And thus, finally, by setting

$$Q_{n-k+2} = \hat{Q}_{n-k+2} \cdot \tilde{Q}_{n-k+2},$$

we may conclude that this latter matrix solves (2.5) for the extended set of input data D_{n-k+2} and $\boldsymbol{w}^{(n-k+2)}$.

The complete reduction scheme is stated below in algorithmic form. For the sake of notational convenience we assume to do nothing whenever index exceeds matrix dimensions.

Procedure CMV_Reduce
Input: $n = 2k$, $n2 = n+2$, $\boldsymbol{w}, \boldsymbol{f} \in \mathbb{C}^n$;
Output: $Q, F \in \mathbb{C}^{n \times n}$ as in (2.5);
  for $k = n2\colon -2\colon 6$
    $E = \widehat{A}(k-3\colon k, 1\colon 2)$; $[Q, R] = \mathtt{qr}(E)$;
    $\widehat{A}(k-3\colon k, \colon) = Q^H \widehat{A}(k-3\colon k, \colon)$; $\widehat{A}(\colon, k-3\colon k) = \widehat{A}(\colon, k-3\colon k)Q$;
    for $j = k\colon 2\colon n2$
      $E = \widehat{A}(j-3\colon j-2, j\colon j+1)$; $E = E^H$; $[Q, R] = \mathtt{qr}(E)$;
      $\widehat{A}(\colon, j\colon j+1) = \widehat{A}(\colon, j\colon j+1)Q$; $\widehat{A}(j\colon j+1, \colon) = Q^H \widehat{A}(j\colon j+1, \colon)$;
      $E = \widehat{A}(j+1\colon j+2, j-2)$; $[Q, R] = \mathtt{qr}(E)$; $\tilde{j} = j+1$
      $\widehat{A}(\colon, \tilde{j}\colon \tilde{j}+1) = \widehat{A}(\colon, \tilde{j}\colon \tilde{j}+1)Q$; $\widehat{A}(\tilde{j}\colon \tilde{j}+1, \colon) = Q^H \widehat{A}(\tilde{j}\colon \tilde{j}+1, \colon)$;
      $E = \widehat{A}(j-3\colon j-2, j-1\colon j)$; $E = E^H$; $[Q, R] = \mathtt{qr}(E)$;
      $\widehat{A}(\colon, j-1\colon j) = \widehat{A}(\colon, j-1\colon j)Q$; $\widehat{A}(j-1\colon j, \colon) = Q^H \widehat{A}(j-1\colon j, \colon)$;
      $E = \widehat{A}(j\colon j+1, j-2)$; $[Q, R] = \mathtt{qr}(E)$;
      $\widehat{A}(\colon, j\colon j+1) = \widehat{A}(\colon, j\colon j+1)Q$; $\widehat{A}(j\colon j+1, \colon) = Q^H \widehat{A}(j\colon j+1, \colon)$;
    end
  end

Since each computation involves unitary matrices of size $m \leq 4$ and the sparse form of $\widehat{A}$ is restored at each step it is easy to show that **CMV_Reduce** has a cost of $O(n^2)$ flops. For the sake of illustration in Figure 1 we visualize the sparsity pattern of the matrix $\widehat{A}$ generated by the **CMV_Reduce** procedure at different times.

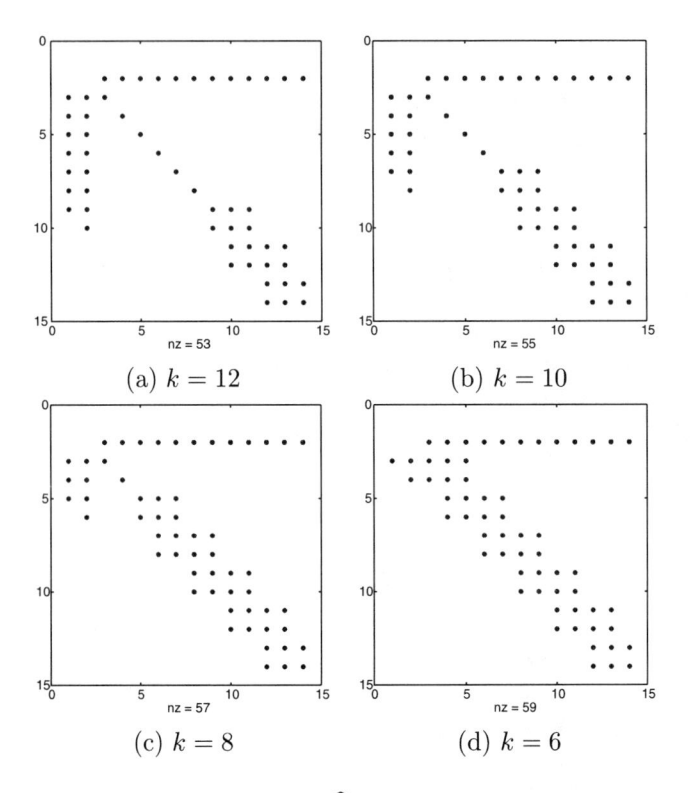

FIGURE 1. Sparsity pattern of $\widehat{A}$ of size $n2 = 14$ at the end of the for cycle with $k = n2 - 2, n2 - 4, n2 - 6$ and $n2 - 8$.

We conclude this section with two remarks concerning the reduction of (2.2) into a CMV–like pencil.

REMARK 2.4. It should be worth noticing that if $\xi_1 = z_1, \ldots, \xi_n = z_n$ and $\boldsymbol{w} = (1/n)\left[1, \ldots, 1\right]^T$ then an alternative reduction scheme of (2.2) into a CMV–like pencil follows from the properties of the Fourier matrix $\Omega = (1/\sqrt{n})(\omega^{(i-1)(j-1)})$ with $\omega = z_2$. In fact it is well known that

$$\Omega^H D \Omega = \begin{bmatrix} 0 & & & 1 \\ 1 & 0 & & 0 \\ & \ddots & \ddots & \vdots \\ & & 1 & 0 \end{bmatrix}, \quad \Omega^H \boldsymbol{w} = \alpha \boldsymbol{e}_1.$$

Then by making use of the permutation matrix P defined in [7] it can easily be seen that $Q = \Omega P$ solves (2.5) and, hence, $Q = 1 \oplus \Omega P$ converts the input arrowhead pencil (2.1), (2.2), into the permuted companion form considered in [7].

REMARK 2.5. The Lanczos methods proposed in [6] and [25] for the reduction of a unitary matrix to CMV–like form can break down for suitable choices of the vector $\boldsymbol{w}$ or when the starting matrix has multiple eigenvalues. It is clear that the possibly coalescence of nodes produces degeneracies also in previous algorithm **CMV_Reduce**.

2.2. Deflation Technique.

Once the modified matrix pair (A_1, B_1), $A_1 = \widehat{Q}^H A \widehat{Q}$, $B_1 = B$, has been constructed then a procedure like that one proposed in [27] can be carried out in order to deflate the two spurious infinite eigenvalues. Specifically we proceed in the following steps.

(1) At first, we swap the first two rows of A_1 and B_1 by using a permutation matrix P_1 on the left. Since $\widehat{A}_1 = P_1 A_1$ is block upper triangular and $\widehat{B}_1 = P_1 B_1$ is upper triangular we can delete the first row and column of $\widehat{A}_1$ and $\widehat{B}_1$ by obtaining the novel matrix pair (A_2, B_2) with $A_2 = \widehat{A}_1(2\colon n+1; 2\colon n+1)$ and $B_2 = \widehat{B}_1(2\colon n+1; 2\colon n+1)$. Observe that A_2 can be expressed as a CMV–like unitary matrix perturbed by a rank–one correction located in its first row.

(2) Then we determine a 2×2 Givens rotation matrix which acts on the first two rows of A_2 by annihilating the first subdiagonal entry. By applying the same transformation to B_2 we find that the modified matrices $\widehat{A}_2$ and $\widehat{B}_2$ are in block upper triangular and upper triangular form, respectively. Thus we can again perform a deflation step by removing the second spurious infinite eigenvalue and reducing the size of A_2 and B_2 by one. The final matrix pair is (A_3, B_3), where $A_3, B_3 \in \mathbb{C}^{(n-1)\times(n-1)}$ and

$$B_3 = \begin{bmatrix} \sigma & & & \\ & 1 & & \\ & & \ddots & \\ & & & 1 \end{bmatrix},$$

for a possibly nonzero scalar σ. Also it should be noticed that A_3 can be further written as a rank–one modification of a CMV–like unitary matrix so that at the very end of this process whenever $\sigma \neq 0$ we obtain that the spectrum of the structured matrix $B_3^{-1} A_3$ gives suitable approximations of the zeros of $p(z)$.

Summing up, if the deflation procedure succeeds then the generalized eigenproblem $A_1 \boldsymbol{x} = \lambda B_1 \boldsymbol{x}$ is reduced to the classical eigenvalue problem $A \boldsymbol{x} = \lambda \boldsymbol{x}$, where $A\colon = B_3^{-1} A_3$ is a CMV–like unitary matrix perturbed by a rank–one correction located in its first row. The structure of this latter matrix will be exploited in the next section for the design of an efficient eigensolver. It is worth pointing out that theoretically, if $\sigma = 0$ then the deflation procedure may be continued by preserving the structural properties of the pencil. However, in practice, from a numerical point of view the critical issue to be addressed is the occurrence of a quite small but nonzero value of σ. We will go back on this later in Section 4.

3. A Fast Structured QR Algorithm

In this section we are going to develop a fast variant of the customary QR eigenvalue algorithm for an input matrix $A = F - \boldsymbol{e}_1 \boldsymbol{w}^H$ which can be represented as a CMV–like unitary matrix plus a rank–one correction located in the first row. The derivation follows largely that in [7], where the interested reader is directed for more details. Our approach is able to exploit both the staircase form of A and its perturbed CMV–like representation.

Staircase matrix patterns can be exploited for eigenvalue computation [2]. The shifted QR algorithm

(3.1) $$\begin{cases} A_s - \rho_s I_n = Q_s R_s \\ A_{s+1} = Q_s^H A_s Q_s, \quad s \geq 0, \end{cases}$$

is the standard algorithm for computing the Schur form of a general matrix $A = A_0 \in \mathbb{C}^{n \times n}$ [23]. The matrix A is said to be staircase if $m_j(A) \geq m_{j-1}(A)$, $2 \leq j \leq n$, where

$$m_j(A) = \max\{j, \max_{i>j}\{i \colon a_{i,j} \neq 0\}\}.$$

The staircase form is preserved under the QR iteration (3.1) in the sense that [2]

$$m_j(A_{s+1}) \leq m_j(A_s), \quad 1 \leq j \leq n.$$

For Hermitian and unitary matrices the staircase form also implies a zero pattern or a rank structure in the upper triangular part. The invariance of this pattern by the QR algorithm is proved in [2] for Hermitian matrices and in [14] for unitary CMV–shaped matrices. An alternative proof for the unitary case that is suitable for generalizations is given in [6] by relying upon the classical nullity theorem [22].

It has already been noticed above that the staircase form of $A_0 =: A$ is preserved under the shifted QR iteration (3.1). This means that each unitary matrix Q_s is also in staircase form. From

$$A_0 = F - e_1 w^H = F_0 - z_0 w_0^H$$

it follows that

(3.2) $\quad A_{s+1} = Q_s^H A_s Q_s = Q_s^H (F_s - z_s w_s^H) Q_s = F_{s+1} - z_{s+1} w_{s+1}^H, \quad s \geq 0,$

where

(3.3) $\quad F_{s+1} := Q_s^H F_s Q_s, \quad z_{s+1} := Q_s^H z_s, \quad w_{s+1} := Q_s^H w_s.$

These relations enable the entries of each A_s to be represented in terms of a linear number of parameters. Hereafter we recall the main result in [7].

THEOREM 3.1. *For any $s \geq 0$ the unitary matrix F_s satisfies*

$$\mathrm{rank}\,(F_s(1:2j, 2(j+1)+1:n)) \leq 1, \quad 1 \leq j \leq \lfloor \frac{n+1}{2} \rfloor - 2, \ s \geq 0.$$

Moreover, if A_0 is invertible then

$$F_s(1:2j, 2(j+1)+1:n) = B_s(1:2j, 2(j+1)+1:n), \quad 1 \leq j \leq \lfloor \frac{n+1}{2} \rfloor - 2, \ s \geq 0,$$

where

(3.4) $$B_s = \frac{F_s w_s z_s^H F_s}{z_s^H F_s w_s - 1} = Q_s^H B_{s-1} Q_s, \quad s \geq 1,$$

is a rank one matrix.

From the previous theorem we derive a structural representation of each matrix A_s, $s \geq 0$, generated under the QR process (3.1) applied to A_0. Let us start by observing that each matrix A_s, $s \geq 0$, generated by (3.1) can be represented by means of the following sparse data set of size $O(n)$:

(1) the nonzero entries of the banded matrix $\widehat{A}_s \in \mathbb{C}^{n \times n}$ obtained from A_s according to

$$\widehat{A}_s = (\hat{a}_{i,j}^{(s)}), \quad \hat{a}_{i,j}^{(s)} = \begin{cases} 0, & \text{if } j \geq 2\lfloor \frac{i+1}{2} \rfloor + 3,\ 1 \leq i \leq 2\lfloor \frac{n+1}{2} \rfloor - 4; \\ a_{i,j}^{(s)}, & \text{elsewhere;} \end{cases}$$

(2) the vectors $\mathbf{z}_s = (z_i^{(s)}), \mathbf{w}_s = (w_i^{(s)}) \in \mathbb{C}^n$ and $\mathbf{f}_s := F_s \mathbf{w}_s, \mathbf{f}_s = (f_i^{(s)})$, and $\mathbf{g}_s := F_s^H \mathbf{z}_s, \mathbf{g}_s = (g_i^{(s)})$.

The nonzero pattern of the matrix $\widehat{A}_s$ looks as below:

$$\widehat{A}_s = \begin{bmatrix} \star & \star & \star & \star & & & & & \\ \star & \star & \star & \star & & & & & \\ & & \star & \star & \star & \star & \star & & \\ & & \star & \star & \star & \star & \star & & \\ & & & & \star & \star & \star & \star & \star \\ & & & & \star & \star & \star & \star & \star \\ & & & & & & \star & \star & \star \\ & & & & & & \star & \star & \star \end{bmatrix}, \quad (n = 2k),$$

or

$$\widehat{A}_s = \begin{bmatrix} \star & \star & \star & \star & & & \\ \star & \star & \star & \star & & & \\ & & \star & \star & \star & \star & \star \\ & & \star & \star & \star & \star & \star \\ & & & & \star & \star & \star \\ & & & & \star & \star & \star \\ & & & & & \star & \star \end{bmatrix}, \quad (n = 2k - 1).$$

From (3.2) and (3.4) we find that the entries of the matrix $A_s = (a_{i,j}^{(s)})$ can be expressed in terms of elements of this data set as follows:
(3.5)
$$a_{i,j}^{(s)} = \begin{cases} -\sigma^{-1} f_i^{(s)} \bar{g}_j^{(s)} - z_i^{(s)} \bar{w}_j^{(s)}, & \text{if } j \geq 2\lfloor \frac{i+1}{2} \rfloor + 3,\ 1 \leq i \leq 2\lfloor \frac{n+1}{2} \rfloor - 4; \\ \hat{a}_{i,j}^{(s)}, & \text{elsewhere;} \end{cases}$$

where $\sigma = 1 - \mathbf{z}_s^H F_s \mathbf{w}_s = 1 - \mathbf{z}_0^H F_0 \mathbf{w}_0$.

A fast adaptation of the QR iteration (3.1) applied to a starting invertible matrix $A_0 = F - \mathbf{e}_1 \mathbf{w}^H \in \mathbb{C}^{n \times n}$ by using the structural properties described above is devised in [7]. At each step this method works on a condensed entrywise data–sparse representation of the matrix using $O(n)$ flops and $O(n)$ memory storage. An efficient MatLab[1] implementation of the fast iteration called **Fast_QR** is also provided in [7]. This implementation is one main building block of our composite algorithm for the approximation of the zeros of an analytic function presented in the next section.

4. Numerical Results

Based on the results stated in the previous sections we propose the following composite algorithm for the computation of the zeros of an analytic function $f : \mathbb{C} \to \mathbb{C}$ inside the unit circle in the complex plane.

---

[1]Matlab is a registered trademark of The MathWorks, Inc..

pencil_eig	poly_roots
4.7e-08+3.4e-08i	-1.2e-03 -1.8e-03i
4.7e-08+3.4e-08i	1.2e-03 +1.8e-03i
4.7e-08+3.4e-08i	-2.2e-03 +1.6e-04i
4.7e-08+3.4e-08i	2.2e-03-1.6e-04i
-1.1e-08 -1.8e-07i	9.7e-04-2.0e-03i
-1.8e-07+4.8e-08i	-9.7e-04+2.0e-03i

TABLE 1. Approximations of the multiple zero at the origin returned by `eig` applied to the arrowhead pencil and `roots` applied to the interpolating polynomial.

(1) For a given $n \in \mathbb{N}$ form the arrowhead pair (A, B) defined in (2.1), (2.2), where for balancing issues the additional parameters ξ_j are set to $\xi_j = 1/\sqrt{|f(z_j)|}$ for $f(z_j) \neq 0$.
(2) Apply the procedure **CMV_Reduce** by transforming the input arrowhead pair (A, B) into (A_1, B) where A_1 is a CMV–like matrix modified by a rank–one correction in its first row.
(3) Apply the **deflation procedure** in order to reduce the generalized eigenvalue problem for (A_1, B) to a classical eigenvalue problem for a matrix A which is still a CMV–like matrix modified by a rank–one correction in its first row.
(4) Perform the **Fast_QR** procedure applied to the input matrix A for the computation of the desired approximations of the zeros of $f(z)$.

The complexity of the algorithm is $O(n^2)$ flops using $O(n)$ memory storage. A Matlab implementation of our composite algorithm has been realized for testing and experimental purposes. The following examples indicate the efficiency and the robustness of the algorithm. In each experiment we compute the finite spectrum of the arrowhead pencil using the Matlab function `eig`, we match these entries with the approximations returned as output by the **Fast_QR** procedure and compute the maximum absolute error.

EXAMPLE 4.1. The first numerical example is preparatory. Let us consider the nonlinear equation
$$f(z) = z^3 \sin(5 (z/2)^3) = 0.$$
In table 1 we show the approximations of the multiple zero at the origin obtained by computing the eigenvalues of the arrowhead pencil defined in (2.1), (2.2) with $n = 64$ and by the command `roots` applied to the coefficients of the interpolating polynomial at the same roots of unity. The results confirm some experimental observations in [11] about the better accuracy of the arrowhead linearization compared with the classical companion form for zeros clustered around the origin in the complex plane.

The second preliminary test concerns the robustness of our approach and specifically the accuracy of the **CMV_Reduce** procedure. In particular, recall that Remark 2.5 indicates that numerical difficulties can arise for large n when the

n	256	512	1024
$err(\boldsymbol{z})$	6.7e-14	1.11e-11	2.9e-06
$err(\hat{\boldsymbol{z}})$	1.4e-14	7.8e-14	9.4e-12

TABLE 2. Error in matching the finite spectra of the pencil in arrow form and the modified pencil under different orderings.

n	128	256	512	1024
err	1.2e-14	4.9e-14	1.3e-13	6.7e-11

TABLE 3. Error in matching the spectra of the pencil in arrow form and the perturbed CMV–like matrix.

separation of the nodes approaches zero. We consider the polynomial function
$$f_1(z) = \sum_{j=0}^{n-1}(j+1)z^j.$$
From the Eneström-Kakeya theorem [24] it follows that the roots of $f_1(z)$ live in the interior of the unit disc. In table 2 we show the errors between the finite spectrum of (A, B) and (A_1, B), where A_1 is generated by means of **CMV_Reduce** applied with the same starting vector $\boldsymbol{w}$ respectively to $D = \mathtt{diag}(\boldsymbol{z})$, $z_k = e^{2\pi(k-1)/n}$, $1 \leq k \leq n$, and to $D = \mathtt{diag}(\hat{\boldsymbol{z}})$, where $\hat{\boldsymbol{z}}$ is the vector produced by the following reordering of the nodes aimed to distance contiguous points:
$$z_{2j-1} = e^{2\pi(j-1)/n}, \quad z_{2j} = -z_{2j-1}, \quad 1 \leq j \leq n/2.$$

Finally, the next table 3 shows the overall errors generated for different values of n starting with the permuted vector of nodes. The condition numbers of the eigenvalues of the perturbed CMV–like matrix of order $n = 1024$ are of order $1.0e+2$ and, therefore, the absolute errors are within the theoretical estimates.

EXAMPLE 4.2. In [19] the problem of computing the zeros of the following function is considered
$$f_2(z) = z^{50} + z^{12} - 5\sin(20z)\cos(5z) - 1.$$
In Figure 2 we plot the the spectra of the pencil in arrow form and the perturbed CMV–like matrix for different values of n.

The test is very hard. The infinity norm of the vector $f(\boldsymbol{z}) = (f(z_j))$ is of order $9.8e+13$, the maximum condition number of the eigenvalues of $A^{-1}B$ is of order $1.0e+8$ and the value of σ is of order $1.0e-08$. It can be observed that for $n = 256$ there is a large absolute error in one computed eigenvalue. The condition numbers of the eigenvalues of the perturbed CMV–like matrix range from $1.0e+8$ to $1.0e+15$ due to the quite small (but nonzero) value of the σ parameter occurring in the deflation procedure. Notwithstanding that, the relevant part of the spectrum, that is, the approximations approaching the zeros of $f_2(z)$ are computed quite accurately. By evaluating the residuals we find that there are 21 "good"

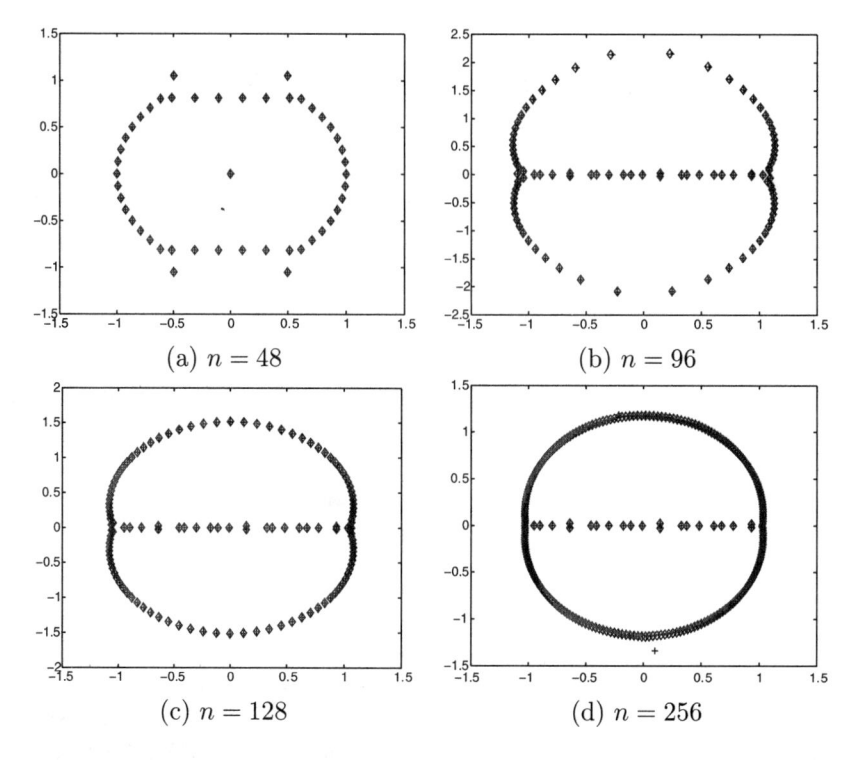

FIGURE 2. Plot of the spectra of the arrowhead pencil (diamond symbol) and the perturbed CMV–like matrix (plus symbol).

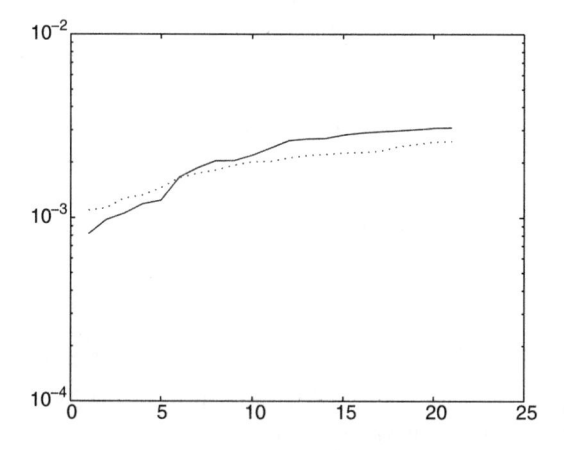

FIGURE 3. Comparison of the residuals generated by the arrowhead pencil (continuous line) and the perturbed CMV–like matrix (dotted line).

approximations. In the next figure 3 we illustrate the semi–log scale plot of the residual for the approximations generated by the arrowhead pencil (continuous line) and the perturbed CMV–like matrix (dotted line).

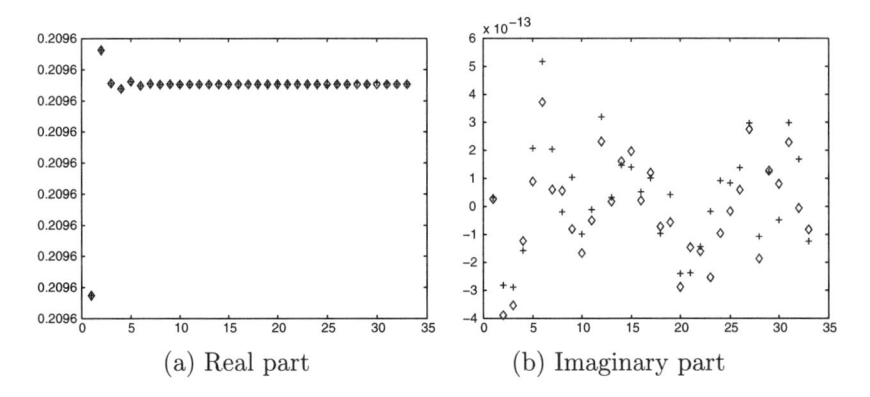

FIGURE 4. Approximations of λ generated from the spectra of the arrowhead pencil (diamond symbol) and the perturbed CMV–like matrix (plus symbol).

EXAMPLE 4.3. To further investigate the approximation of the spectrum of nonlinear eigenproblems around the origin in the complex plane we consider the scaled Hadeler problem in the NLEVP library [5], that is,

$$\det(T(z)/\gamma(z)) = 0, \quad T(z) = (e^z - 1)B_1 + z^2 B_2 - B_0,$$

where $B_i \in \mathbb{R}^{8\times 8}$ are defined as

$$B_0 = 100 I_8, \quad (B_1)_{i,j} = (9 - \max(i,j))ij, \quad (B_2)_{i,j} = 1/(i+j) + 8\delta_{i,j},$$

with $\delta_{i,j}$ the Kronecker delta symbol and, moreover,

$$\gamma(z) = (\det(B_1))^{1/8}(e^{z/2} - 1) + (\det(B_2))^{1/8}(z/2)^2 + (\det(B_0))^{1/8}.$$

It is known that there is a real zero $\lambda \simeq 0.21$ close to the origin [29]. We compute the eigenvalues of the arrowhead pencil and the perturbed CMV–like matrix for any even n ranging from 64 to 128. From these spectra we determine the best approximations of λ. The condition numbers of the eigenvalues of the CMV–like matrix range from $1.0e+07$ to $1.0e+15$. The parameter σ occurring in the deflation procedure is generally small in magnitude (less than $1.0e-07$). For instance, for $n = 86$ we find that the infinity norm of $f(z)$ is of order $1.1e+4$ and `polyeig` applied to (A, B) returns an eigenvalue $\lambda = 2.095592506490536e-01 + 2.317181655677081e-13i$ with estimated condition number of order $1.0e+5$. The deflation procedure finds a value of σ of order $1.0e-12$. Notwithstanding that, the fast QR algorithm applied to the deflated matrix computes $\lambda = 2.095592506545526e-01 + 8.022746309423756e-13i$.

In the next figure 4 we plot the real and the imaginary parts of the approximations of λ computed from the spectra for different values of $n = 64 + 2(j-1)$, $1 \leq j \leq 33$.

We see that the computed approximations of λ persist to be accurate independently of the conditioning upper bounds.

EXAMPLE 4.4. The location of the zeros of the holomorphic function

$$f_3(z) = a + bz + z^2 - hz^2 e^{-\tau z},$$

determines the stability of a steady state solution of a neutral functional differential equation [21]. Similar models are also used to study the stability of a flow

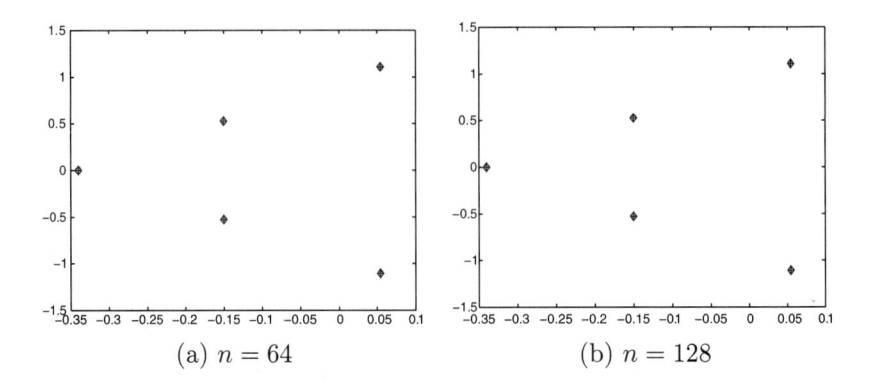

FIGURE 5. Approximations of the zeros of $f_3(z)$ generated from the spectra of the arrowhead pencil (diamond symbol) and the perturbed CMV–like matrix (plus symbol).

$n = 64$	3.7e-15	3.7e-15	5.6e-15	1.6e-12	6.4e-13
$n = 128$	7.8e-15	5.0e-15	5.4e-15	1.3e-09	1.8e-09

TABLE 4. Errors for the approximations of the five zeros generated starting from the initial pencil in arrow form and the perturbed CMV–like matrix

inside an annular combustion chamber [19]. The case where $a = 1, b = 0.5, h = -0.82465048736655, \tau = 6.74469732735569$ is analyzed in [26] corresponding to a Hopf bifurcation point. In figure 5 we compare the approximations of the zeros of $f_3(z)$ generated from the spectra of the arrowhead pencil and the perturbed CMV–like matrix.

For $n = 128$ we find that the infinity norm of $f(z)$ is $7.0e + 02$, the magnitude of the parameter σ arising in the deflation procedure of order $1.0e-13$ and among the considered approximations three eigenvalues of (A, B) have condition numbers of order $1.0e + 03$ while the remaining two eigenvalues have condition numbers of order $1.0e + 08$. In table 4 for $n = 64, 128$ we report the absolute errors of the five approximations displayed in the figure generated starting from the initial pencil in arrow form and the perturbed CMV–like matrix.

EXAMPLE 4.5. In [4] a mathematical model of cancer growth is proposed which consists of three linear delay differential equations. The stability of the model reduces to investigate the root distribution around the origin in the complex plane of the function
$$f_4(z) = \det(zI_3 - A_0 - A_1 e^{-rz}),$$
with
$$A_0 = \begin{bmatrix} -\mu_1 & 0 & 0 \\ 2b_1 & -(\mu_0 + \mu_Q) & b_Q \\ 0 & \mu_Q & -(b_Q + \mu_{G_0}) \end{bmatrix},$$

$$A_1 = e^{-(\mu_0 + \mu_Q)r} \begin{bmatrix} 2b_1 & 0 & b_Q \\ -2b_1 & 0 & -b_Q \\ 0 & 0 & 0 \end{bmatrix},$$

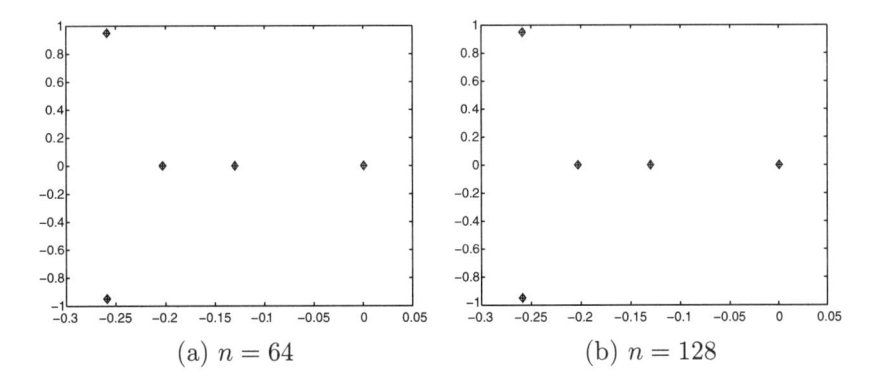

FIGURE 6. Approximations of the zeros of $f_4(z)$ with $b_1 = 0.25$ generated from the spectra of the arrowhead pencil (diamond symbol) and the perturbed CMV–like matrix (plus symbol).

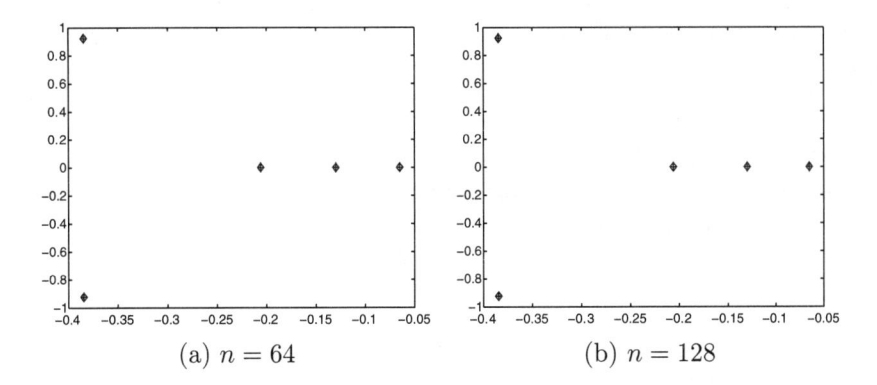

FIGURE 7. Approximations of the zeros of $f_4(z)$ with $b_1 = 0.13$ generated from the spectra of the arrowhead pencil (diamond symbol) and the perturbed CMV–like matrix (plus symbol).

and

$$r = 5, \ b_1 = 0.25, \ b_Q = 0.2, \ \mu_1 = 0.28, \ \mu_0 = 0.11, \ \mu_Q = 0.02, \ \mu_{G_0} = 0.0001.$$

In figure 6 we plot the approximations of the zeros computed by solving the matrix eigenvalue problems.

The approximations returned for the rightmost zero are $\lambda_\diamond = 6.73e-04 +$ i $9.0e-15$ and $\lambda_+ = 6.73e-04 +$ i $2.6e-13$ which means that the system is unstable. A different selection of the system parameters where we decrease the value of b_1 corresponding to the duplication rate of tumor cells from $b_1 = 0.25$ to $b_1 = 0.13$ modifies the stability of the system. In the next figure 7 we show the approximated zeros for the novel set of parameters.

The rightmost approximation is $\lambda_\diamond = -6.49e-02 +$ i $2.3e-14$ and $\lambda_+ = -6.49e-02 -$ i $1.1e-09$ and the system is stable. The conditioning of the five eigenvalues range from $1.4e+01$ and $3.1e+04$. In the next table 5 for $n = 64, 128$ we show the absolute errors of the five approximations displayed in the figure generated

$n = 64$	6.5e-14	1.4e-13	8.0e-14	2.0e-15	1.6e-15
$n = 128$	1.6e-13	3.0e-13	1.5e-13	2.0e-14	1.0e-14

TABLE 5. Errors for the approximations of the five zeros generated starting from the initial pencil in arrow form and the perturbed CMV–like matrix

starting from the initial pencil in arrow form and the perturbed CMV–like matrix.

5. Conclusion and Future Work

In this paper we have presented a matrix algorithm for the computation of the zeros of an analytic function inside the unit circle in the complex plane. At the heart of the proposed approach there is a fast reduction of the initial generalized eigenproblem in arrow form into a perturbed CMV–like form combined with a fast adaptation of the QR eigenvalue algorithm which exploits the structural properties of this latter formulation. The overall complexity is $O(n^2)$ using $O(n)$ memory storage, where n is the number of interpolation nodes on the unit circle used to discretize the original zerofinding computation. The numerical experience is promising and confirms that the proposed approach performs rather accurately in general. In principle numerical difficulties can be encountered if the nodes are quite close or the deflation procedure provides a small but nonzero cutting parameter. In the first case we think that the problem can be alleviated by a careful analysis of the unitary transformation matrices used in the reduction procedure in order to guarantee the invariance of the rank properties of the transformed matrix. Future work is also concerned with the design of more robust deflation procedures taking into account for the occurrence of tiny but nonzero parameters.

References

[1] G.S. Ammar, W.B. Gragg, and L. Reichel, *On the eigenproblem for orthogonal matrices*, Decision and Control, 1986 25th IEEE Conference on, Dec 1986, pp. 1963–1966.

[2] Peter Arbenz and Gene H. Golub, *Matrix shapes invariant under the symmetric QR algorithm*, Numer. Linear Algebra Appl. **2** (1995), no. 2, 87–93, DOI 10.1002/nla.1680020203. MR1323814 (95m:65060)

[3] Anthony P. Austin, Peter Kravanja, and Lloyd N. Trefethen, *Numerical algorithms based on analytic function values at roots of unity*, SIAM J. Numer. Anal. **52** (2014), no. 4, 1795–1821, DOI 10.1137/130931035. MR3240851

[4] Maria Vittoria Barbarossa, Christina Kuttler, and Jonathan Zinsl, *Delay equations modeling the effects of phase-specific drugs and immunotherapy on proliferating tumor cells*, Math. Biosci. Eng. **9** (2012), no. 2, 241–257, DOI 10.3934/mbe.2012.9.241. MR2897077 (2012m:92001)

[5] Timo Betcke, Nicholas J. Higham, Volker Mehrmann, Christian Schröder, and Françoise Tisseur, *NLEVP: a collection of nonlinear eigenvalue problems*, ACM Trans. Math. Software **39** (2013), no. 2, Art. 7, 28, DOI 10.1145/2427023.2427024. MR3031626

[6] R. Bevilacqua, G. M. Del Corso, and L. Gemignani, *Compression of unitary rank–structured matrices to CMV-like shape with an application to polynomial rootfinding*, ArXiv e-prints (2013).

[7] ———, *A CMV-based eigensolver for companion matrices*, ArXiv e-prints (2014), Submitted to SIAM J.Matrix Anal. Appl.

[8] D. A. Bini, P. Boito, Y. Eidelman, L. Gemignani, and I. Gohberg, *A fast implicit QR eigenvalue algorithm for companion matrices*, Linear Algebra Appl. **432** (2010), no. 8, 2006–2031, DOI 10.1016/j.laa.2009.08.003. MR2599839 (2011g:65056)

[9] D. A. Bini, Y. Eidelman, L. Gemignani, and I. Gohberg, *Fast QR eigenvalue algorithms for Hessenberg matrices which are rank-one perturbations of unitary matrices*, SIAM J. Matrix Anal. Appl. **29** (2007), no. 2, 566–585, DOI 10.1137/050627563. MR2318363 (2008f:65072)

[10] D. A. Bini, Y. Eidelman, L. Gemignani, and I. Gohberg, *The unitary completion and QR iterations for a class of structured matrices*, Math. Comp. **77** (2008), no. 261, 353–378, DOI 10.1090/S0025-5718-07-02004-2. MR2353957 (2008j:65054)

[11] Dario A. Bini and Leonardo Robol, *Solving secular and polynomial equations: a multiprecision algorithm*, J. Comput. Appl. Math. **272** (2014), 276–292, DOI 10.1016/j.cam.2013.04.037. MR3227386

[12] P. Boito, Y. Eidelman, and L. Gemignani, *Implicit QR for rank-structured matrix pencils*, BIT **54** (2014), no. 1, 85–111, DOI 10.1007/s10543-014-0478-0. MR3177956

[13] Adhemar Bultheel and Marc Van Barel, *Vector orthogonal polynomials and least squares approximation*, SIAM J. Matrix Anal. Appl. **16** (1995), no. 3, 863–885, DOI 10.1137/S0895479893244572. MR1337650 (96h:65060)

[14] Angelika Bunse-Gerstner and Ludwig Elsner, *Schur parameter pencils for the solution of the unitary eigenproblem*, Linear Algebra Appl. **154/156** (1991), 741–778, DOI 10.1016/0024-3795(91)90402-I. MR1113168 (92c:65048)

[15] M. J. Cantero, L. Moral, and L. Velázquez, *Five-diagonal matrices and zeros of orthogonal polynomials on the unit circle*, Linear Algebra Appl. **362** (2003), 29–56, DOI 10.1016/S0024-3795(02)00457-3. MR1955452 (2003k:42046)

[16] M. J. Cantero, L. Moral, and L. Velázquez, *Minimal representations of unitary operators and orthogonal polynomials on the unit circle*, Linear Algebra Appl. **408** (2005), 40–65, DOI 10.1016/j.laa.2005.04.025. MR2166854 (2007m:47068)

[17] Shiv Chandrasekaran, Ming Gu, Jianlin Xia, and Jiang Zhu, *A fast QR algorithm for companion matrices*, Recent advances in matrix and operator theory, Oper. Theory Adv. Appl., vol. 179, Birkhäuser, Basel, 2008, pp. 111–143, DOI 10.1007/978-3-7643-8539-2_7. MR2397851 (2009a:65087)

[18] R. Corless, *Generalized companion matrices for the Lagrange basis*, Proceedings EACA, 2004.

[19] Michael Dellnitz, Oliver Schütze, and Qinghua Zheng, *Locating all the zeros of an analytic function in one complex variable*, J. Comput. Appl. Math. **138** (2002), no. 2, 325–333, DOI 10.1016/S0377-0427(01)00371-5. MR1876239 (2002k:65038)

[20] Yuli Eidelman, Luca Gemignani, and Israel Gohberg, *Efficient eigenvalue computation for quasiseparable Hermitian matrices under low rank perturbations*, Numer. Algorithms **47** (2008), no. 3, 253–273, DOI 10.1007/s11075-008-9172-0. MR2385737 (2009a:65090)

[21] Koen Engelborghs, Dirk Roose, and Tatyana Luzyanina, *Bifurcation analysis of periodic solutions of neutral functional-differential equations: a case study*, Internat. J. Bifur. Chaos Appl. Sci. Engrg. **8** (1998), no. 10, 1889–1905, DOI 10.1142/S0218127498001595. MR1670631 (99j:34097)

[22] Miroslav Fiedler and Thomas L. Markham, *Completing a matrix when certain entries of its inverse are specified*, Linear Algebra Appl. **74** (1986), 225–237, DOI 10.1016/0024-3795(86)90125-4. MR822149 (87g:15005)

[23] Gene H. Golub and Charles F. Van Loan, *Matrix computations*, 3rd ed., Johns Hopkins Studies in the Mathematical Sciences, Johns Hopkins University Press, Baltimore, MD, 1996. MR1417720 (97g:65006)

[24] Peter Henrici, *Applied and computational complex analysis. Vol. 1*, Wiley Classics Library, John Wiley & Sons, Inc., New York, 1988. Power series—integration—conformal mapping—location of zeros; Reprint of the 1974 original; A Wiley-Interscience Publication. MR1008928 (90d:30002)

[25] Rowan Killip and Irina Nenciu, *CMV: the unitary analogue of Jacobi matrices*, Comm. Pure Appl. Math. **60** (2007), no. 8, 1148–1188, DOI 10.1002/cpa.20160. MR2330626 (2008m:47042)

[26] Peter Kravanja and Marc Van Barel, *Computing the zeros of analytic functions*, Lecture Notes in Mathematics, vol. 1727, Springer-Verlag, Berlin, 2000. MR1754959 (2001c:65004)

[27] Piers W. Lawrence, *Fast reduction of generalized companion matrix pairs for barycentric Lagrange interpolants*, SIAM J. Matrix Anal. Appl. **34** (2013), no. 3, 1277–1300, DOI 10.1137/130904508. MR3095477

[28] Piers W. Lawrence and Robert M. Corless, *Stability of rootfinding for barycentric Lagrange interpolants*, Numer. Algorithms **65** (2014), no. 3, 447–464, DOI 10.1007/s11075-013-9770-3. MR3172327

[29] Axel Ruhe, *Algorithms for the nonlinear eigenvalue problem*, SIAM J. Numer. Anal. **10** (1973), 674–689. MR0329231 (48 #7573)

[30] Barry Simon, *CMV matrices: five years after*, J. Comput. Appl. Math. **208** (2007), no. 1, 120–154, DOI 10.1016/j.cam.2006.10.033. MR2347741 (2009g:47085)

[31] M. Van Barel and A. Bultheel, *Orthonormal polynomial vectors and least squares approximation for a discrete inner product*, Electron. Trans. Numer. Anal. **3** (1995), no. Mar., 1–23 (electronic). MR1320600 (96d:42041)

[32] Marc Van Barel, Raf Vandebril, Paul Van Dooren, and Katrijn Frederix, *Implicit double shift QR-algorithm for companion matrices*, Numer. Math. **116** (2010), no. 2, 177–212, DOI 10.1007/s00211-010-0302-y. MR2672262 (2011i:65062)

[33] Raf Vandebril and Gianna M. Del Corso, *An implicit multishift QR-algorithm for Hermitian plus low rank matrices*, SIAM J. Sci. Comput. **32** (2010), no. 4, 2190–2212, DOI 10.1137/090754522. MR2678097 (2011g:65062)

Dipartimento di Informatica, Università di Pisa, Largo B. Pontecorvo 3, 56127, Pisa, Italy

E-mail address: l.gemignani@di.unipi.it

Multi-frequency acousto-electromagnetic tomography

Giovanni S. Alberti, Habib Ammari, and Kaixi Ruan

ABSTRACT. This paper focuses on the acousto-electromagnetic tomography, a recently introduced hybrid imaging technique. In a previous work, the reconstruction of the electric permittivity of the medium from internal data was achieved under the Born approximation assumption. In this work, we tackle the general problem by a Landweber iteration algorithm. The convergence of such scheme is guaranteed with the use of a multiple frequency approach, that ensures uniqueness and stability for the corresponding linearized inverse problem. Numerical simulations are presented.

1. Introduction

In hybrid imaging inverse problems, two different techniques are combined to obtain high resolution and high contrast images. More precisely, two types of waves are coupled simultaneously: one gives high resolution, and the other one high contrast. Much research has been done in the last decade to develop and study several new methods; the reader is referred to [6, 10, 12, 15, 21] for a review on hybrid techniques. A typical combination is between ultrasonic waves and a high contrast wave, such as light or microwaves. The high resolution of ultrasounds can be used to perturb the medium, thereby changing the electromagnetic properties, and cross-correlating electromagnetic boundary measurements lead to internal data (see e.g. [7, 8, 13, 14, 17, 18]).

This paper focuses on the technique introduced in [14], so called acousto-electromagnetic tomography. Spherical ultrasonic waves are sent from sources around the domain under investigation. The pressure variations create a displacement in the tissue, thereby modifying the electrical properties. Microwave boundary measurements are taken in the unperturbed and in the perturbed situation (see Figure 1). In a first step, the cross-correlation of all the boundary values, after the inversion of a spherical mean Radon transform, gives the internal data of the form

$$|u_\omega(x)|^2 \nabla q(x),$$

where q is the spatially varying electric permittivity of the body $\Omega \subset \mathbb{R}^d$ for $d = 2, 3$, $\omega > 0$ is the frequency and u_ω satisfies the Helmholtz equation with Robin boundary

---

2010 *Mathematics Subject Classification.* Primary 35R30, 35B30.

Key words and phrases. Acousto-electromagnetic tomography, multi-frequency measurements, optimal control, Landweber scheme, convergence, hybrid imaging.

This work was supported by the ERC Advanced Grant Project MULTIMOD–267184.

conditions

(1) $$\begin{cases} \Delta u_\omega + \omega^2 q u_\omega = 0 \text{ in } \Omega, \\ \frac{\partial u_\omega}{\partial \nu} - i\omega u_\omega = -i\omega\varphi \text{ on } \partial\Omega. \end{cases}$$

(In fact, only the gradient part ψ_ω of $|u_\omega|^2 \nabla q = \nabla \psi_\omega + \text{curl}\Phi$ is measured.) The second step of this hybrid methodology consists in recovering q from the knowledge of ψ_ω. In [14] an algorithm based on the inverse Radon transform was considered, but it works only under the Born approximation, namely under the assumption that q has small variations around a certain constant value q_0.

The purpose of this work is to discuss a reconstruction algorithm valid for general values of q. Denoting the measured datum by ψ_ω^*, we propose to minimize the energy functional

$$J_\omega(q) = \frac{1}{2} \int_\Omega |\psi_\omega(q) - \psi_\omega^*|^2 dx$$

with a gradient descent method. In this case, this is equivalent to a Landweber iteration scheme. The convergence of such algorithm [20] is guaranteed provided that $\|D\psi_\omega[q](\rho)\| \geq C \|\rho\|$. This condition represents the uniqueness and stability for the linearized inverse problem $D\psi_\omega[q](\rho) \mapsto \rho$. This problem has been studied for certain classes of internal functionals in [23] by looking at the ellipticity of the associated pseudo-differential operator. Using these techniques, stability up to a finite dimensional kernel could be established. However, uniqueness is a harder issue [22], and in general only generic injectivity can be proved. Indeed, the kernel of $\rho \mapsto D\psi_\omega[q](\rho)$ may well be non-trivial.

In order to obtain an injective problem, we propose here to use a multiple frequency approach. If the boundary condition φ is suitably chosen (e.g. $\varphi = 1$), the kernels of the operators $\rho \mapsto D\psi_\omega[q](\rho)$ "move" as ω changes, and by choosing a finite number of frequencies K in a fixed range, determined a priori, it is possible to show that the intersection becomes empty. In particular, there holds $\sum_{\omega \in K} \|D\psi_\omega[q](\rho)\| \geq C \|\rho\|$ and the convergence of an optimal control algorithm for the functional $J = \sum_{\omega \in K} J_\omega$ follows [9] (see Theorem 1).

The reader is referred to [11, 16, 25, 27] and references therein for recent works on uniqueness and stability results on inverse problems from internal data. The use of multiple frequencies to enforce non-zero constraints in PDE, and to obtain well-posedness for several hybrid problems, has been discussed in [1–5, 9].

This paper is structured as follows. Section 2 describes the physical model and the proposed optimization approach. In Section 3 we prove the convergence of the multi-frequency Landweber scheme. Some numerical simulations are discussed in Section 4. Finally, Section 5 is devoted to some concluding remarks.

2. Acousto-Electromagnetic Tomography

In this section we recall the coupled physics inverse problem introduced in [14] and discuss the proposed Landweber scheme.

2.1. Physical Model. We now briefly describe how to measure the internal data in the hybrid problem under consideration. The reader is referred to [14] for full details.

Let $\Omega \subset \mathbb{R}^d$ be a bounded and smooth domain, for $d = 2$ or $d = 3$ and $q \in L^\infty(\Omega; \mathbb{R}) \cap H^1(\Omega; \mathbb{R})$ be the electric permittivity of the medium. We assume

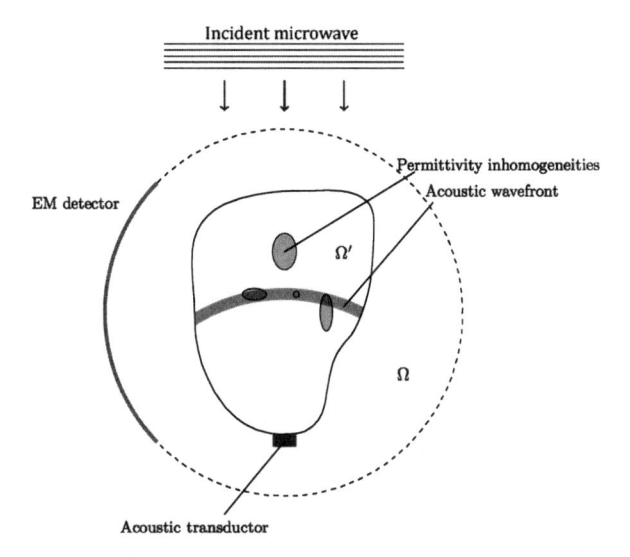

FIGURE 1. The acousto-electromagnetic tomography experiment.

that q is known and constant near the boundary $\partial\Omega$, namely $q = 1$ in $\Omega \setminus \Omega'$, for some $\Omega' \Subset \Omega$. More precisely, suppose that $q \in Q$, where for some $\Lambda > 0$

(2) $\quad Q := \{q \in H^1(\Omega; \mathbb{R}) : \Lambda^{-1} \leq q \leq \Lambda \text{ in } \Omega, \|q\|_{H^1(\Omega)} \leq \Lambda \text{ and } q = 1 \text{ in } \Omega \setminus \Omega'\}.$

In this paper, we model electromagnetic propagation in Ω at frequency $\omega \in \mathcal{A} = [K_{min}, K_{max}] \subset \mathbb{R}_+$ by (1). The boundary value problem model allows us to consider arbitrary q beyond the Born approximation, and so it is used here instead of the free propagation model, which was originally considered in [14]. Problem (1) admits a unique solution $u_\omega \in H^1(\Omega; \mathbb{C})$ for a fixed boundary condition $\varphi \in H^1(\Omega; \mathbb{C})$ (see Lemma 3).

Let us discuss how microwaves are combined with acoustic waves. A short acoustic wave creates a displacement field v in Ω (whose support is the blue area in Figure 1), which we suppose continuous and bijective. Then, the permittivity distribution q becomes q_v defined by

$$q_v(x + v(x)) = q(x), \qquad x \in \Omega,$$

and the complex amplitude u_ω^v of the electric wave in the perturbed medium satisfies

(3) $\quad\quad\quad\quad\quad\quad \Delta u_\omega^v + \omega^2 q_v u_\omega^v = 0 \text{ in } \Omega.$

Using (1) and (3), for v small enough we obtain the cross-correlation formula

$$\int_{\partial\Omega} (\frac{\partial u_\omega}{\partial n} \overline{u_\omega^v} - \frac{\partial \overline{v}}{\partial n} u_\omega) \, d\sigma = \omega^2 \int_\Omega (q_v - q) u_\omega \overline{u_\omega^v} \, dx \approx \omega^2 \int_\Omega |u_\omega|^2 \nabla q \cdot v \, dx,$$

By boundary measurements, the left hand side of this equality is known. Thus, we have measurements of the form

$$\int_\Omega |u_\omega|^2 \nabla q \cdot v \, dx,$$

for all perturbations v. It is shown in [13, 14] that choosing radial displacements v allows to recover the gradient part of $|u_\omega|^2 \nabla q$ by using the inversion for the spherical mean Radon transform. Namely, writing the Helmholtz decomposition of $|u_\omega|^2 \nabla q$

$$|u_\omega|^2 \nabla q = \nabla \psi_\omega + \text{curl} \Phi_\omega,$$

for $\psi_\omega \in H^1(\Omega; \mathbb{R})$ and $\Phi_\omega \in H^1(\Omega; \mathbb{R}^{2d-3})$, the potential ψ_ω can be measured. Moreover, ψ_ω is the unique solution to [19, Chapter I, Theorem 3.2 and Corollary 3.4]

(4)
$$\begin{cases} \Delta \psi_\omega = \text{div}(|u_\omega|^2 \nabla q) & \text{in } \Omega, \\ \frac{\partial \psi_\omega}{\partial \nu} = 0 & \text{on } \partial \Omega, \\ \int_\Omega \psi_\omega \, dx = 0. \end{cases}$$

In this paper, we assume that the inversion of the spherical mean Radon transform has been performed and that we have access to ψ_ω. In the following, we shall deal with the second step of this hybrid imaging problem: recovering the map q from the knowledge of ψ_ω.

2.2. The Landweber Iteration. Let q^* be the real permittivity with corresponding measurements ψ_ω^*. Let $K \subset \mathcal{A}$ be a finite set of admissible frequencies for which we have the measurements ψ_ω^*, $\omega \in K$. The set K will be determined later. Let us denote the error map by

(5)
$$F_\omega : Q \to H^1_\nu(\Omega; \mathbb{R}), \qquad q \mapsto \psi_\omega(q) - \psi_\omega^*,$$

where $\psi_\omega(q)$ is the unique solution to (4), and $H^1_\nu(\Omega; \mathbb{R}) = \{u \in H^1(\Omega; \mathbb{R}) : \frac{\partial u}{\partial \nu} = 0 \text{ on } \partial \Omega\}$.

A natural approach to recover the real conductivity is to minimize the discrepancy functional J defined as

(6)
$$J(q) = \frac{1}{2} \sum_{\omega \in K} \int_\Omega |F_\omega(q)|^2 dx, \qquad q \in Q.$$

The gradient descent method can be employed to minimize J. At each iteration we compute

$$q_{n+1} = T(q_n - h DJ[q_n]),$$

where $h > 0$ is the step size and $T : H^1(\Omega; \mathbb{R}) \to Q$ is the Hilbert projection onto the convex closed set Q, which guarantees that at each iteration q_n belongs to the admissible set Q. Since $DJ[q] = \sum_\omega DF_\omega[q]^*(F_\omega(q))$, this algorithm is equivalent to the Landweber scheme [20] given by

(7)
$$q_{n+1} = T\Big(q_n - h \sum_{\omega \in K} DF_\omega(q_n)^*(F_\omega(q_n))\Big).$$

(For the Fréchet differentiability of the map F_ω, see Lemma 5.)

The main result of this paper states that the Landweber scheme defined above converges to the real unknown q^*, provided that K is suitably chosen and that h and $\|q_0 - q^*\|_{H^1(\Omega)}$ are small enough. The most natural choice for the set of frequencies K is as a uniform sample of $\mathcal{A}$, namely let

$$K^{(m)} = \{\omega_1^{(m)}, \ldots, \omega_m^{(m)}\}, \qquad \omega_i^{(m)} = K_{min} + \frac{(i-1)}{(m-1)}(K_{max} - K_{min}).$$

THEOREM 1. *Set $\varphi = 1$. There exist $C > 0$ and $m \in \mathbb{N}^*$ depending only on Ω, Λ and $\mathcal{A}$ such that for any $q \in Q$ and $\rho \in H^1_\nu(\Omega; \mathbb{R})$*

$$\sum_{\omega \in K^{(m)}} \|DF_\omega[q](\rho)\|_{H^1(\Omega;\mathbb{R})} d\omega \geq C \|\rho\|_{H^1(\Omega;\mathbb{R})}. \tag{8}$$

As a consequence, the sequence defined in (7) converges to q^* provided that h and $\|q_0 - q^*\|_{H^1(\Omega)}$ are small enough.

The proof of this theorem is presented in Section 3. In view of the results in [9, 20], the convergence of the Landweber iteration follows from the Lipschitz continuity of F_ω and from inequality (8). The Lipschitz continuity of F_ω is a simple consequence of the elliptic theory.

On the other hand, the lower bound given in (8) is non-trivial, since it represents the uniqueness and stability of the multi-frequency linearized inverse problem

$$(DF_\omega[q](\rho))_{\omega \in K^{(m)}} \longmapsto \rho.$$

As it has been discussed in the Introduction, the kernels of the operators $\rho \mapsto DF_\omega[q](\rho)$ "move" as ω changes. More precisely, the intersection of the kernels corresponding to the a priori determined finite set of frequencies $K^{(m)}$ is empty. Moreover, the argument automatically gives an a priori constant C in (8).

The multi-frequency method is based on the analytic dependence of the problem with respect to the frequency ω, and on the fact that in $\omega = 0$ the problem is well posed. Indeed, when $\omega \to 0$ it is easy to see that $u_\omega \to 1$ in (1), so that $u_0 = 1$. Thus, looking at (4), the measurement datum ψ_0 is nothing else than q^* (up to a constant). Therefore, q^* could be easily determined when $\omega = 0$ since q^* is known on the boundary $\partial \Omega$. As we show in the following section, the analyticity of the problem with respect to ω allows to "transfer" this property to the desired range of frequencies $\mathcal{A}$.

3. Convergence of the Landweber Iteration

In order to use the well-posedness of the problem in $\omega = 0$ we shall need the following result on quantitative unique continuation for vector-valued holomorphic functions.

LEMMA 2. *Let V be a complex Banach space, $\mathcal{A} = [K_{min}, K_{max}] \subset \mathbb{R}_+$, C_0, and $D > 0$. Let*

$$\mathcal{S}_{C_0, D, K_{max}} := \Big\{ g \colon B(0, K_{max}) \subset \mathbb{C} \to V, g \text{ is holomorphic}, \|g(0)\| \geq C_0,$$
$$\sup_{\omega \in B(0, K_{max})} \|g(\omega)\| \leq D \Big\}.$$

Then there exists $C > 0$, depending only on $\mathcal{A}$, C_0 and D, such that for any $g \in \mathcal{S}_{C_0, D, K_{max}}$

$$\max_{\omega \in \mathcal{A}} \|g(\omega)\| \geq C.$$

PROOF. By contradiction, assume that there exists a sequence of functions $g_n \in \mathcal{S}_{C_0, D, K_{max}}$ such that $\max_{\omega \in \mathcal{A}} \|g_n(\omega)\| \to 0$. By Hahn Banach theorem, for any n there exists $T_n \in V'$ such that $\|T_n\| \leq 1$ and $T_n(g_n(0)) = \|g_n(0)\|$. Set

$f_n := T_n \circ g_n \colon B(0, K_{max}) \to \mathbb{C}$. Thus (f_n) is a family of complex-valued uniformly bounded holomorphic functions, since
$$|f_n(\omega)| \leq \|T_n\| \, \|g_n(\omega)\| \leq D, \qquad \omega \in B(0, K_{max}).$$
As a consequence, by standard complex analysis (see, for instance, [**26**, Theorem 3.3, p.225], there exists a holomorphic function $f \colon B(0, K_{max}) \to \mathbb{C}$ such that $f_n \to f$ uniformly. We readily observe that for any $\omega \in \mathcal{A}$ there holds
$$|f(\omega)| = \lim_n |f_n(\omega)| \leq \lim_n \|T_n\| \, \|g_n(\omega)\| = 0,$$
since $\max_{\omega \in \mathcal{A}} \|g_n(\omega)\| \to 0$. By the unique continuation theorem $f(0) = 0$. On the other hand, as $T_n(g_n(0)) = \|g_n(0)\|$,
$$f(0) = \lim_n f_n(0) = \lim \|g_n(0)\| \geq C_0 > 0,$$
which yields a contradiction. $\square$

In view of (8), we need to study the Fréchet differentiability of the map F_ω and characterize its derivative. Before doing this, we study the well-posedness of (1) with $q \in Q$. The result is classical; for a proof, see [**24**, Section 8.1].

LEMMA 3. *Let $\Omega \subset \mathbb{R}^d$ be a bounded and smooth domain for $d = 2, 3$, $\omega \in B(0, K_{max})$ and $q \in Q$. For any $f \in L^2(\Omega; \mathbb{C})$ and $\varphi \in H^1(\Omega; \mathbb{C})$ the problem*

(9)
$$\begin{cases} \Delta u + \omega^2 q u = \omega f & \text{in } \Omega, \\ \frac{\partial u}{\partial \nu} - i\omega u = -i\omega\varphi & \text{on } \partial\Omega, \end{cases}$$

augmented with the condition

(10)
$$\int_{\partial\Omega} u \, d\sigma = \int_{\partial\Omega} \varphi \, d\sigma - i \int_\Omega f \, dx$$

if $\omega = 0$ admits a unique solution $u \in H^2(\Omega; \mathbb{C})$. Moreover
$$\|u\|_{H^2(\Omega;\mathbb{C})} \leq C(\|f\|_{L^2(\Omega;\mathbb{C})} + \|\varphi\|_{H^1(\Omega;\mathbb{C})})$$
for some $C > 0$ depending only on Ω, Λ and K_{max}.

Since for $\omega = 0$ the solution to (9) is unique up to a constant, condition (10) is needed to have uniqueness. Even though it may seem mysterious, this condition is natural in order to ensure continuity of u with respect to ω. Indeed an integration by parts gives
$$\omega \int_\Omega f \, dx = \int_{\partial\Omega} \frac{\partial u}{\partial \nu} \, d\sigma + \omega^2 \int_\Omega q u \, dx = i\omega \int_{\partial\Omega} (u - \varphi) \, d\sigma + \omega^2 \int_\Omega q u \, dx,$$
whence for $\omega \neq 0$ we obtain
$$\int_{\partial\Omega} u \, d\sigma = \int_{\partial\Omega} \varphi \, d\sigma - i \int_\Omega f \, dx + \omega i \int_\Omega q u \, dx,$$
and so for $\omega = 0$ we are left with (10). The above condition is a consequence of (9) for $\omega \neq 0$, but needs to be added in the case $\omega = 0$ to guarantee uniqueness.

Let us go back to (1). Fix $\varphi \in H^1(\Omega; \mathbb{C})$ and $\omega \in B(0, K_{max})$. By Lemma 3 the problem

(11)
$$\begin{cases} \Delta u_\omega + \omega^2 q u_\omega = 0 & \text{in } \Omega, \\ \frac{\partial u_\omega}{\partial n} - i\omega u_\omega = -i\omega\varphi & \text{on } \partial\Omega, \end{cases}$$

together with condition

$$\tag{12} \int_{\partial\Omega} u_\omega \, d\sigma = \int_{\partial\Omega} \varphi \, d\sigma + \omega i \int_\Omega q u_\omega \, dx$$

admits a unique solution $u_\omega \in H^2(\Omega; \mathbb{C})$ such that

$$\tag{13} \|u_\omega\|_{H^2(\Omega;\mathbb{C})} \leq C \|\varphi\|_{H^1(\Omega;\mathbb{C})}$$

for some $C > 0$ depending only on Ω, Λ and K_{max}. As above, (12) guarantees uniqueness and continuity in $\omega = 0$ and is implicit in (11) if $\omega \neq 0$.

Next, we study the dependence of u_ω on ω.

LEMMA 4. *Let $\Omega \subset \mathbb{R}^d$ be a bounded and smooth domain for $d = 2, 3$, $q \in Q$ and $\varphi \in H^1(\Omega; \mathbb{C})$. The map*

$$\mathcal{F}(q) \colon B(0, K_{max}) \subset \mathbb{C} \longrightarrow H^2(\Omega; \mathbb{C}), \qquad \omega \longmapsto u_\omega$$

is holomorphic. Moreover, the derivative $\partial_\omega u_\omega \in H^2(\Omega; \mathbb{C})$ is the unique solution to

$$\tag{14} \begin{cases} \Delta \partial_\omega u_\omega + \omega^2 q \partial_\omega u_\omega = -2\omega q u_\omega & \text{in } \Omega, \\ \frac{\partial(\partial_\omega u_\omega)}{\partial \nu} - i\omega \partial_\omega u_\omega = i u_\omega - i\varphi & \text{on } \partial\Omega, \end{cases}$$

together with condition

$$\tag{15} \int_{\partial\Omega} \partial_\omega u_\omega \, d\sigma = \omega i \int_\Omega q \partial_\omega u_\omega \, dx + i \int_\Omega q u_\omega \, dx,$$

and satisfies

$$\|\partial_\omega u_\omega\|_{H^2(\Omega;\mathbb{C})} \leq C \|\varphi\|_{H^1(\Omega;\mathbb{C})}$$

for some $C > 0$ depending only on Ω, Λ and K_{max}.

PROOF. The proof of this result is completely analogous to the ones given in [1, 2, 9] in similar situations. Here only a sketch will be presented.

Fix $\omega \in B(0, K_{max})$: we shall prove that $\mathcal{F}(q)$ is holomorphic in ω and that the derivative is $\partial_\omega u_\omega$, i.e., the unique solution to (14)-(15). For $h \in \mathbb{C}$ let $v_h = (u_{\omega+h} - u_\omega)/h$. We need to prove that $v_h \to \partial_\omega u_\omega$ in $H^2(\Omega)$ as $h \to 0$. Suppose first $\omega \neq 0$. A direct calculation shows that

$$\begin{cases} \Delta v_h + \omega^2 q v_h = -2\omega q u_{\omega+h} - h q u_{\omega+h} & \text{in } \Omega, \\ \frac{\partial v_h}{\partial \nu} - i\omega v_h = i(u_{\omega+h} - \varphi) & \text{on } \partial\Omega. \end{cases}$$

Arguing as in Lemma 3, we obtain $u_{\omega+h} \to u_\omega$ as $h \to 0$ in $H^2(\Omega)$, whence $v_h \to \partial_\omega u_\omega$ in $H^2(\Omega)$, as desired.

When $\omega = 0$, the above system must be augmented with the condition

$$\int_{\partial\Omega} v_h \, d\sigma = i \int_\Omega q u_0 \, dx,$$

which is a simple consequence of (12), and the result follows. $\square$

We now study the Fréchet differentiability of the map F_ω defined in (5). The proof of this result is trivial, and the details are left to the reader.

LEMMA 5. *Let $\Omega \subset \mathbb{R}^d$ be a bounded and smooth domain for $d = 2, 3$, $q \in Q$, $\omega \in B(0, K_{max})$ and $\varphi \in H^1(\Omega; \mathbb{C})$. The map F_ω is Fréchet differentiable and for $\rho \in H^1_\nu(\Omega; \mathbb{R})$, the derivative $\xi_\omega(\rho) := DF_\omega[q](\rho)$ is the unique solution to the problem*

$$\begin{cases} \Delta \xi_\omega(\rho) = \operatorname{div}\left(|u_\omega|^2 \nabla \rho + (\overline{u}_\omega v_\omega(\rho) + u_\omega \overline{v}_\omega(\rho)) \nabla q\right) & \text{in } \Omega, \\ \frac{\partial \xi_\omega(\rho)}{\partial \nu} = 0 & \text{on } \partial\Omega, \\ \int_\Omega \xi_\omega(\rho) \, dx = 0, \end{cases}$$

where $v_\omega(\rho) \in H^2(\Omega; \mathbb{C})$ is the unique solution to

(16) $$\begin{cases} \Delta v_\omega(\rho) + \omega^2 q v_\omega(\rho) = -\omega^2 \rho \, u_\omega & \text{in } \Omega, \\ \frac{\partial v_\omega(\rho)}{\partial \nu} - i\omega v_\omega(\rho) = 0 & \text{on } \partial\Omega, \end{cases}$$

together with $\int_{\partial\Omega} v_0(\rho) \, d\sigma = 0$ if $\omega = 0$. In particular, F_ω is Lipschitz continuous, namely

$$\|\xi_\omega(\rho)\|_{H^1(\Omega; \mathbb{R})} \le C(\Omega, \Lambda, K_{max}, \|\varphi\|_{H^1(\Omega; \mathbb{C})}) \|\rho\|_{H^1(\Omega; \mathbb{R})}.$$

The main step in the proof of Theorem 1 is inequality (8), which we now prove. The argument in the proof clarifies the multi-frequency method illustrated in the previous section. The proof is structured as the proof of [**3**, Theorem 1].

PROPOSITION 6. *Set $\varphi = 1$. There exist $C > 0$ and $m \in \mathbb{N}^*$ depending on Ω, Λ and $\mathcal{A}$ such that for any $q \in Q$ and $\rho \in H^1_\nu(\Omega; \mathbb{R})$*

$$\sum_{\omega \in K^{(m)}} \|DF_\omega[q](\rho)\|_{H^1(\Omega; \mathbb{R})} d\omega \ge C \|\rho\|_{H^1(\Omega; \mathbb{R})}.$$

PROOF. In the proof, several positive constants depending only on Ω, Λ and $\mathcal{A}$ will be denoted by C or Z.

Fix $q \in Q$. For $\rho \in H^1_\nu(\Omega; \mathbb{R})$ such that $\|\rho\|_{H^1(\Omega; \mathbb{R})} = 1$ define the map

$$g_\rho(\omega) = \operatorname{div}\left(u_\omega \overline{u}_{\overline{\omega}} \nabla \rho + (\overline{u}_{\overline{\omega}} v_{\overline{\omega}}(\rho) + u_\omega \overline{v}_{\overline{\omega}}(\rho)) \nabla q\right), \qquad \omega \in B(0, K_{max}).$$

Hence $g_\rho \colon B(0, K_{max}) \to H^1_\nu(\Omega; \mathbb{C})'$ is holomorphic. We shall apply Lemma 2 to g_ρ, and so we now verify the hypotheses.

Since $\varphi = 1$, by (11)-(12) we have $u_0 = 1$ and by (16) we have $v_0(\rho) = 0$, whence $g_\rho(0) = \operatorname{div}(\nabla \rho)$. Since $\frac{\partial \rho}{\partial \nu} = 0$ on $\partial\Omega$ there holds

$$\|g_\rho(0)\|_{H^1_\nu(\Omega; \mathbb{C})'} = \|\operatorname{div}(\nabla \rho)\|_{H^1_\nu(\Omega; \mathbb{C})'} \ge C \|\nabla \rho\|_{L^2(\Omega)} \ge C > 0,$$

since $\|\rho\|_{H^1(\Omega; \mathbb{R})} = 1$. For $\omega \in B(0, K_{max})$ we readily derive

$$\|g_\rho(\omega)\|_{H^1_\nu(\Omega; \mathbb{C})'} \le C \|u_\omega \overline{u}_{\overline{\omega}} \nabla \rho + (\overline{u}_{\overline{\omega}} v_{\overline{\omega}}(\rho) + u_\omega \overline{v}_{\overline{\omega}}(\rho)) \nabla q\|_{L^2(\Omega)}$$
$$\le C \left(\|\rho\|_{H^1(\Omega)} + \|q\|_{H^1(\Omega)}\right)$$
$$\le C,$$

where the second inequality follows from (13), Lemma 3 applied to $v_\omega(\rho)$ and the Sobolev embedding $H^2 \hookrightarrow L^\infty$. Therefore, by Lemma 2 there exists $\omega_\rho \in \mathcal{A}$ such that

(17) $$\|g_\rho(\omega_\rho)\|_{H^1_\nu(\Omega; \mathbb{C})'} \ge C.$$

Consider now for $\omega \in B(0, K_{max})$

$$g'_\rho(\omega) = \operatorname{div}\left((u'_\omega \overline{u}_{\overline{\omega}} + u_\omega \overline{u}'_{\overline{\omega}}) \nabla \rho + (\overline{u}'_{\overline{\omega}} v_{\overline{\omega}}(\rho) + \overline{u}_{\overline{\omega}} v'_{\overline{\omega}}(\rho) + u'_\omega \overline{v}_{\overline{\omega}}(\rho) + u_\omega \overline{v}'_{\overline{\omega}}(\rho)) \nabla q\right)$$

where for simplicity the partial derivative ∂_ω is replaced by $'$. Arguing as before, and using Lemma 4 we obtain
$$\|g'_\rho(\omega)\|_{H^1_\nu(\Omega;\mathbb{C})'} \leq C, \qquad \omega \in B(0, K_{max}).$$

As a consequence, by (17) we obtain

(18) $$\|g_\rho(\omega)\|_{H^1_\nu(\Omega;\mathbb{C})'} \geq C, \qquad \omega \in [\omega_\rho - Z, \omega_\rho + Z] \cap \mathcal{A}.$$

Since $\mathcal{A} = [K_{min}, K_{max}]$ there exists $P = P(Z, \mathcal{A}) \in \mathbb{N}$ such that

(19) $$\mathcal{A} \subseteq \bigcup_{p=1}^{P} I_p, \qquad I_p = [K_{min} + (p-1)Z, K_{min} + pZ].$$

Choose now $m \in \mathbb{N}^*$ big enough so that for every $p = 1, \ldots, P$ there exists $i_p = 1, \ldots, m$ such that $\omega(p) := \omega_{i_p}^{(m)} \in I_p$ (recall that $\omega_i^{(m)} = K_{min} + \frac{(i-1)}{(m-1)}(K_{max} - K_{min})$). Note that m depends only on Z and $|\mathcal{A}|$.

Since $|[\omega_\rho - Z, \omega_\rho + Z]| = 2Z$ and $|I_p| = Z$, in view of (19) there exists $p_\rho = 1, \ldots, P$ such that $I_{p_\rho} \subseteq [\omega_\rho - Z, \omega_\rho + Z]$. Therefore $\omega(p_\rho) \in [\omega_\rho - Z, \omega_\rho + Z] \cap \mathcal{A}$, whence by (18) there holds $\|g_\rho(\omega(p_\rho))\|_{H^1_\nu(\Omega;\mathbb{C})'} \geq C$. Since $\omega(p_\rho) \in \mathbb{R}$ this implies

$$\left\|\operatorname{div}\left(|u_{\omega(p_\rho)}|^2 \nabla \rho + (\overline{u}_{\omega(p_\rho)} v_\omega(\rho) + u_\omega \overline{v}_{\omega(p_\rho)}(\rho))\nabla q\right)\right\|_{H^1_\nu(\Omega;\mathbb{C})'} \geq C,$$

which by Lemma 5 yields $\|\Delta \xi_{\omega(p_\rho)}(\rho)\|_{H^1_\nu(\Omega;\mathbb{C})'} \geq C$. Hence, since $\frac{\partial \xi_{\omega(p_\rho)}(\rho)}{\partial \nu} = 0$ on $\partial\Omega$, there holds $\|\xi_{\omega(p_\rho)}(\rho)\|_{H^1(\Omega)} \geq C$. Thus, since $\omega(p_\rho) \in K^{(m)}$

$$\sum_{\omega \in K^{(m)}} \|\xi_\omega(\rho)\|_{H^1(\Omega)} \geq C.$$

We have proved this inequality only for $\rho \in H^1_\nu(\Omega; \mathbb{R})$ with unitary norm. By using the linearity of $\xi_\omega(\rho)$ with respect to ρ we immediately obtain

$$\sum_{\omega \in K^{(m)}} \|\xi_\omega(\rho)\|_{H^1(\Omega)} \geq C \|\rho\|_{H^1(\Omega)}, \qquad \rho \in H^1_\nu(\Omega; \mathbb{R}),$$

as desired. $\square$

We are now ready to prove Theorem 1.

PROOF OF THEOREM 1. Inequality (8) follows from Proposition 6. Moreover, F_ω is Lipschitz continuous by Lemma 5. Therefore, the convergence of the Landweber iteration is a consequence of the results in [9, 20], provided that $\|q_0 - q^*\|_{H^1(\Omega)}$ and h are small enough. $\square$

4. Numerical Results

In this section we present some numerical results. Let Ω be the unit square $[0,1] \times [0,1]$. We set the mesh size to be 0.01. A phantom image is used for the true permittivity distribution q^* (see Figure 2). According to Theorem 1 we set the Robin boundary condition to be a constant function $\varphi = 1$. Let K be the set of frequencies for which we have measurements ψ^*_ω, $\omega \in K$. As discussed in § 2.2, we minimize the functional J in (6) with the Landweber iteration scheme given in (7). The initial guess is $q_0 = 1$.

We start with the imaging problem at a single frequency. In Figure 3 we display the findings for the case $K = \{3\}$. Figure 3a shows the reconstructed distribution

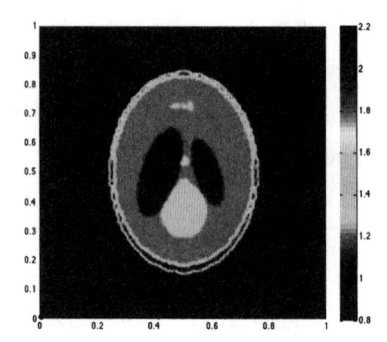

FIGURE 2. The true permittivity distribution q^*.

(A) Reconstructed distribution after 100 iterations.

(B) Relative error depending on the number of iterations.

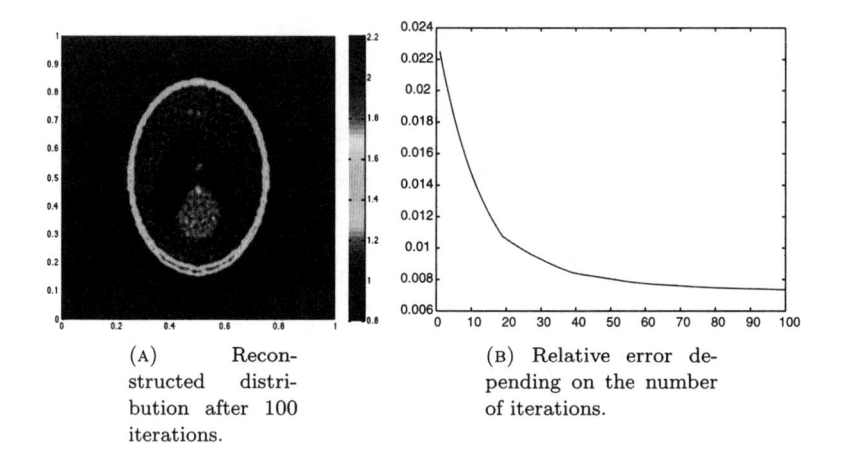

FIGURE 3. Reconstruction of q for the set of frequencies $K = \{3\}$.

after 100 iterations. Figure 3b shows the relative error as a function of the number of iterations. This suggests the convergence of the iterative algorithm, even though the algorithm is proved to be convergent only in the multi-frequency case. It is possible that for small frequencies ω (with respect to the domain size) we are still in the coercive case, i.e. the kernel R_ω of $\rho \mapsto DF_\omega[q](\rho)$ is trivial and a single frequency is sufficient.

However, this does not work at higher frequencies (with respect to the domain size). Figure 4 shows some reconstructed maps for $\omega = 10$, $\omega = 15$ and $\omega = 20$, which suggest that the algorithm may not converge numerically for high frequencies. In each case, there are areas that remain invisible. This may be an indication that $R_\omega \neq \{0\}$ for these values of the frequency.

The invisible areas in Figure 4 are different for different frequencies, and so combining these measurements may give a satisfactory reconstruction. More precisely, according to Theorem 1, by using multiple frequencies it is possible to make the problem injective, namely $\cap_\omega R_\omega = \{0\}$, since the kernels R_ω change as ω varies. Figure 5 shows the results for the case $K = \{10, 15, 20\}$. (According to the notation introduced in Section 2, this choice of frequencies corresponds to $\mathcal{A} = [10, 20]$ and $m = 3$.) These findings suggest the convergence of the multi-frequency Landweber

(A) $K = \{10\}$

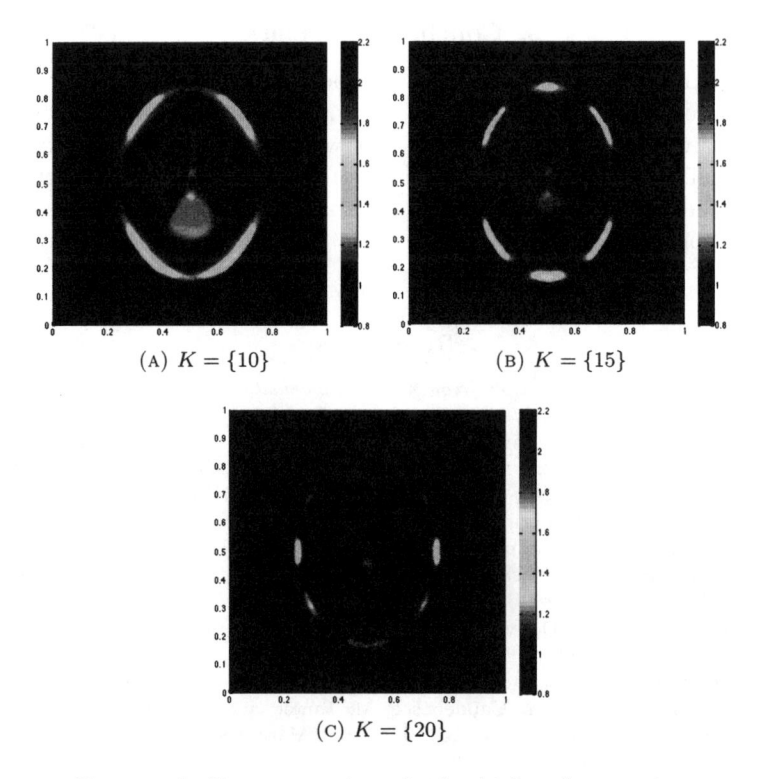

(B) $K = \{15\}$

(C) $K = \{20\}$

FIGURE 4. Reconstruction of q for higher frequencies.

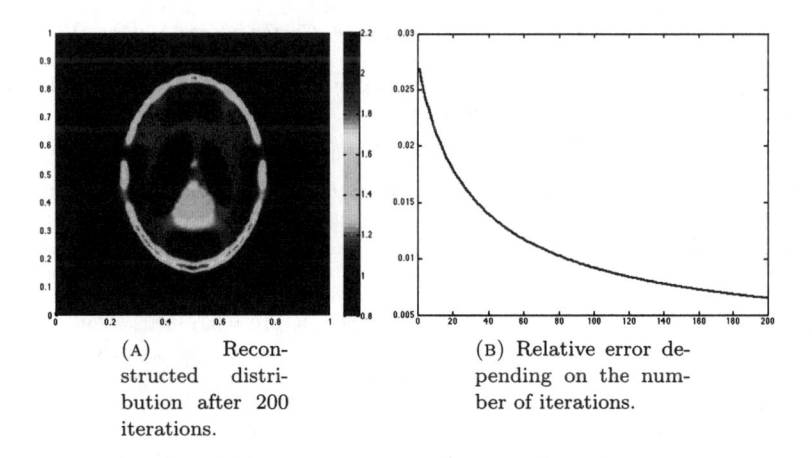

(A) Reconstructed distribution after 200 iterations.

(B) Relative error depending on the number of iterations.

FIGURE 5. Reconstruction of q for $K = \{10, 15, 20\}$.

iteration, even though it was not convergent in each single-frequency case. Since we chose higher frequencies, the convergence is slower.

5. Concluding Remarks

In this paper, we proved that the Landweber scheme in acousto-electromagnetic tomography converges to the true solution provided that multi-frequency measurements are used. We illustrated this result with several numerical examples. It would be challenging to estimate the robustness of the proposed algorithm with respect to random fluctuations in the electromagnetic parameters. This will be the subject of a forthcoming work.

References

[1] Giovanni S. Alberti, *On multiple frequency power density measurements*, Inverse Problems **29** (2013), no. 11, 115007, 25, DOI 10.1088/0266-5611/29/11/115007. MR3116343

[2] Giovanni S. Alberti, *On multiple frequency power density measurements*, Inverse Problems **29** (2013), no. 11, 115007, 25, DOI 10.1088/0266-5611/29/11/115007. MR3116343

[3] Giovanni S. Alberti, *Enforcing local non-zero constraints in PDEs and applications to hybrid imaging problems*, ArXiv e-prints (2014).

[4] Giovanni S. Alberti, *On local constraints and regularity of PDE in electromagnetics. Applications to hybrid imaging inverse problems*, Ph.D. thesis, University of Oxford, 2014.

[5] Giovanni S. Alberti and Yves Capdeboscq, *A propos de certains problèmes inverses hybrides*, Seminaire: Equations aux Dérivées Partielles. 2013–2014, Sémin. Équ. Dériv. Partielles, École Polytech., Palaiseau, 2014, p. Exp. No. II.

[6] ———, *Lectures on elliptic methods for hybrid inverse problems*, in preparation.

[7] H. Ammari, E. Bonnetier, Y. Capdeboscq, M. Tanter, and M. Fink, *Electrical impedance tomography by elastic deformation*, SIAM J. Appl. Math. **68** (2008), no. 6, 1557–1573, DOI 10.1137/070686408. MR2424952 (2009h:35439)

[8] Habib Ammari, Yves Capdeboscq, Frédéric de Gournay, Anna Rozanova-Pierrat, and Faouzi Triki, *Microwave imaging by elastic deformation*, SIAM J. Appl. Math. **71** (2011), no. 6, 2112–2130, DOI 10.1137/110828241. MR2873260

[9] H. Ammari, L. Giovangigli, L. Hoang Nguyen, and J.-K. Seo, *Admittivity imaging from multifrequency micro-electrical impedance tomography*, ArXiv e-prints (2014).

[10] *Mathematical and statistical methods for imaging*, Contemporary Mathematics, vol. 548, American Mathematical Society, Providence, RI, 2011. Papers from the NIMS Thematic Workshop held at Inha University, Incheon, August 10–13, 2010; Edited by Habib Ammari, Josselin Garnier, Hyeonbae Kang and Knut Sølna. MR2868483 (2012h:65006)

[11] H. Ammari, A. Waters, and H. Zhang, *Stability Analysis for Magnetic Resonance Elastography*, ArXiv e-prints (2014).

[12] Habib Ammari, *An introduction to mathematics of emerging biomedical imaging*, Mathématiques & Applications (Berlin) [Mathematics & Applications], vol. 62, Springer, Berlin, 2008. MR2440857 (2010j:44002)

[13] Habib Ammari, Emmanuel Bossy, Josselin Garnier, Loc Hoang Nguyen, and Laurent Seppecher, *A reconstruction algorithm for ultrasound-modulated diffuse optical tomography*, Proc. Amer. Math. Soc. **142** (2014), no. 9, 3221–3236, DOI 10.1090/S0002-9939-2014-12090-9. MR3223378

[14] Habib Ammari, Emmanuel Bossy, Josselin Garnier, and Laurent Seppecher, *Acousto-electromagnetic tomography*, SIAM J. Appl. Math. **72** (2012), no. 5, 1592–1617, DOI 10.1137/120863654. MR3022278

[15] Guillaume Bal, *Hybrid inverse problems and internal functionals*, Inverse problems and applications: inside out. II, Math. Sci. Res. Inst. Publ., vol. 60, Cambridge Univ. Press, Cambridge, 2013, pp. 325–368. MR3098661

[16] *Inverse problems and applications*, Contemporary Mathematics, vol. 615, American Mathematical Society, Providence, RI, 2014. Papers from the conference in honor of Gunther Uhlmann held at the University of California, Irvine, CA, June 18–22, 2012 and the International Conference in Honor of Gunther Uhlmann's 60th Birthday held at Zhejiang University, Hangzhou, September 17–21, 2012; Edited by Plamen Stefanov, András Vasy and Maciej Zworski. MR3236876

[17] Guillaume Bal and Shari Moskow, *Local inversions in ultrasound-modulated optical tomography*, Inverse Problems **30** (2014), no. 2, 025005, 17, DOI 10.1088/0266-5611/30/2/025005. MR3162107
[18] Guillaume Bal and John C. Schotland, *Ultrasound-modulated bioluminescence tomography*, Phys. Rev. E **89** (2014), 031201.
[19] Vivette Girault and Pierre-Arnaud Raviart, *Finite element methods for Navier-Stokes equations*, Springer Series in Computational Mathematics, vol. 5, Springer-Verlag, Berlin, 1986. Theory and algorithms. MR851383 (88b:65129)
[20] Martin Hanke, Andreas Neubauer, and Otmar Scherzer, *A convergence analysis of the Landweber iteration for nonlinear ill-posed problems*, Numer. Math. **72** (1995), no. 1, 21–37, DOI 10.1007/s002110050158. MR1359706 (96i:65046)
[21] Peter Kuchment, *Mathematics of hybrid imaging: a brief review*, The mathematical legacy of Leon Ehrenpreis, Springer Proc. Math., vol. 16, Springer, Milan, 2012, pp. 183–208, DOI 10.1007/978-88-470-1947-8_12. MR3289684
[22] Peter Kuchment and Dustin Steinhauer, *Stabilizing inverse problems by internal data*, Inverse Problems **28** (2012), no. 8, 084007, 20, DOI 10.1088/0266-5611/28/8/084007. MR2956563
[23] Peter Kuchment and Dustin Steinhauer, *Stabilizing inverse problems by internal data*, Inverse Problems **28** (2012), no. 8, 084007, 20, DOI 10.1088/0266-5611/28/8/084007. MR2956563
[24] Jens Markus Melenk, *On generalized finite-element methods*, ProQuest LLC, Ann Arbor, MI, 1995. Thesis (Ph.D.)–University of Maryland, College Park. MR2692949
[25] Carlos Montalto and Plamen Stefanov, *Stability of coupled-physics inverse problems with one internal measurement*, Inverse Problems **29** (2013), no. 12, 125004, 13, DOI 10.1088/0266-5611/29/12/125004. MR3129114
[26] Elias M. Stein and Rami Shakarchi, *Complex analysis*, Princeton Lectures in Analysis, II, Princeton University Press, Princeton, NJ, 2003. MR1976398 (2004d:30002)
[27] T. Widlak and O. Scherzer, *Stability in the linearized problem of quantitative elastography*, ArXiv e-prints (2014).

DEPARTMENT OF MATHEMATICS AND APPLICATIONS, ECOLE NORMALE SUPÉRIEURE, 45 RUE D'ULM, 75005 PARIS, FRANCE
E-mail address: giovanni.alberti@ens.fr

DEPARTMENT OF MATHEMATICS AND APPLICATIONS, ECOLE NORMALE SUPÉRIEURE, 45 RUE D'ULM, 75005 PARIS, FRANCE
E-mail address: habib.ammari@ens.fr

INSTITUTE FOR COMPUTATIONAL & MATHEMATICAL ENGINEERING, STANFORD, CALIFORNIA 94305-2215
E-mail address: kaixi@stanford.edu

On the space-time fractional Schrödinger equation with time independent potentials

Saleh Baqer and Lyubomir Boyadjiev

ABSTRACT. This paper is about the fractional Schrödinger equation (FSE) expressed in terms of the quantum Riesz-Feller space fractional and the Caputo time fractional derivatives. The main focus is on the case of time independent potential fields as a Dirac-delta potential and a linear potential. For such type of potential fields the separation of variables method allows to split the FSE into space fractional equation and time fractional one. The results obtained in this paper contain as particular cases already known results for FSE in terms of the quantum Riesz space fractional derivative and standard Laplace operator.

1. Problem Formulation

The particle motion in quantum mechanics is governed by the Schrödinger equation (SE) and the particle motion state is determined by the wave function. Feynman and Hibbs ([1]) deduced the standard SE by means of path integrals over the Brownian trajectories. This idea was generalized by Laskin ([2], [3]) who introduced the principles of the fractional quantum mechanics through the path integral approach over the Lévy trajectories. For a particle with a mass m moving in a potential field $V(x,t)$ Laskin introduced the FSE of the form ($1 < \alpha \leq 2$),

$$-D_\alpha (\hbar \nabla)^\alpha \psi(x,t) + V(x,t)\psi(x,t) = i\hbar \frac{\partial}{\partial t}\psi(x,t). \tag{1}$$

In (1), $\hbar$ is the reduced Planck constant, D_α is the generalized fractional diffusion coefficient with physical dimension $[D_\alpha] = \text{erg}^{1-\alpha} \times \text{cm}^\alpha \times \text{sec}^{-\alpha}$ ($D_\alpha = 1/2m$ for $\alpha = 2$), $\psi(x,t)$ is the wave function and $-(\hbar\nabla)^\alpha$ is the quantum Riesz space fractional derivative which is defined by ([3])

$$\mathcal{F}\{-(\hbar\nabla)^\alpha \psi(x,t); p\} = |p|^\alpha \hat{\psi}(p,t), \tag{2}$$

where

$$\hat{\psi}(p,t) = \mathcal{F}\{\psi(x,t); p\} = \frac{1}{2\pi\hbar} \int_{-\infty}^{\infty} e^{-ipx/\hbar} \psi(x,t) \, dx \tag{3}$$

2010 *Mathematics Subject Classification.* Primary 26A33, 35Q40, 42A38, 44A10, 44A99, 33B15, 33E12, 33B10.

Key words and phrases. Fractional calculus, fractional Schrödinger equation, quantum Riesz-Feller space fractional derivative, Caputo time fractional derivative, Fourier transform in momentum representation.

S. Baqer: Corresponding author.

is the Fourier transform in momentum representation of a function $\psi \in S$, where S is the space of rapidly decreasing functions, and

$$\psi(x,t) = \mathcal{F}^{-1}\left\{\hat{\psi}(p,t); x\right\} = \frac{1}{2\pi\hbar}\int_{-\infty}^{\infty} e^{ipx/\hbar}\, \hat{\psi}(p,t)\, dp \tag{4}$$

is the inverse Fourier transform. The adopted Fourier transform (3) and its inverse (4) satisfy the Plancherel theorem (see e.g. [15, Sec. 2.7]).

The FSE (1) is solved in case of the infinite potential well, a Dirac-delta potential, a linear potential and a Coulomb potential (see [3], [4]). For $\alpha = 2$, the equation (1) reduces to the standard SE.

The generalization of (1) when the time derivative is replaced by the Caputo fractional derivative ([12], [17]) of order β,

$$_a^C D_t^\beta f(t) = \begin{cases} \frac{1}{\Gamma(n-\beta)}\int_a^t \frac{f^{(n)}(\tau)}{(t-\tau)^{\beta+1-n}}\, d\tau, & n-1 < \beta < n,\ n \in \mathbb{N} \\ \frac{d^n}{dt^n} f(t), & \beta = n \in \mathbb{N} \end{cases}, \tag{5}$$

was studied recently by many authors (see e.g. [5], [7]). The FSE with the quantum Riesz-Feller space fractional derivative D_θ^α of order α, skewness θ and the Caputo time fractional derivative $_0^C D_t^\beta$ of order β has the form

$$C_\alpha D_\theta^\alpha \psi(x,t) + V(x,t)\psi(x,t) = (i\hbar)^\beta\, _0^C D_t^\beta \psi(x,t), \tag{6}$$

where C_α is a positive constant $(C_\alpha = \hbar^2/2m$ for $\alpha = 2)$. The Riesz-Feller space fractional derivative $_x D_\theta^\alpha$, which is the generalization of the Riesz space fractional derivative, of order α and skewness θ is defined through its Fourier transform as ([10, Def. 6.7.])

$$\mathcal{F}\{_x D_\theta^\alpha \psi(x,t); p\} = -\eta_\alpha^\theta \hat{\psi}(p,t), \tag{7}$$

where

$$\eta_\alpha^\theta = |p|^\alpha e^{iSgn(p)\theta\pi/2},\ 0 < \alpha \leq 2,\ |\theta| \leq \min\{\alpha, 2-\alpha\}. \tag{8}$$

The quantum Riesz-Feller space fractional derivative D_θ^α is defined to be the Riesz-Feller space fractional derivative timed by a minus sign (See [6]), that is,

$$D_\theta^\alpha = -\,_x D_\theta^\alpha.$$

Thus for the quantum Riesz-Feller space fractional derivative

$$\mathcal{F}\{D_\theta^\alpha \psi(x,t); p\} = \eta_\alpha^\theta \hat{\psi}(p,t). \tag{9}$$

When $\theta = 0$, (9) reduces to

$$\mathcal{F}\{D_0^\alpha \psi(x,t); p\} = |p|^\alpha \hat{\psi}(p,t),$$

where ([10, (6.149)])

$$D_0^\alpha = -\,_x D_0^\alpha = -\left[-(-\Delta)^{\alpha/2}\right] = (-\Delta)^{\alpha/2}.$$

Thus, the operator $(-\Delta)^{\alpha/2}$ is the quantum Riesz space fractional derivative in case of generalizing the SE with the quantum Riesz-Feller space fractional derivative. For $\theta = 0$, $\alpha = 2$ and $\beta = 1$ (6) reduces to the standard SE. The interested readers in some new results related to the Riesz-Feller space fractional derivative, like its possible geometrical and physical meanings, can refer to the recent papers ([18]-[20]).

When $\theta = 0$, the following FSE
$$C_\alpha (-\Delta)^{\alpha/2} \psi(x,t) + V(x,t)\psi(x,t) = E\psi(x,t) \tag{10}$$
is equivalent to the one in terms of the fractional diffusion coefficient D_α, that is,
$$-D_\alpha (\hbar \nabla)^\alpha \psi(x,t) + V(x,t)\psi(x,t) = E\psi(x,t). \tag{11}$$
Indeed, when $\alpha = 2$ the equations (10) and (11) reduce to the standard SE.

For a free particle the equation (6) with the standard first order time derivative is solved and the solution is obtained ([6]) in terms of the Fox H-function $H_{p,q}^{m,n}(z)$ ([10, Sec. 1.2]).

We consider in this paper two cases of time independent potentials: a Dirac-delta potential and a linear potential. The method of separation of variables can be effectively used and under the assumption $\psi(x,t) = f(t)\phi(x)$, to reduce equation (6) to the time fractional equation
$$(i\hbar)^\beta \, {}^C_0 D_t^\beta f(t) = E f(t), \tag{12}$$
and the space fractional equation
$$C_\alpha D_\theta^\alpha \phi(x) + V(x)\phi(x) = E\phi(x), \tag{13}$$
where E refers to the energy.

2. Time Fractional Equation

THEOREM 2.1. ([5]) *If $0 < \beta \leq 1$, the solution of the time fractional equation (12) is of the form*
$$f(t) = f(0) H_{1,2}^{1,1} \left[-\left(\frac{t}{i\hbar}\right)^\beta E \, \middle| \, \begin{matrix} (0,1) \\ (0,1), (0,\beta) \end{matrix} \right].$$

The proof of Theorem 2.1. is based on the usage of the Laplace transform applied to (12) which implies ([12, (2.253)])
$$F(s) = f(0) \frac{s^{\beta-1}}{s^\beta - (i\hbar)^{-\beta} E}. \tag{14}$$
Since
$$\frac{1}{1 - (i\hbar)^{-\beta} E s^{-\beta}} = \sum_{k=0}^\infty E^k (i\hbar)^{-\beta k} s^{-\beta k},$$
then (14) becomes
$$F(s) = f(0) \sum_{k=0}^\infty \frac{E^k (i\hbar)^{-\beta k}}{s^{\beta k + 1}}. \tag{15}$$
By the inverse Laplace transform applied to (15) it can be seen that
$$f(t) = f(0) \sum_{k=0}^\infty \frac{\left(E (i\hbar)^{-\beta} t^\beta\right)^k}{\Gamma(\beta k + 1)} = f(0) E_\beta \left(\left(\frac{t}{i\hbar}\right)^\beta E\right), \tag{16}$$
where $E_\beta(z)$ is the one-parameter Mittag-Leffler function ([12, (1.55)]). It is known that ([10, (1.135)])
$$E_\beta(z) = H_{1,2}^{1,1}\left(-z \, \middle| \, \begin{matrix} (0,1) \\ (0,1), (0,\beta) \end{matrix}\right),$$

and thus the validity of the theorem is proved. □

COROLLARY 2.1. ([8]) For $\beta = 1$, the solution of the time fractional equation (12) is of the form

$$f(t) = f(0) e^{-iEt/\hbar}.$$

3. Space Fractional Equation

LEMMA 3.1. For any $x, \alpha > 0, \rho \geq 0$, and $b \in \mathbb{C} - \{0\}$ with $|\arg(b)| < \pi$, the following identical formula of the Fox H-function holds,

$$\frac{x^\rho}{1 + bx^\alpha} = b^{-\frac{\rho}{\alpha}} H_{1,1}^{1,1} \left(bx^\alpha \,\middle|\, \begin{array}{c} (\frac{\rho}{\alpha}, 1) \\ (\frac{\rho}{\alpha}, 1) \end{array} \right).$$

PROOF. According to ([14, (C.12)]),

(17) $$\frac{x^\rho}{1 + bx^\alpha} = b^{-\frac{\rho}{\alpha}} G_{1,1}^{1,1} \left(bx^\alpha \,\middle|\, \begin{array}{c} \frac{\rho}{\alpha} \\ \frac{\rho}{\alpha} \end{array} \right),$$

where $G_{p,q}^{m,n}(z)$ is the G-function. Since a G-function and a Fox H-function are related by

$$G_{p,q}^{m,n}\left(z \,\middle|\, \begin{array}{c} a_p \\ b_q \end{array} \right) = H_{p,q}^{m,n}\left(z \,\middle|\, \begin{array}{c} (a_p, 1) \\ (b_q, 1) \end{array} \right),$$

then the identical formula follows. The condition on the argument b, $|\arg(b)| < \pi$, provides the existence of the Fox H-function ([10, (1.20)]) which is

if $\sigma > 0$ and $|\arg(z)| < \dfrac{\pi\sigma}{2}$, then the Fox H – function exists for all $z \neq 0$

where

$$\sigma = \sum_{j=1}^{n} A_j - \sum_{j=n+1}^{p} A_j + \sum_{j=1}^{m} B_j - \sum_{j=m+1}^{q} B_j.$$

If $z = bx^\alpha$, $z \neq 0$ as $x \in (0, \infty)$ and $b \in \mathbb{C} - \{0\}$, then $|\arg(z)| = |\arg(bx^\alpha)| = |\arg(b)|$. In our case $\sigma = 2 > 0$, and therefore $|\arg(b)| < \pi$. □

3.1. Dirac-delta Potential.
Suppose a particle is moving in a Dirac-delta potential $V(x) = -\gamma\delta(x)(\gamma > 0)$. Then the space fractional equation (13) takes the form

(18) $$C_\alpha D_\theta^\alpha \phi(x) - \gamma\delta(x)\phi(x) = E\phi(x).$$

THEOREM 3.1. *If $1 < \alpha \leq 2$ and $|\theta| \leq \min\{\alpha, 2-\alpha\}$, then for $x \neq 0$ the solution of the equation (18) has the form*

$$\phi(x) = \frac{-\pi\gamma k}{(2\pi\hbar)^2 \alpha E} \left(\frac{C_\alpha}{-E}\right)^{\frac{-1}{\alpha}}$$

$$\times \left\{ e^{\frac{-i\theta\pi}{2\alpha}} H_{2,3}^{2,1} \left[|x| \left(\frac{\hbar^\alpha e^{i\theta\pi/2} C_\alpha}{-E}\right)^{-1/\alpha} \middle| \begin{array}{c} \left(\frac{\alpha-1}{\alpha}, \frac{1}{\alpha}\right), \left(\frac{1}{2}, \frac{1}{2}\right) \\ (0,1), \left(\frac{\alpha-1}{\alpha}, \frac{1}{\alpha}\right), \left(\frac{1}{2}, \frac{1}{2}\right) \end{array} \right] \right.$$

$$+ e^{\frac{i\theta\pi}{2\alpha}} H_{2,3}^{2,1} \left[|x| \left(\frac{\hbar^\alpha e^{-i\theta\pi/2} C_\alpha}{-E}\right)^{-1/\alpha} \middle| \begin{array}{c} \left(\frac{\alpha-1}{\alpha}, \frac{1}{\alpha}\right), \left(\frac{1}{2}, \frac{1}{2}\right) \\ (0,1), \left(\frac{\alpha-1}{\alpha}, \frac{1}{\alpha}\right), \left(\frac{1}{2}, \frac{1}{2}\right) \end{array} \right] \right\} - i \frac{\gamma k \sqrt{\pi}}{2(2\pi\hbar)^2 \alpha E}$$

$$\times \left(\frac{C_\alpha}{-E}\right)^{\frac{-1}{\alpha}} \left\{ e^{\frac{-i\theta\pi}{2\alpha}} H_{1,3}^{2,1} \left[|x| \left(\frac{2^\alpha \hbar^\alpha C_\alpha e^{i\theta\pi/2}}{-E}\right)^{-1/\alpha} \middle| \begin{array}{c} \left(\frac{\alpha-1}{\alpha}, \frac{1}{\alpha}\right) \\ \left(\frac{1}{2}, \frac{1}{2}\right), \left(\frac{\alpha-1}{\alpha}, \frac{1}{\alpha}\right), \left(0, \frac{1}{2}\right) \end{array} \right] \right.$$

$$\left. - e^{\frac{i\theta\pi}{2\alpha}} H_{1,3}^{2,1} \left[|x| \left(\frac{2^\alpha \hbar^\alpha C_\alpha e^{-i\theta\pi/2}}{-E}\right)^{-1/\alpha} \middle| \begin{array}{c} \left(\frac{\alpha-1}{\alpha}, \frac{1}{\alpha}\right) \\ \left(\frac{1}{2}, \frac{1}{2}\right), \left(\frac{\alpha-1}{\alpha}, \frac{1}{\alpha}\right), \left(0, \frac{1}{2}\right) \end{array} \right] \right\},$$

where

$$k = 2\pi\hbar \, \mathcal{F}\{\delta(x)\phi(x); p\}.$$

PROOF. Taking into account (9), the application of the Fourier transform (3) to (18) leads to

$$\hat{\phi}(p) = \frac{\gamma k}{2\pi\hbar} \frac{1}{(C_\alpha |p|^\alpha e^{iSgn(p)\theta\pi/2} - E)}, \tag{19}$$

and thus

$$\phi(x) = \frac{\gamma k}{(2\pi\hbar)^2} (I_1 + iI_2), \tag{20}$$

where

$$I_1 = \int_{-\infty}^{\infty} \frac{\cos(px/\hbar)}{(C_\alpha |p|^\alpha e^{iSgn(p)\theta\pi/2} - E)} \, dp,$$

$$I_2 = \int_{-\infty}^{\infty} \frac{\sin(px/\hbar)}{(C_\alpha |p|^\alpha e^{iSgn(p)\theta\pi/2} - E)} \, dp.$$

If $x > 0$, by Lemma 3.1. it is possible to see that

$$I_1 = \frac{-1}{E} \left\{ \int_0^\infty \cos(px/\hbar) H_{1,1}^{1,1} \left(-p^\alpha \frac{C_\alpha e^{-i\theta\pi/2}}{E} \middle| \begin{array}{c} (0,1) \\ (0,1) \end{array} \right) dp \right.$$

$$\left. + \int_0^\infty \cos(px/\hbar) H_{1,1}^{1,1} \left(-p^\alpha \frac{C_\alpha e^{i\theta\pi/2}}{E} \middle| \begin{array}{c} (0,1) \\ (0,1) \end{array} \right) dp \right\},$$

$$I_2 = \frac{-1}{E} \left\{ \int_0^\infty \sin(px/\hbar) H_{1,1}^{1,1} \left(-p^\alpha \frac{C_\alpha e^{i\theta\pi/2}}{E} \middle| \begin{array}{c} (0,1) \\ (0,1) \end{array} \right) dp \right.$$

$$\left. - \int_0^\infty \sin(px/\hbar) H_{1,1}^{1,1} \left(-p^\alpha \frac{C_\alpha e^{-i\theta\pi/2}}{E} \middle| \begin{array}{c} (0,1) \\ (0,1) \end{array} \right) dp \right\}.$$

According to the formulas for cosine and sine transforms of the Fox H-funcion ([**11**, (17)]) and ([**10**, (2.49)]), respectively, and by using the following properties of the Fox H-function ([**10**]):

$$H_{p,q}^{m,n}\left(z \left|\begin{array}{c}(a_p, A_p)\\(b_q, B_q)\end{array}\right.\right) = kH_{p,q}^{m,n}\left(z^k \left|\begin{array}{c}(a_p, kA_p)\\(b_q, kB_q)\end{array}\right.\right); \ k > 0,$$

$$H_{p,q}^{m,n}\left(z \left|\begin{array}{c}(a_p, A_p)\\(b_q, B_q)\end{array}\right.\right) = H_{q,p}^{n,m}\left(\frac{1}{z} \left|\begin{array}{c}(1-b_q, B_q)\\(1-a_p, A_p)\end{array}\right.\right),$$

$$z^\sigma H_{p,q}^{m,n}\left(z \left|\begin{array}{c}(a_p, A_p)\\(b_q, B_q)\end{array}\right.\right) = H_{p,q}^{m,n}\left(z \left|\begin{array}{c}(a_p + \sigma A_p, A_p)\\(b_q + \sigma B_q, B_q)\end{array}\right.\right); \ \sigma \in \mathbb{C},$$

we can get

$$I_1 = \frac{-\pi}{\alpha E}\left(\frac{C_\alpha}{-E}\right)^{\frac{-1}{\alpha}} \times$$

$$\left\{e^{\frac{-i\theta\pi}{2\alpha}} H_{2,3}^{2,1}\left[x\left(\frac{\hbar^\alpha e^{i\theta\pi/2} C_\alpha}{-E}\right)^{-1/\alpha} \left|\begin{array}{c}(1-(1/\alpha), 1/\alpha), (1/2, 1/2)\\(0,1), (1-(1/\alpha), 1/\alpha), (1/2, 1/2)\end{array}\right.\right]\right.$$

$$\left.+ e^{\frac{i\theta\pi}{2\alpha}} H_{2,3}^{2,1}\left[x\left(\frac{\hbar^\alpha e^{-i\theta\pi/2} C_\alpha}{-E}\right)^{-1/\alpha} \left|\begin{array}{c}(1-(1/\alpha), 1/\alpha), (1/2, 1/2)\\(0,1), (1-(1/\alpha), 1/\alpha), (1/2, 1/2)\end{array}\right.\right]\right\},$$

and

$$I_2 = \frac{-\sqrt{\pi}}{2\alpha E}\left(\frac{C_\alpha}{-E}\right)^{\frac{-1}{\alpha}} \times$$

$$\left\{e^{\frac{-i\theta\pi}{2\alpha}} H_{1,3}^{2,1}\left[x\left(\frac{2^\alpha \hbar^\alpha C_\alpha e^{i\theta\pi/2}}{-E}\right)^{-1/\alpha} \left|\begin{array}{c}(1-(1/\alpha), 1/\alpha)\\(1/2, 1/2), (1-(1/\alpha), 1/\alpha), (0, 1/2)\end{array}\right.\right]\right.$$

$$\left.- e^{\frac{i\theta\pi}{2\alpha}} H_{1,3}^{2,1}\left[x\left(\frac{2^\alpha \hbar^\alpha C_\alpha e^{-i\theta\pi/2}}{-E}\right)^{-1/\alpha} \left|\begin{array}{c}(1-(1/\alpha), 1/\alpha)\\(1/2, 1/2), (1-(1/\alpha), 1/\alpha), (0, 1/2)\end{array}\right.\right]\right\}.$$

The substitution of these expressions for I_1 and I_2 into (20) confirms the validity of the theorem.

The case $x < 0$ can be considered similarly in order to accomplish the proof. $\square$

Taking into account (10) and (11) the following assertions follow.

COROLLARY 3.1. ([**4**]) *If* $1 < \alpha \leq 2$ *and* $\theta = 0$, *the solution of the equation* (18) *for* $x \neq 0$ *has the form*

$$(21) \qquad \phi(x) = \xi_0 H_{2,3}^{2,1}\left[|x|\left(\frac{\hbar^\alpha D_\alpha}{-E}\right)^{-1/\alpha} \left|\begin{array}{c}\left(\frac{\alpha-1}{\alpha}, \frac{1}{\alpha}\right), \left(\frac{1}{2}, \frac{1}{2}\right)\\(0,1), \left(\frac{\alpha-1}{\alpha}, \frac{1}{\alpha}\right), \left(\frac{1}{2}, \frac{1}{2}\right)\end{array}\right.\right],$$

where

$$\xi_0 = \left(\frac{-\gamma k}{2\pi\hbar^2 E\alpha}\right)\left(\frac{D_\alpha}{-E}\right)^{\frac{-1}{\alpha}}.$$

COROLLARY 3.2. ([**8**]) *If* $\alpha = 2$ *and* $\theta = 0$, *the standard wave function solution of the equation* (18) *for* $x \neq 0$ *has the form*

$$\phi(x) = \lambda H_{0,1}^{1,0}\left(|x|\frac{\sqrt{-2mE}}{\hbar} \left|\begin{array}{c}\\(0,1)\end{array}\right.\right) = \lambda e^{-|x|\sqrt{-2mE}/\hbar} \quad (\lambda \text{ is a constant}).$$

3.2. Linear Potential.
Consider a particle in a linear potential field

$$V(x) = \begin{cases} Ax, & x \geq 0 \, (A > 0) \\ \infty, & x < 0 \end{cases}.$$

Then the space fractional equation (13) becomes

(22) $$C_\alpha D_\theta^\alpha \phi(x) + Ax\phi(x) = E\phi(x), \; x \geq 0.$$

THEOREM 3.2. *If $1 < \alpha \leq 2$ and $|\theta| \leq \min\{\alpha, 2 - \alpha\}$, then the solution of the equation (22) has the form*

$$\phi(x) = \frac{2\pi N}{(\alpha+1)} H_{2,2}^{1,1}\left[\left(x - \frac{E}{A}\right)\frac{1}{\hbar}\left(\frac{C_\alpha}{\hbar A(\alpha+1)}\right)^{\frac{-1}{\alpha+1}} \middle| \begin{array}{c} \left(\frac{\alpha}{\alpha+1}, \frac{1}{1+\alpha}\right), \left(\frac{2+\alpha-\theta}{2(\alpha+1)}, \frac{\alpha+\theta}{2(\alpha+1)}\right) \\ (0,1), \left(\frac{2+\alpha-\theta}{2(\alpha+1)}, \frac{\alpha+\theta}{2(\alpha+1)}\right) \end{array}\right],$$

where

$$N = \frac{1}{2\pi\hbar}\left(\frac{C_\alpha}{A\hbar(\alpha+1)}\right)^{\frac{-1}{(\alpha+1)}}.$$

PROOF. According to (9) and the formula ([4])

$$\mathcal{F}\{x\phi(x); p\} = i\hbar \frac{d}{dp}\hat{\phi}(p),$$

the application of the Fourier transform (3) to the equation (22) leads to

$$\frac{d\hat{\phi}(p)}{\hat{\phi}(p)} = \frac{1}{Ai\hbar}\left(E - C_\alpha |p|^\alpha e^{iSgn(p)\theta\pi/2}\right),$$

from where it follows readily that (omitting the constant of the integration)

$$\hat{\phi}(p) = \begin{cases} \exp\left[\frac{-i}{A\hbar}\left(Ep - \frac{C_\alpha}{\alpha+1}p^{\alpha+1}e^{i\theta\pi/2}\right)\right]; & p > 0 \\ \exp\left[\frac{-i}{A\hbar}\left(Ep + \frac{C_\alpha}{\alpha+1}|p|^{\alpha+1}e^{-i\theta\pi/2}\right)\right]; & p < 0 \end{cases}.$$

Setting

(23) $$w = p\left(\frac{C_\alpha}{A\hbar(\alpha+1)}\right)^{\frac{1}{(\alpha+1)}}, \; y = \frac{1}{\hbar}\left(x - \frac{E}{A}\right)\left(\frac{C_\alpha}{A\hbar(\alpha+1)}\right)^{\frac{-1}{(\alpha+1)}},$$

it is possible by the application of the inverse Fourier transform (4) to get that

(24) $$\phi(y) = N\{\phi_1(y) + \phi_2(y)\},$$

where

$$\phi_1(y) = \int_0^\infty e^{iyw} e^{ie^{i\theta\pi/2} w^{\alpha+1}} dw,$$

and

$$\phi_2(y) = \int_{-\infty}^0 e^{iyw} e^{-ie^{-i\theta\pi/2}|w|^{\alpha+1}} dw.$$

Denote by $\hat{\phi}(s) = \mathcal{M}\{\phi(y); s\}$ the Mellin transform of $\phi(y)$. From the formula ([16, Ch.8])

(25) $$\mathcal{M}\{e^{-i\rho x}; s\} = (i\rho)^{-s}\Gamma(s) = \rho^{-s}\Gamma(s)\left(\cos\left(\frac{\pi s}{2}\right) - i\sin\left(\frac{\pi s}{2}\right)\right); \; \rho \in \mathbb{C},$$

it follows that

(26)
$$\tilde{\phi}_1(s) = (-i)^{-s}\Gamma(s)\int_0^\infty e^{ie^{i\theta\pi/2}w^{\alpha+1}} w^{-s}dw,$$

and

(27)
$$\tilde{\phi}_2(s) = (-i)^{-s}\Gamma(s)\int_{-\infty}^0 e^{-ie^{-i\theta\pi/2}|w|^{\alpha+1}} w^{-s}dw.$$

Now by the formula (25), the substitution $u = w^{\alpha+1}$ in (26) and the substitutions $u = -w$, $\xi = u^{\alpha+1}$ in (27), it can be seen that

$$\tilde{\phi}_1(s) = \frac{1}{\alpha+1}\Gamma\left(\frac{1-s}{1+\alpha}\right)\Gamma(s)\exp\left(\frac{i\pi}{2}\left[\frac{1-\theta}{\alpha+1} + \frac{(\alpha+\theta)s}{\alpha+1}\right]\right),$$

and

$$\tilde{\phi}_2(s) = \frac{1}{\alpha+1}\Gamma\left(\frac{1-s}{1+\alpha}\right)\Gamma(s)\exp\left(\frac{-i\pi}{2}\left[\frac{1-\theta}{\alpha+1} + \frac{(\alpha+\theta)s}{\alpha+1}\right]\right).$$

These representations and the formula

$$\cos(\pi z/2) = \frac{\pi}{\Gamma\left(\frac{1+z}{2}\right)\Gamma\left(\frac{1-z}{2}\right)}$$

allow from (24) to be obtained that

(28)
$$\tilde{\phi}(s) = \frac{2\pi N}{(\alpha+1)}\frac{\Gamma(s)\Gamma\left(\frac{1-s}{1+\alpha}\right)}{\Gamma\left(\frac{\alpha+\theta-\alpha s-\theta s}{2(\alpha+1)}\right)\Gamma\left(\frac{2+\alpha-\theta+\alpha s+\theta s}{2(\alpha+1)}\right)}.$$

Therefore

(29)
$$\phi(y) = \frac{2\pi N}{(\alpha+1)}\frac{1}{2\pi i}\int_{\gamma-i\infty}^{\gamma+i\infty}\frac{\Gamma(s)\Gamma\left(\frac{1-s}{1+\alpha}\right)}{\Gamma\left(\frac{\alpha+\theta-\alpha s-\theta s}{2(\alpha+1)}\right)\Gamma\left(\frac{2+\alpha-\theta+\alpha s+\theta s}{2(\alpha+1)}\right)}y^{-s}ds,$$

that is ([**10**, Sec. 1.2]),

$$\phi(y) = \frac{2\pi N}{(\alpha+1)}H_{2,2}^{1,1}\left(y\left|\begin{array}{c}\left(\frac{\alpha}{\alpha+1},\frac{1}{1+\alpha}\right),\left(\frac{2+\alpha-\theta}{2(\alpha+1)},\frac{\alpha+\theta}{2(\alpha+1)}\right)\\(0,1),\left(\frac{2+\alpha-\theta}{2(\alpha+1)},\frac{\alpha+\theta}{2(\alpha+1)}\right)\end{array}\right.\right).$$

Taking into account the substitutions (23), the validity of the theorem follows directly. □

Having in mind (10) and (11) the following assertions follow.

COROLLARY 3.3. ([4]) *If $1 < \alpha \le 2$ and $\theta = 0$, the solution of the equation (22) has the form*

(30) $\phi(x) =$

$$\frac{2\pi N}{(\alpha+1)}H_{2,2}^{1,1}\left[\left(x-\frac{E}{A}\right)\frac{1}{\hbar}\left(\frac{D_\alpha}{\hbar A(\alpha+1)}\right)^{\frac{-1}{\alpha+1}}\left|\begin{array}{c}\left(\frac{\alpha}{\alpha+1},\frac{1}{1+\alpha}\right),\left(\frac{2+\alpha}{2(\alpha+1)},\frac{\alpha}{2(\alpha+1)}\right)\\(0,1),\left(\frac{2+\alpha}{2(\alpha+1)},\frac{\alpha}{2(\alpha+1)}\right)\end{array}\right.\right].$$

By the power series representation of the Fox H-function ([10, (A.69)]) and the formula
$$\sin(\pi z) = \frac{\pi}{\Gamma(z)\Gamma(1-z)},$$
we can write (30) as

(31) $\phi(x) = \frac{2N}{(\alpha+1)}$
$$\times \sum_{k=0}^{\infty} \Gamma\left(\frac{k+1}{\alpha+1}\right) \sin\left(\frac{(\alpha+2)(k+1)\pi}{2(\alpha+1)}\right) \frac{1}{k!} \left[\left(x - \frac{E}{A}\right) \frac{1}{\hbar} \left(\frac{D_\alpha}{A\hbar(\alpha+1)}\right)^{\frac{-1}{(\alpha+1)}}\right]^k.$$

This enables us to formulate the following statement.

COROLLARY 3.4. ([9]) If $\alpha = 2$ and $\theta = 0$, the standard wave function solution of the equation (22) has the form
$$\phi(x) = \frac{\lambda}{\pi} \sum_{k=0}^{\infty} \Gamma\left(\frac{k+1}{3}\right) \sin\left(\frac{2(k+1)\pi}{3}\right) \frac{1}{k!} \left(3^{\frac{1}{3}} u\right)^k,$$
where λ is a constant and $u = \left(x - \left(\frac{E}{A}\right)\right) \left(\frac{2mA}{\hbar^2}\right)^{\frac{1}{3}}$.

References

[1] Feynman, R. P. and Hibbs, A. R. (1965). *Quantum Mechanics and Path Integrals*. McGraw-Hill, New York.
[2] Nikolai Laskin, *Fractional quantum mechanics and Lévy path integrals*, Phys. Lett. A **268** (2000), no. 4-6, 298–305, DOI 10.1016/S0375-9601(00)00201-2. MR1755089 (2000m:81097)
[3] Nick Laskin, *Principles of fractional quantum mechanics*, Fractional dynamics, World Sci. Publ., Hackensack, NJ, 2012, pp. 393–427. MR2932616
[4] Jianping Dong and Mingyu Xu, *Some solutions to the space fractional Schrödinger equation using momentum representation method*, J. Math. Phys. **48** (2007), no. 7, 072105, 14, DOI 10.1063/1.2749172. MR2337665 (2008h:81040)
[5] Jianping Dong and Mingyu Xu, *Space-time fractional Schrödinger equation with time-independent potentials*, J. Math. Anal. Appl. **344** (2008), no. 2, 1005–1017, DOI 10.1016/j.jmaa.2008.03.061. MR2426328 (2010e:35291)
[6] Boyadjiev, L., Lucko, Yu. and Al-Saqabi, B. (2013). Comments on employing the Riesz-Feller derivative in the Schrödinger equation. *Eur. Phys. J. Special Topics* 222, 1779-1794.
[7] Selçuk Ş. Bayın, *Time fractional Schrödinger equation: Fox's H-functions and the effective potential*, J. Math. Phys. **54** (2013), no. 1, 012103, 18, DOI 10.1063/1.4773100. MR3059867
[8] Griffiths, D. J. (2004). *Introduction to Quantum Mechanics*, 2nd ed. Prentice-Hall, Englewood Cliffs NJ.
[9] L. D. Landau and E. M. Lifshitz, *A shorter course of theoretical physics. Vol. 2*, Pergamon Press, Oxford-New York-Toronto, Ont., 1974. Quantum mechanics; Translated from the Russian by J. B. Sykes and J. S. Bell. MR0400931 (53 #4761)
[10] A. M. Mathai, Ram Kishore Saxena, and Hans J. Haubold, *The H-function*, Springer, New York, 2010. Theory and applications. MR2562766 (2011j:33002)
[11] Lyubomir Boyadjiev and Bader Al-Saqabi, *On the fractional Schrödinger equation*, Int. J. Appl. Math. **24** (2011), no. 6, 873–884. MR2933732
[12] Igor Podlubny, *Fractional differential equations*, Mathematics in Science and Engineering, vol. 198, Academic Press, Inc., San Diego, CA, 1999. An introduction to fractional derivatives, fractional differential equations, to methods of their solution and some of their applications. MR1658022 (99m:26009)
[13] Larry C. Andrews, *Special functions of mathematics for engineers*, 2nd ed., SPIE Optical Engineering Press, Bellingham, WA; Oxford University Press, Oxford, 1998. MR1492179 (99a:33001)

[14] Virginia Kiryakova, *Generalized fractional calculus and applications*, Pitman Research Notes in Mathematics Series, vol. 301, Longman Scientific & Technical, Harlow; copublished in the United States with John Wiley & Sons, Inc., New York, 1994. MR1265940 (95d:26010)
[15] Gasiorowicz, S. (2003). *Quantum Physics*, 3rd ed. Wiley.
[16] Lokenath Debnath and Dambaru Bhatta, *Integral transforms and their applications*, 2nd ed., Chapman & Hall/CRC, Boca Raton, FL, 2007. MR2253985 (2007e:44001)
[17] Michele Caputo, *Linear models of dissipation whose Q is almost frequency independent. II*, Fract. Calc. Appl. Anal. **11** (2008), no. 1, 4–14. Reprinted from Geophys. J. R. Astr. Soc. **13** (1967), no. 5, 529–539. MR2379269 (2009b:86009)
[18] Richard Herrmann, *Covariant fractional extension of the modified Laplace-operator used in 3D-shape recovery*, Fract. Calc. Appl. Anal. **15** (2012), no. 2, 332–343, DOI 10.2478/s13540-012-0024-1. MR2897783
[19] Richard Herrmann, *Towards a geometric interpretation of generalized fractional integrals—Erdélyi-Kober type integrals on R^N, as an example*, Fract. Calc. Appl. Anal. **17** (2014), no. 2, 361–370, DOI 10.2478/s13540-014-0174-4. MR3181060
[20] Richard Herrmann, *Reflection symmetric Erdélyi-Kober type operators—a quasi-particle interpretation*, Fract. Calc. Appl. Anal. **17** (2014), no. 4, 1215–1228, DOI 10.2478/s13540-014-0221-1. MR3254688

DEPARTMENT OF MATHEMATICS, FACULTY OF SCIENCE, KUWAIT UNIVERSITY
E-mail address: mpcosmo57@gmail.com

DEPARTMENT OF MATHEMATICS, FACULTY OF SCIENCE, KUWAIT UNIVERSITY
E-mail address: boyadjievl@yahoo.com

A note on companion pencils

Jared L. Aurentz, Thomas Mach, Raf Vandebril, and David S. Watkins

ABSTRACT. Various generalizations of companion matrices to companion pencils are presented. Companion matrices link to monic polynomials, whereas companion pencils do not require monicity of the corresponding polynomial. In the classical companion pencil case (A, B) only the coefficient of the highest degree appears in B's lower right corner. We will show, however, that all coefficients of the polynomial can be distributed over both A and B creating additional flexibility.

Companion matrices admit a so-called Fiedler factorization into essentially 2×2 matrices. These Fiedler factors can be reordered without affecting the eigenvalues (which equal the polynomial roots) of the assembled matrix. We will propose a generalization of this factorization and the reshuffling property for companion pencils.

Special examples of the factorizations and extensions to matrix polynomials and product eigenvalue problems are included.

1. Introduction

It is well known that the polynomial

(1) $$p(z) = c_n z^n + c_{n-1} z^{n-1} + \cdots + c_1 z^1 + c_0$$

2000 *Mathematics Subject Classification.* Primary 65F15; Secondary 15A23.

Key words and phrases. Companion pencil, polynomial, matrix polynomial, rootfinding, product eigenvalue problems.

The research was partially supported by the Research Council KU Leuven, projects CREA-13-012 Can Unconventional Eigenvalue Algorithms Supersede the State of the Art, OT/11/055 Spectral Properties of Perturbed Normal Matrices and their Applications, and CoE EF/05/006 Optimization in Engineering (OPTEC); by the Fund for Scientific Research–Flanders (Belgium) project G034212N Reestablishing Smoothness for Matrix Manifold Optimization via Resolution of Singularities; by the Interuniversity Attraction Poles Programme, initiated by the Belgian State, Science Policy Office, Belgian Network DYSCO (Dynamical Systems, Control, and Optimization); and by the European Research Council under the European Union's Seventh Framework Programme (FP7/20072013)/ERC grant agreement no. 291068. The views expressed in this article are not those of the ERC or the European Commission, and the European Union is not liable for any use that may be made of the information contained here.

©2016 American Mathematical Society

equals the determinant $\det(zB - A)$, with

(2) $\quad A = \begin{bmatrix} 0 & & & & -c_0 \\ 1 & 0 & & & -c_1 \\ & \ddots & \ddots & & \vdots \\ & & 1 & 0 & -c_{n-2} \\ & & & 1 & -c_{n-1} \end{bmatrix} \quad \text{and} \quad B = \begin{bmatrix} 1 & & & & \\ & 1 & & & \\ & & \ddots & & \\ & & & 1 & \\ & & & & c_n \end{bmatrix}.$

The matrix pencil (A, B) is called the *companion pencil* of $p(z)$ and can be used to compute the roots of $p(z)$ as the roots of $p(z)$ coincide with the eigenvalues of the pencil (A, B).

At the moment, generalizing Frobenius companion matrices and/or pencils [3, 7, 9] is an active research topic as new matrix forms and/or factorizations open up possibilities to develop new, possibly faster or more accurate algorithms for both the classical rootfinding problem as well as matrix polynomial eigenvalue problems [6, 8, 12]. Even though the companion pencil exhibits a favorable numerical behavior [10] with respect to the companion matrix, there are only a few structured QZ algorithms for companion pencils available [4, 5].

The backward error bound for polynomial rootfinding based on companion matrices is

$$\|c - \tilde{c}\| < k_1 \frac{c_{\max}}{c_n} \|c\| \epsilon_m + \mathcal{O}(\epsilon_m^2),$$

with $\tilde{c}$ the coefficients of a polynomial with the computed roots, k_2 a constant, $c_{\max}$ the coefficient with the largest absolute value, and ϵ_m the machine precision. However, the backward error for polynomial rootfinding based on companion pencils is only

$$\|c - \tilde{c}\| < k_2 \|c\| \epsilon_m,$$

where a scaling to $\|c\| = 1$ is possible [8, 10, 12]. Thus polynomial rootfinding with companion pencils is advantageous for polynomials with large $\frac{c_{\max}}{c_n}$.

In this article no new rootfinders will be proposed, only theoretical extensions of the existing companion pencil are introduced. First, in Section 2 we will generalize the companion pencil (1) by distributing not only the highest order, but all polynomial coefficients over both A and B. Moreover, one does not have to choose whether to put a specific coefficient c_i in A or B, one can also write $c_i = v_{i+1} + w_i$, where the term v_i will appear in A, and w_i in B. For matrix polynomials a related approach has been investigated in [11]. Second, we will adjust Fiedler's factorization of a companion matrix to obtain a similar factorization of the companion pencil and we will prove that also in this case one can reorder all the factors without altering the pencil's eigenvalues. Finally, in Sections 4 and 5 we link some of the results to matrix polynomials and product eigenvalue problems.

2. Generalized companion pencils

The following theorem shows that the coefficients can be distributed over the last column of A and the last column of B.

THEOREM 2.1. *Consider $p(z)$, a polynomial with complex coefficients $p(z) = c_n z^n + c_{n-1} z^{n-1} + \cdots + c_1 z + c_0$. Let v, w be vectors of length n with*

$$\begin{aligned} v_1 &= c_0, \\ v_{i+1} + w_i &= c_i, \text{ for } i = 1, \ldots, n-1, \text{ and} \\ w_n &= c_n. \end{aligned} \tag{3}$$

Then the determinant $\det(zB - A)$ of the pencil (A, B), with

$$A = Z - v e_n^T, \quad B = Z^H Z + w e_n^T, \tag{4}$$

e_n the nth standard unit vector, and Z the downshift matrix

$$Z = \begin{bmatrix} 0 & & & \\ 1 & 0 & & \\ & \ddots & \ddots & \\ & & 1 & 0 \end{bmatrix},$$

equals $p(z)$.

PROOF. We have

$$\det(zB - A) = \det \begin{pmatrix} z & & & & zw_1 + v_1 \\ -1 & z & & & zw_2 + v_2 \\ & \ddots & \ddots & & \vdots \\ & & -1 & z & zw_{n-1} + v_{n-1} \\ & & & -1 & zw_n + v_n \end{pmatrix}.$$

Applying Laplace's formula to the first row yields

$$\det(zB - A) = z \det \begin{pmatrix} z & & & & zw_2 + v_2 \\ -1 & z & & & zw_3 + v_3 \\ & \ddots & \ddots & & \vdots \\ & & -1 & z & zw_{n-1} + v_{n-1} \\ & & & -1 & zw_n + v_n \end{pmatrix}$$

$$+ (-1)^{n+1} (zw_1 + v_1) \underbrace{\det \begin{pmatrix} -1 & z & & & \\ & -1 & z & & \\ & & \ddots & \ddots & \\ & & & -1 & z \\ & & & & -1 \end{pmatrix}}_{=(-1)^{n-1}}.$$

The proof is completed by applying Laplace's formula recursively. □

The coefficients c_0 in A and c_n in B are fixed. All other coefficients can be freely distributed between parts in A and parts in B as long as v_{i+1} and w_i sum up to c_i. The following example demonstrates that the additional freedom can be used to create additional structure in A and B.

EXAMPLE 2.2. Let $p(z) = 5z^3 - z^2 + 2z - 1$. Then both A and B can be chosen to consist solely of non-negative entries, e.g.,

$$A = \begin{bmatrix} 0 & & 1 \\ 1 & 0 & 1 \\ & 1 & 2 \end{bmatrix} \quad \text{and} \quad B = \begin{bmatrix} 1 & & 3 \\ & 1 & 1 \\ & & 5 \end{bmatrix}.$$

Another option is to choose always either $w_i = 0$ or $v_{i+1} = 0$:

COROLLARY 2.3. Consider again $p(z)$ from (1). Take two disjunct subsets $\mathcal{I} \subset \{0, \ldots, n-1\}$ and $\mathcal{J} \subset \{1, \ldots, n\}$, $\mathcal{I} \cap \mathcal{J} = \emptyset$ such that their union $\mathcal{I} \cup \mathcal{J} = \{0, \ldots, n\}$.

Define two vectors v and w as

$$v_{i+1} = \begin{cases} c_i & \text{if } i \in \mathcal{I}, \\ 0 & \text{if } i \notin \mathcal{I}, \end{cases} \quad \text{and} \quad w_j = \begin{cases} c_j & \text{if } j \in \mathcal{J}, \\ 0 & \text{if } j \notin \mathcal{J}. \end{cases}$$

Then the eigenvalues of the pencil (A, B) with

$$A = Z - v e_n^T, \quad B = Z^H Z + w e_n^T,$$

and Z be the downshift matrix coincide with the polynomial roots of $p(z)$.

EXAMPLE 2.4. Let $p(z) = c_5 z^5 + c_4 z^4 + \cdots + c_1 z + c_0$. Then (A, B), with

$$A = \begin{bmatrix} 0 & & & & -c_0 \\ 1 & 0 & & & 0 \\ & 1 & 0 & & -c_2 \\ & & 1 & 0 & 0 \\ & & & 1 & -c_4 \end{bmatrix} \quad \text{and} \quad B = \begin{bmatrix} 1 & & & & c_1 \\ & 1 & & & 0 \\ & & 1 & & c_3 \\ & & & 1 & 0 \\ & & & & c_5 \end{bmatrix}$$

is a companion pencil of $p(z)$.

REMARK 2.5. When considering the existing QZ-algorithms by Boito, Eidelman, and Gemignani [4, 5] we notice that the algorithms are based on the property that both A and B in (2) are of unitary-plus-rank-one form, a structure maintained under QZ-iterates. This property also holds for the new pencil matrices in Theorem 2.1 implying that with modest modifications the existing algorithms could compute the eigenvalues of the new pencils as well.

3. Fiedler companion factorization

The matrix A of the pencil (A, B) is the companion matrix of the monic polynomial whose coefficients are defined by v. As shown by Fiedler [9], companion matrices can be factored in n core transformations. A *core transformation* X_i is a modified identity matrix, where a 2×2 submatrix, for $i = 0$ and $i = n$ one entry, on the diagonal is replaced by an arbitrary matrix; for all core transformations the subindex i links to the submatrix $X_i(i : i+1, i : i+1)$ differing from the identity. For the factorization of the companion matrix the core transformations $A_0, A_i, \ldots, A_{n-1}$, typically named Fiedler factors, have the form[1] $A_0(1,1) = -v_1$

---

[1] Note that A_0's active part is restricted to the upper left matrix entry. Later B_n's active part is restricted to the lower right matrix entry.

and $A_{i-1}(i-1:i, i-1:i) = \begin{bmatrix} 0 & 1 \\ 1 & -v_i \end{bmatrix}$ for $i = 2, \ldots, n$. For example, the factorization of A, with $n = 5$ equals

$$A = \begin{bmatrix} 0 & & & & -v_1 \\ 1 & 0 & & & -v_2 \\ & 1 & 0 & & -v_3 \\ & & 1 & 0 & -v_4 \\ & & & 1 & -v_5 \end{bmatrix} = \begin{bmatrix} -v_1 \end{bmatrix} \begin{bmatrix} 0 & 1 \\ 1 & -v_2 \end{bmatrix} \begin{bmatrix} 0 & 1 \\ 1 & -v_3 \end{bmatrix} \begin{bmatrix} 0 & 1 \\ 1 & -v_4 \end{bmatrix} \begin{bmatrix} 0 & 1 \\ 1 & -v_5 \end{bmatrix}$$
$$= A_0 A_1 A_2 A_3 A_4,$$

where only the essential parts of the core transformations are shown. Moreover, Fiedler also proved [9] that reordering the Fiedler factors arbitrarily has no effect on the eigenvalues of their product matrix. Thus for all permutations σ of $\{0, \ldots, n-1\}$ the eigenvalues of $A_{\sigma(0)} \cdots A_{\sigma(n-1)}$ equal those of $A_0 \cdots A_{n-1}$. The product $A_{\sigma(0)} \cdots A_{\sigma(n-1)}$ is often referred to as a *Fiedler companion matrix*. In the setting of companion pencils Fiedler factorizations have been used, e.g., in [1, 2].

A similar factorization exists for B, but since B is upper triangular two sequences of core transformations are required. For $n = 5$ we have

$$B = \begin{bmatrix} w_4 & 1 \\ 1 & 0 \end{bmatrix} \begin{bmatrix} w_3 & 1 \\ 1 & 0 \end{bmatrix} \begin{bmatrix} w_2 & 1 \\ 1 & 0 \end{bmatrix} \begin{bmatrix} w_1 & 1 \\ 1 & 0 \end{bmatrix} \begin{bmatrix} 0 & 1 \\ 1 & 0 \end{bmatrix} \begin{bmatrix} 0 & 1 \\ 1 & 0 \end{bmatrix} \begin{bmatrix} 0 & 1 \\ 1 & 0 \end{bmatrix} \begin{bmatrix} 0 & 1 \\ 1 & 0 \end{bmatrix}$$
$$[w_5] \begin{bmatrix} 1 & 0 \end{bmatrix}$$
$$= B_5 B_4 B_3 B_2 B_1 G_1 G_2 G_3 G_4,$$

with $B_i(i:i+1, i:i+1) = \begin{bmatrix} w_i & 1 \\ 1 & 0 \end{bmatrix}$ and $G_i(i:i+1, i:i+1) = \begin{bmatrix} 0 & 1 \\ 1 & 0 \end{bmatrix}$.

We will now generalize Fiedler's reordering identity of the companion matrix' factorization to the companion pencil case.

THEOREM 3.1. *Let (A, B) be a companion pencil of $p(z)$, with A and B factored in core transformations $A = A_0 \cdots A_{n-1}$ and $B = B_n \cdots B_1 G_1 \cdots G_{n-1}$ as in Theorem 2.1. Then, for every permutation $\sigma : \{1, \ldots, n-1\} \to \{1, \ldots, n-1\}$ the pencil (A_σ, B_σ), with*

$$A_\sigma = A_0 A_{\sigma(1)} \cdots A_{\sigma(n-1)},$$
$$B_\sigma = B_n B_{\sigma(n-1)} \cdots B_{\sigma(1)} G_{\sigma(1)} \cdots G_{\sigma(n-1)}$$

satisfies $p(z) = \det(zB_\sigma - A_\sigma)$.

PROOF. We will form $F_\sigma = A_\sigma B_{\sigma(1)}^{-1} \cdots B_{\sigma(n-1)}^{-1} = A_\sigma B_n^{-1} B_n$ and show that F_σ is a Fiedler companion matrix determined by σ and related to the monic polynomial $\tilde{p}(z) = z^n + c_{n-1} z^{n-1} + \cdots + c_0$, missing coefficient c_n. Note that all the B_i for $i < n$ are invertible as is B_n, as otherwise the original polynomial would have had a leading coefficient 0.[2] The resulting pencil (F_σ, B_n) is equivalent to the companion pencil (2) as one can prove by using the similarity transformations described in [9, Lemma 2.2] and altering the reasoning slightly not to touch transformation F_{n-1}. The Fiedler companion matrix F_σ can be factored into core transformations, the Fiedler factors, $F_0 F_{\sigma(1)} \cdots F_{\sigma(n-1)}$, with $F_0(1,1) = -c_0$ and $F_i(i:i+1, i:$

---

[2]In fact, B_n does not need to be inverted, so the proof still holds even for singular B_n.

$i+1) = \begin{bmatrix} 0 & 1 \\ 1 & -c_i \end{bmatrix}$. The matrix F_{n-1} does not commute with B_n. Consequently the necessary transformations involve only $F_1, \ldots, F_{n-2}$.

To construct $F_\sigma = A_\sigma B_\sigma^{-1} B_n$ we will multiply the pencil (A_σ, B_σ) on the right with factors to eliminate parts in B_σ and to move them to A_σ's side. Core transformations X_i and X_j commute if $|i-j| > 1$, so we can reorder $A_\sigma = A_0 A_{\sigma(1)} \cdots A_{\sigma(n-1)}$ as $A_\sigma = A_{\tilde\sigma} = A_0(A_{j_1-1}A_{j_1-2}\cdots A_1)(A_{j_2-1}A_{j_2-2}\cdots A_{j_1})\cdots (A_{n-1}A_{n-2}\cdots A_{j_s})$, We define $j_0 = 1$. This reordering appeared also in [9], and contains factors $(A_{j_{\ell+1}-1}A_{j_{\ell+1}-2}\cdots A_{j_\ell+1}A_{j_\ell})$ in which each core transformation A_{j-1} is positioned to the right of A_j, where $j = j_{\ell+1}-1, j_{\ell+1}-2, \ldots, j_\ell + 1$. The ordering between these larger factors is, however, different $A_{j_\ell-1}$ is positioned to the left of A_{j_ℓ}, with $\ell = 0, \ldots, s$.

We can also reorder the factors B_i and G_i to get

$$B_\sigma = B_{\tilde\sigma} = B_n(B_{j_s}\cdots B_{n-1})\cdots(B_1\cdots B_{j_1-1})(G_{j_1-1}\cdots G_1)\cdots(G_{n-1}\cdots G_{j_s}),$$

Let us now form the product of the two inner factors

$$S = (B_1\cdots B_{j_1-1})(G_{j_1-1}\cdots G_1) = \begin{bmatrix} 1 & w_1 & \cdots & w_{j_1-1} \\ & 1 & & \\ & & \ddots & \\ & & & 1 \end{bmatrix}.$$

Let r be the largest index with $j_r = j_1 + r - 1$, which is in fact the number of factors $(A_{j_{\ell+1}-1}A_{j_{\ell+1}-2}\cdots A_{j_\ell})$ containing only a single matrix. This means that the matrix $A_{\tilde\sigma}$ actually equals

$$A_{\tilde\sigma} = A_0(A_{j_1-1}A_{j_1-2}\cdots A_1)A_{j_1}\cdots A_{j_{r-1}}(A_{j_r+1-1}\cdots A_{j_r})\cdots(A_{n-1}\cdots A_{j_s}),$$

where there are now r factors consisting of a single matrix.

One can show that S commutes with $(G_{j_2-1}\cdots G_{j_1+1})\cdots(G_{n-1}\cdots G_{j_s})$ but not with $G_{j_1},\ldots,G_{j_r}$. These factors $G_{j_1},\ldots,G_{j_r}$, when applied from the right on S, will each move w_{j_1-1} one column to the right. The matrix

$$\tilde S = ((G_{j_2-1}\cdots G_{j_1+1})\cdots(G_{n-1}\cdots G_{j_s}))^{-1} S (G_{j_2-1}\cdots G_{j_1+1})\cdots(G_{n-1}\cdots G_{j_s}),$$

is an elimination matrix. Thus inverting $\tilde S$ is equivalent to changing all w_i into $-w_i$. By multiplying the pencil with $\tilde S^{-1}$ from the right, we can change it to

$$\left(A_{\tilde\sigma}\tilde S^{-1}, B_n(B_{j_s}\cdots B_{n-1})\cdots(B_{j_1}\cdots B_{j_2-1})(G_{j_2-1}\cdots G_{j_1})\cdots(G_{n-1}\cdots G_{j_s})\right)$$

Again one can show that $\tilde S^{-1}$ commutes with $(A_{j_2-1}\cdots A_{j_1+1})\cdots(A_{n-1}\cdots A_{j_s})$.

Further $A_{j_1},\ldots,A_{j_r}$ will each move w_{j_1-1} one column back to the left, so that

$$(A_{j_2-1}\cdots A_{j_1+1})\cdots(A_{n-1}\cdots A_{j_s})\tilde S((A_{j_2-1}\cdots A_{j_1+1})\cdots(A_{n-1}\cdots A_{j_s}))^{-1} = S.$$

We can now combine $(A_{j_1-1}A_{j_1-2}\cdots A_1)S^{-1}$, yielding

$$(A_{j_1-1}\cdots A_1)S^{-1} = \begin{bmatrix} & & & 1 & \\ & & \ddots & & \\ & & & & 1 \\ 1 & -v_2-w_1 & \cdots & -v_{j_1}-w_{j_1-1} \end{bmatrix} = (F_{j_1-1}\cdots F_1).$$

By repeating this procedure the remaining factors in B_σ can be moved to the left and $F_\sigma = A_0(F_{j_1-1}F_{j_1-2}\cdots F_1)(F_{j_2-1}F_{j_2-2}\cdots F_{j_1})\cdots(F_{n-1}F_{n-2}\cdots F_{j_s})$. □

REMARK 3.2. We would like to emphasize that in Fiedler factorizations it is possible to reposition the A_0 term, so that it can appear everywhere in the factorization. In the pencil case, however, this is in general not possible. Neither A_0 nor B_n can be moved freely, unless in some specific configurations, of which the Fiedler factorization with all $B'_i s$ identities is one.

Let us now look at some examples of different shapes for companion pencils.

EXAMPLE 3.3. Let $p(z) = 2z^4 + (\sin(t) + \cos(t))z^3 + (3+i)z^2 - 4z - 2$. A standard Hessenberg-upper triangular pencil (A, B) could look like

$$A = A_0 A_1 A_2 A_3 = \begin{bmatrix} 0 & & & 2 \\ 1 & 0 & & 5 \\ & 1 & 0 & -3 \\ & & 1 & -\sin(t) \end{bmatrix} \text{ and}$$

$$B = B_1 B_2 B_3 B_4 = \begin{bmatrix} 1 & & & 1 \\ & 1 & & i \\ & & 1 & \cos(t) \\ & & & 2 \end{bmatrix}.$$

Another option would be a CMV-like ordering

$$A = A_0 A_2 A_1 A_3 = \begin{bmatrix} 0 & 2 & & \\ & 0 & & 1 \\ 1 & 5 & 0 & -3 \\ & & 1 & -\sin(t) \end{bmatrix} \text{ and}$$

$$B = B_2 B_1 B_3 B_4 = \begin{bmatrix} 1 & 1 & i & \\ & 1 & 0 & \\ & & 1 & \cos(t) \\ & & & 2 \end{bmatrix} = \begin{bmatrix} 1 & i & 1 & \\ 1 & & & \\ & \cos(t) & 1 & \\ & 2 & & \end{bmatrix} \begin{bmatrix} & & & 1 \\ 1 & & & \\ & 1 & & \\ & & 1 & \end{bmatrix} \begin{bmatrix} & 1 & & \\ 1 & & & \\ & & & 1 \\ & & 1 & \end{bmatrix}.$$

The advantage is that A is pentadiagonal and B the product of a pentadiagonal and a permutation matrix, properties that can be exploited when developing QR or QZ algorithms.

EXAMPLE 3.4. In this example we will demonstrate the effect of the permutation σ. Let $p(z) = c_8 z^8 + \cdots c_1 z^1 + c_0$ and $\sigma = \begin{pmatrix} 1 & 2 & 3 & 4 & 5 & 6 & 7 \\ 1 & 5 & 4 & 3 & 2 & 6 & 7 \end{pmatrix}$. Then we have

$$\underbrace{\begin{bmatrix} 0 & 1 \\ 1 & -v_6 \end{bmatrix} \begin{bmatrix} 0 & 1 \\ 1 & -v_5 \end{bmatrix} \begin{bmatrix} [-v_1] \\ \begin{bmatrix} 0 & 1 \\ 1 & -v_2 \end{bmatrix} \\ \begin{bmatrix} 0 & 1 \\ 1 & -v_4 \end{bmatrix} \begin{bmatrix} 0 & 1 \\ 1 & -v_3 \end{bmatrix} \\ \begin{bmatrix} 0 & 1 \\ 1 & -v_7 \end{bmatrix} \begin{bmatrix} 0 & 1 \\ 1 & -v_8 \end{bmatrix} \end{bmatrix}}_{A_0 A_{\sigma(1)} \cdots A_{\sigma(7)}}, \quad \underbrace{\begin{bmatrix} \begin{bmatrix} w_1 & 1 \\ w_2 & 1 \\ 1 & 0 \\ w_3 & 1 \\ 1 & 0 \\ w_4 & 1 \\ 1 & 0 \\ w_5 & 1 \\ 1 & 0 \\ w_6 & 1 \\ 1 & 0 \\ [w_8] \begin{bmatrix} w_7 & 1 \\ 1 & 0 \end{bmatrix} \end{bmatrix}}_{B_8 B_{\sigma(7)} \cdots B_{\sigma(1)}}, \quad \underbrace{\begin{bmatrix} 0 & 1 \\ 1 & 0 \\ 0 & 1 \\ 1 & 0 \\ 0 & 1 \\ 1 & 0 \\ 0 & 1 \\ 1 & 0 \\ 0 & 1 \\ 1 & 0 \\ 0 & 1 \\ 1 & 0 \\ 0 & 1 \\ 1 & 0 \end{bmatrix}}_{G_{\sigma(1)} \cdots G_{\sigma(7)}},$$

where only the active parts of the Fiedler factors are shown. We see that the shape is the same in A and G but mirrored in B. In the stacking notation above, one should replace each *active matrix part* by a full matrix; the notation is unambiguous as two blocks positioned on top of each other commute so their ordering does not play any role.

REMARK 3.5. We focused on a specific factorization linked to the B-matrix. However, the matrix B could also be factored as (e.g., let $n = 5$)

$$B = \begin{bmatrix} 0 & 1 \\ 1 & 0 \end{bmatrix} \begin{bmatrix} 0 & 1 \\ 1 & 0 \end{bmatrix} \begin{bmatrix} 0 & 1 \\ 1 & 0 \end{bmatrix} \begin{bmatrix} 0 & 1 \\ 1 & 0 \end{bmatrix} \begin{bmatrix} 0 & 1 \\ 1 & w_1 \end{bmatrix} \begin{bmatrix} 0 & 1 \\ 1 & w_2 \end{bmatrix} \begin{bmatrix} 0 & 1 \\ 1 & w_3 \end{bmatrix} \begin{bmatrix} 0 & 1 \\ 1 & w_4 \end{bmatrix} \begin{bmatrix} w_5 \end{bmatrix}.$$

The factorization differs, but results similar to Theorem 3.1 can be proved.

4. Factoring companion pencils and product eigenvalue problems

Even though Theorem 2.1 is probably the most useful in practice, it is still only a special case of a general product setting, where the polynomial rootfinding problem is written as a product eigenvalue problem [13].

THEOREM 4.1. *Consider $p(z)$, a polynomial with complex coefficients $p(z) = c_n z^n + c_{n-1} z^{n-1} + \cdots + c_1 z + c_0$. Define A, B, and $F^{(k)}$ (for $k = 1, \ldots, m$) based on the n-tuples v, w, and $f^{(k)}$*

$$v_1 = c_0, \ v_i = 0, \ for \ i = 2, \ldots, n,$$
$$w_n = c_n, \ w_i = 0, \ for \ i = 1, \ldots, n-1,$$
$$f_n^{(k)} = 0, \ and \ \sum_{k=1}^{m} f_i^{(k)} = c_i, \ for \ i = 1, \ldots, n-1,$$

as follows

$$A = Z - v e_n^T = \begin{bmatrix} 0 & & & -c_0 \\ 1 & 0 & & 0 \\ & \ddots & \ddots & \vdots \\ & & 1 & 0 \\ & & & 1 & 0 \end{bmatrix}, \quad B = Z^H Z + w e_n^T = \begin{bmatrix} 1 & & & \\ & 1 & & \\ & & \ddots & \\ & & & 1 & \\ & & & & c_n \end{bmatrix}$$

and $F^{(k)} = I - f^{(k)} e_n^T = \begin{bmatrix} 1 & & & -f_1^{(k)} \\ & 1 & & \vdots \\ & & \ddots & \\ & & & 1 & -f_{n-1}^{(k)} \\ & & & & 1 \end{bmatrix},$

with Z the downshift matrix. Then the product $A(\prod_{k=1}^{m} F^{(k)}) B_n^{-1}$ equals the companion matrix in (2).

PROOF. Straightforward multiplication of the elimination matrices provides the result. The factors $F^{(k)}$ commute and thus the product notation is unambiguous. □

Theorem 2.1 is a simple consequence of Theorem 4.1. The inverse of $F^{(k)}$ is again an elimination matrix, where only a sign change of the elements determined by $f^{(k)}$ needs to be affected; as a result, we have for all $1 \leq \ell < m$ that the pencil $(A \prod_{k=1}^{\ell} F^{(k)}, B_n^{-1} \prod_m^{\ell+1} (F^{(k)})^{-1})$ shares the eigenvalues of the companion matrix. Moreover, instead of explicitly computing the product one could also use the factorization to retrieve the eigenvalues of the pencil.

REMARK 4.2. Each of the factors $F^{(k)}$ can be written in terms of his Fiedler factorization as proposed in Section 3. Moreover, Theorem 3.1 can be generalized to the product setting, where each of the factors $F^{(k)}$ obeys the ordering imposed by the permutation σ. The proof proceeds similarly to Theorem 3.1; instead of having to move only one matrix B from the right to the left entry in the pencil, one has to move now each of the factors, one by one to the left.

5. Matrix Polynomials

In this section we will generalize the results for classical polynomials to matrix polynomials. Let $p(z) = C_n z^n + C_{n-1} z^{n-1} + \cdots + C_1 z + C_0$ be a matrix polynomial having coefficients $C_i \in \mathbb{C}^{m \times m}$. Then $p(z)$ matches the determinant $\det(zB - A)$, with

$$(5) \quad A = \begin{bmatrix} 0_m & & & & -V_1 \\ I_m & 0_m & & & -V_2 \\ & \ddots & \ddots & & \vdots \\ & & I_m & 0_m & -V_{n-1} \\ & & & I_m & -V_n \end{bmatrix} \text{ and } B = \begin{bmatrix} I_m & & & & W_1 \\ & I_m & & & W_2 \\ & & \ddots & & \vdots \\ & & & I_m & W_{n-1} \\ & & & & W_n \end{bmatrix},$$

where $V_{i+1} + W_i = C_i$, $V_1 = C_0$, $W_n = C_n$, I_m equals the identity matrix of size $m \times m$, and 0_m stands for the zero matrix of size $m \times m$. In the following corollary it is shown that the additional freedom in the splitting can be used to form block-matrices A and B with structured blocks.

COROLLARY 5.1. Let $p(z) = C_n z^n + C_{n-1} z^{n-1} + \cdots + C_1 z + C_0$ be a matrix polynomial with coefficients $C_i \in \mathbb{C}^{m \times m}$. Let further C_0 be skew-Hermitian ($C_0^H = -C_0$) and C_n Hermitian ($C_n = C_n^H$). Then there are $V = \begin{bmatrix} V_1 & V_2 & \ldots & V_n \end{bmatrix}^T$ and $W = \begin{bmatrix} W_1 & W_2 & \ldots & W_n \end{bmatrix}^T$ as in (5), so that all V_i are skew-Hermitian and all W_i are Hermitian.

PROOF. V_1 is skew-Hermitian and W_n is Hermitian. For all other coefficients C_i we can use the splitting $V_{i+1} = \frac{1}{2}(C_i - C_i^H)$ and $W_i = \frac{1}{2}(C_i + C_i^H)$. □

In Section 4 we already showed that we can distribute the coefficients over many more factors to obtain a product eigenvalue problem. If C_0 and C_n are of special form, one can even obtain a matrix product with structured matrices.

COROLLARY 5.2. Let $p(z) = C_n z^n + C_{n-1} z^{n-1} + \cdots + C_1 z + C_0$ be a matrix polynomial with coefficients $C_i \in \mathbb{C}^{m \times m}$, C_0 unitary-plus-rank-1, and $C_n = I_m$.

Then there exist m upper-triangular and unitary-plus-rank-1 matrices $F^{(k)}$ and a unitary-plus-rank-1 matrix A, so that the eigenvalues of the matrix polynomial coincide with those of the product eigenvalue problem $A(\prod_{k=1}^{m} F^{(k)}) B_n^{-1}$.

PROOF. Adapting the notation from Theorem 4.1 to fit matrix polynomials as in (5) and letting V_i and W_i as in (5) and $F_i^{(k)}$ as in Theorem 4.1 be matrices. The desired property holds for the $m(n-1)$ matrices $F_i^{(k)}$, with ($k = 1, \ldots, m$;

$i = 1, \ldots, n-1$; and s and t general indices)

$$F_i^{(k)}(s,t) = \begin{cases} C_i(s,t), & t = k, \\ 0, & t \neq k, \end{cases}$$

as every $F^{(k)}$ takes one column out of the coefficient matrices. $\square$

6. Conclusions & Future work

A generalization of Fiedler's factorization of companion matrices to companion pencils was presented. It was shown that the pencil approach can be seen as a specific case of a product eigenvalue problem, and it was noted that all results are applicable to the matrix polynomial case.

Forthcoming investigations focus on exploiting these splittings in rootsolvers, based on companion pencils to obtain fast reliable solutions.

Acknowledgments

We'd like to thank Vanni Noferini and Leonardo Robol for the fruitful discussions on the conditioning of these pencils during their visit to the Dept. of Computer Science, November 2014. Furthermore we'd like to express our gratitude to the referees, who pinpointed some shortcomings and errors; their careful reading has significantly improved the presentation and readability.

References

[1] E. Antoniou, S. Vologiannidis, and N. Karampetakis, *Linearizations of polynomial matrices with symmetries and their applications.*, Intelligent Control, 2005. Proceedings of the 2005 IEEE International Symposium on, Mediterrean Conference on Control and Automation, June 2005, pp. 159–163.

[2] E. N. Antoniou and S. Vologiannidis, *A new family of companion forms of polynomial matrices*, Electron. J. Linear Algebra **11** (2004), 78–87. MR2111515 (2005h:15069)

[3] R. Bevilacqua, G. M Del Corso, and L. Gemignani, *A CMV-based eigensolver for companion matrices*, arXiv preprint arXiv:1406.2820 (2014).

[4] P. Boito, Y. Eidelman, and L. Gemignani, *Implicit QR for rank-structured matrix pencils*, BIT **54** (2014), no. 1, 85–111, DOI 10.1007/s10543-014-0478-0. MR3177956

[5] P. Boito, Y. Eidelman, and L. Gemignani, *Implicit QR for companion-like pencils*, arXiv preprint arXiv:1401.5606 (2014).

[6] F. de Terán, F. M. Dopico, and J. Pérez, *Backward stability of polynomial root-finding using Fiedler companion matrices*, IMA Journal of Numerical Analysis (2014).

[7] B. Eastman, I.-J. Kim, B. L. Shader, and K. N. Vander Meulen, *Companion matrix patterns*, Linear Algebra Appl. **463** (2014), 255–272, DOI 10.1016/j.laa.2014.09.010. MR3262399

[8] A. Edelman and H. Murakami, *Polynomial roots from companion matrix eigenvalues*, Math. Comp. **64** (1995), no. 210, 763–776, DOI 10.2307/2153450. MR1262279 (95f:65075)

[9] M. Fiedler, *A note on companion matrices*, Linear Algebra Appl. **372** (2003), 325–331, DOI 10.1016/S0024-3795(03)00548-2. MR1999154 (2004g:15017)

[10] G. F. Jónsson and S. Vavasis, *Solving polynomials with small leading coefficients*, SIAM J. Matrix Anal. Appl. **26** (2004/05), no. 2, 400–414, DOI 10.1137/S0895479899365720. MR2124155 (2005k:12010)

[11] D. S. Mackey, N. Mackey, C. Mehl, and V. Mehrmann, *Vector spaces of linearizations for matrix polynomials*, SIAM J. Matrix Anal. Appl. **28** (2006), no. 4, 971–1004 (electronic), DOI 10.1137/050628350. MR2276550 (2008b:65053)

[12] P. Van Dooren and P. Dewilde, *The eigenstructure of an arbitrary polynomial matrix: computational aspects*, Linear Algebra Appl. **50** (1983), 545–579, DOI 10.1016/0024-3795(83)90069-1. MR699575 (84j:15009)

[13] D. S. Watkins, *Product eigenvalue problems*, SIAM Rev. **47** (2005), no. 1, 3–40, DOI 10.1137/S0036144504443110. MR2147197 (2006b:65060)

Mathematical Institute, University of Oxford, Andrew Wiles Building, Woodstock Road, OX2 6GG Oxford, United Kingdom
 E-mail address: aurentz@maths.ox.ac.uk
 URL: http://www.maths.ox.ac.uk/people/jared.aurentz

Department of Computer Science, KU Leuven, Celestijnenlaan 200A, 3001 Leuven (Heverlee), Belgium
 E-mail address: Thomas.Mach@cs.kuleuven.be
 URL: http://people.cs.kuleuven.be/thomas.mach/

Department of Computer Science, KU Leuven, Celestijnenlaan 200A, 3001 Leuven (Heverlee), Belgium
 E-mail address: Raf.Vandebril@cs.kuleuven.be
 URL: http://people.cs.kuleuven.be/raf.vandebril/

Department of Mathematics, Washington State University, Pullman, Washington 99164-3113
 E-mail address: watkins@math.wsu.edu
 URL: http://www.math.wsu.edu/faculty/watkins/

Tame systems of linear and semilinear mappings and representation-tame biquivers

Tatiana Klimchuk, Dmitry Kovalenko, Tetiana Rybalkina, and Vladimir V. Sergeichuk

ABSTRACT. We study systems of linear and semilinear mappings considering them as representations of a directed graph G with full and dashed arrows: a representation of G is given by assigning to each vertex a complex vector space, to each full arrow a linear mapping, and to each dashed arrow a semilinear mapping of the corresponding vector spaces. We extend to such representations the classical theorems by Gabriel about quivers of finite type and by Nazarova, Donovan, and Freislich about quivers of tame types.

1. Introduction

We study systems of linear and semilinear mappings on complex vector spaces. A mapping $\mathcal{A}$ from a complex vector space U to a complex vector space V is called *semilinear* if

$$\mathcal{A}(a+b) = \mathcal{A}a + \mathcal{A}b, \qquad \mathcal{A}(\alpha a) = \bar{\alpha}\mathcal{A}a$$

for all $a, b \in U$ and $\alpha \in \mathbb{C}$. We write $\mathcal{A}: U \to V$ if $\mathcal{A}$ is a linear mapping and $\mathcal{A}: U \dashrightarrow V$ (using a dashed arrow) if $\mathcal{A}$ is a semilinear mapping.

We study systems of linear and semilinear mappings considering them as representations of biquivers introduced by Sergeichuk [12, Section 5] (see also [5]); they generalize the notion of representations of quivers introduced by Gabriel [9].

DEFINITION 1.1.
- A *biquiver* is a directed graph G with vertices $1, 2, \ldots, t$ and with full and dashed arrows; for example,

(1)

- The *category* $\text{rep}_{\mathbb{C}}(G)$ *of representations* of a biquiver G is defined as follows:

2010 *Mathematics Subject Classification*. Primary 15A04, 15A21, 16G60.
Key words and phrases. Linear and semilinear mappings, quivers of finite and tame types, classification.

- A *representation* $\mathcal{A}$ of a biquiver G is given by assigning to each vertex v a complex vector space $\mathcal{A}_v$, to each full arrow $\alpha : u \longrightarrow v$ a linear mapping $\mathcal{A}_\alpha : \mathcal{A}_u \to \mathcal{A}_v$, and to each dashed arrow $\alpha : u \dashrightarrow v$ a semilinear mapping $\mathcal{A}_\alpha : \mathcal{A}_u \dashrightarrow \mathcal{A}_v$. The vector

$$\dim \mathcal{A} := (\dim \mathcal{A}_1, \ldots, \dim \mathcal{A}_t)$$

is called the *dimension* of a representation $\mathcal{A}$. For example, a representation

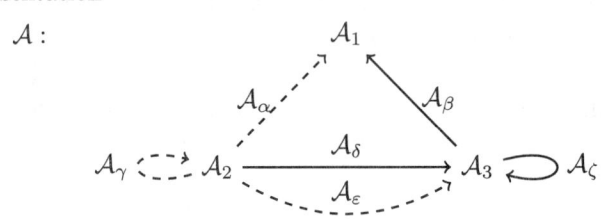

$\mathcal{A}$:

<!-- diagram: triangle with vertices $\mathcal{A}_1$, $\mathcal{A}_2$, $\mathcal{A}_3$; arrows $\mathcal{A}_\alpha$ (dashed) from $\mathcal{A}_2$ to $\mathcal{A}_1$, $\mathcal{A}_\beta$ from $\mathcal{A}_1$ to $\mathcal{A}_3$, $\mathcal{A}_\delta$ from $\mathcal{A}_2$ to $\mathcal{A}_3$, $\mathcal{A}_\varepsilon$ (dashed) from $\mathcal{A}_2$ to $\mathcal{A}_3$, $\mathcal{A}_\gamma$ (dashed) loop/arrow at $\mathcal{A}_2$, $\mathcal{A}_\zeta$ loop at $\mathcal{A}_3$ -->

of (1) is formed by complex spaces $\mathcal{A}_1, \mathcal{A}_2, \mathcal{A}_3$, linear mappings $\mathcal{A}_\beta$, $\mathcal{A}_\delta$, $\mathcal{A}_\zeta$, and semilinear mappings $\mathcal{A}_\alpha$, $\mathcal{A}_\gamma$, $\mathcal{A}_\varepsilon$.

- A *morphism* $\mathcal{F} : \mathcal{A} \to \mathcal{B}$ between representations $\mathcal{A}$ and $\mathcal{B}$ of a biquiver G is a family of linear mappings $\mathcal{F}_1 : \mathcal{A}_1 \to \mathcal{B}_1, \ldots, \mathcal{F}_t : \mathcal{A}_t \to \mathcal{B}_t$ such that for each arrow α from u to v the diagram

(2)
$$\begin{cases} \begin{array}{c} \mathcal{A}_u \xrightarrow{\mathcal{A}_\alpha} \mathcal{A}_v \\ \mathcal{F}_u \downarrow \quad \downarrow \mathcal{F}_v \\ \mathcal{B}_u \xrightarrow{\mathcal{B}_\alpha} \mathcal{B}_v \end{array} & \text{if } u \xrightarrow{\alpha} v, \\[2em] \begin{array}{c} \mathcal{A}_u \dashrightarrow^{\mathcal{A}_\alpha} \mathcal{A}_v \\ \mathcal{F}_u \downarrow \quad \downarrow \mathcal{F}_v \\ \mathcal{B}_u \dashrightarrow^{\mathcal{B}_\alpha} \mathcal{B}_v \end{array} & \text{if } u \dashrightarrow^{\alpha} v \end{cases}$$

is commutative (i.e., $\mathcal{B}_\alpha \mathcal{F}_u = \mathcal{F}_v \mathcal{A}_\alpha$). We write $\mathcal{A} \simeq \mathcal{B}$ if $\mathcal{A}$ and $\mathcal{B}$ are isomorphic; i.e., if all $\mathcal{F}_i$ are bijections.

For example, each cycle of linear and semilinear mappings

$$\mathcal{A}: \quad V_1 \xrightarrow{\mathcal{A}_1} V_2 \xrightarrow{\mathcal{A}_2} \cdots \xrightarrow{\mathcal{A}_{t-2}} V_{t-1} \xrightarrow{\mathcal{A}_{t-1}} V_t \quad \overset{\mathcal{A}_t}{\frown}$$

(in which each edge is a full or dashed arrow $\longrightarrow$, $\longleftarrow$, $\dashrightarrow$, or $\dashleftarrow$) is a representation of the biquiver

$$C: \quad 1 \xrightarrow{\alpha_1} 2 \xrightarrow{\alpha_2} \cdots \xrightarrow{\alpha_{t-2}} (t-1) \xrightarrow{\alpha_{t-1}} t \quad \overset{\alpha_t}{\frown}$$

its representations were classified in [6].

Note that the notion of a biquiver is equivalent to the notion of a *signed digraph*, which is a directed graph in which each arc is labeled by $+$ or $-$; see a huge mathematical bibliography and glossary of signed graphs on Zaslavsky's web page

[16]. A biquiver without dashed arrows is a quiver and its representations are the quiver representations. All quivers, for which the problem of classifying their representations does not contain the problem of classifying pairs of matrices up to similarity (i.e., the problem of classifying representations of the quiver $\circlearrowright 1 \circlearrowleft$), are called *tame* (this definition is informal; formal definitions are given in [10, Section 14.10]). The list of all tame quivers and the classification of their representations were obtained independently by Donovan and Freislich [4] and Nazarova [11]. We extend their results to representations of biquivers.

2. Formulation of the main results

The *direct sum* of representations $\mathcal{A}$ and $\mathcal{B}$ of a biquiver is the representation $\mathcal{A} \oplus \mathcal{B}$ formed by the spaces $\mathcal{A}_v \oplus \mathcal{B}_v$ and the mappings $\mathcal{A}_\alpha \oplus \mathcal{B}_\alpha$. A representation of nonzero dimension is *indecomposable* if it is not isomorphic to a direct sum of representations of nonzero dimensions.

By analogy with quiver representations, we say that a biquiver is *representation-finite* if it has only finitely many nonisomorphic indecomposable representations. A biquiver is *wild* if the problem of classifying its representations contains the problem of classifying matrix pairs up to similarity transformations

$$(A, B) \mapsto (S^{-1}AS, S^{-1}AS), \qquad S \text{ is nonsingular,}$$

otherwise the biquiver is *tame*. Clearly, each representation-finite biquiver is tame. The problem of classifying matrix pairs up to similarity is the problem of classifying representations of the quiver $\circlearrowright 1 \circlearrowleft$; it contains the problem of classifying representations of each quiver (see [2]) and so it is considered as hopeless. An analogous statement for representations of biquivers was proved in [5]: the problem of classifying representations of the biquiver $\dashrightarrow 1 \dashleftarrow$ contains the problem of classifying representations of each biquiver.

The *Tits form* of a biquiver G with vertices $1, \ldots, t$ is the integral quadratic form

$$q_G(x_1, \ldots, x_t) := x_1^2 + \cdots + x_t^2 - \sum x_u x_v,$$

in which the sum $\sum$ is taken over all arrows $u \longrightarrow v$ and $u \dashrightarrow v$ of the biquiver.

The following theorem extends Gabriel's theorem [9] (see also [7, Theorem 2.6.1]) to each biquiver G and coincides with it if G is a quiver.

THEOREM 2.1 (proved in Section 4). *Let G be a connected biquiver with vertices $1, 2, \ldots, t$.*

(a) *G is representation-finite if and only if G can be obtained from one of the Dynkin diagrams*

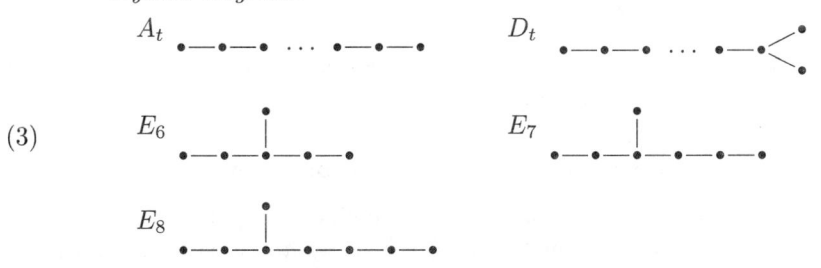

(3)
$$A_t \quad \bullet\!-\!\bullet\!-\!\bullet \ \cdots \ \bullet\!-\!\bullet\!-\!\bullet \qquad D_t \quad \bullet\!-\!\bullet\!-\!\bullet \ \cdots \ \bullet\!-\!\bullet\!\!<^{\bullet}_{\bullet}$$

$$E_6 \quad \bullet\!-\!\bullet\!-\!\overset{\bullet}{\bullet}\!-\!\bullet\!-\!\bullet \qquad E_7 \quad \bullet\!-\!\bullet\!-\!\overset{\bullet}{\bullet}\!-\!\bullet\!-\!\bullet\!-\!\bullet$$

$$E_8 \quad \bullet\!-\!\bullet\!-\!\overset{\bullet}{\bullet}\!-\!\bullet\!-\!\bullet\!-\!\bullet\!-\!\bullet$$

by replacing each edge with a full or dashed arrow, if and only if the Tits form q_G is positive definite.

(b) Let G be representation-finite and let $z = (z_1, \dots, z_t)$ be a nonzero integer vector with nonnegative components. There exists an indecomposable representation of dimension z if and only if $q_G(z) = 1$; this representation is determined by z uniquely up to isomorphism.

Representations of representation-finite quivers were classified by Gabriel [9]; see also [7, Theorem 2.6.1] and [3].

The following theorem extends the Donovan Freislich Nazarova theorem [4,11] (see also [7, Chapter 2]) to each biquiver G and coincides with it if G is a quiver.

THEOREM 2.2 (proved in Section 5). *Let G be a connected biquiver with vertices $1, 2, \dots, t$.*

(a) *G is tame if and only if G can be obtained from one of the Dynkin diagrams (3) or from one of the extended Dynkin diagrams*

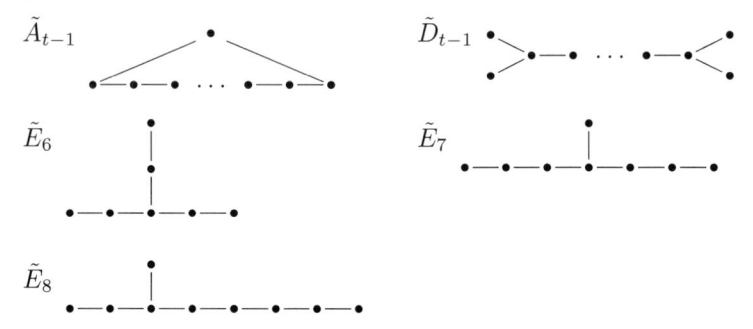

$\tilde{A}_{t-1}$

$\tilde{D}_{t-1}$

$\tilde{E}_6$

$\tilde{E}_7$

$\tilde{E}_8$

by replacing each edge with a full or dashed arrow, if and only if the Tits form q_G is positive semidefinite.

(b) *Let G be tame and let $z = (z_1, \dots, z_t)$ be a nonzero integer vector with nonnegative components. There exists an indecomposable representation of dimension z if and only if $q_G(z) = 0$ or $q_G(z) = 1$.*

Representations of tame quivers were classified by Donovan and Freislich [4] and independently by Nazarova [11].

The following theorem is a special case of the Krull–Schmidt theorem for additive categories [1, Chapter I, Theorem 3.6] (it holds for representations of a biquiver G since $\operatorname{rep}_{\mathbb{C}}(G)$ is an additive category in which all idempotents split).

THEOREM 2.3. *Each representation of a biquiver is isomorphic to a direct sum of indecomposable representations. This direct sum is uniquely determined, up to permutations and isomorphisms of direct summands, which means that if*

$$\mathcal{A}_1 \oplus \cdots \oplus \mathcal{A}_r \simeq \mathcal{B}_1 \oplus \cdots \oplus \mathcal{B}_s,$$

in which all $\mathcal{A}_i$ and $\mathcal{B}_j$ are indecomposable representations, then $r = s$ and all $\mathcal{A}_i \simeq \mathcal{B}_i$ after a suitable renumbering of $\mathcal{A}_1, \dots, \mathcal{A}_r$.

3. Matrix representations of biquivers

Let us recall some elementary facts about semilinear mappings.

We denote by $[a]_e$ the coordinate vector of a vector a in a basis $e_1, \dots, e_n$. For each two bases $e_1, \dots, e_n$ and $e'_1, \dots, e'_n$ of a vector space, we denote by $S_{e \to e'}$ the change of basis matrix; its columns are $[e'_1]_e, \dots, [e'_n]_e$. If $A = [\alpha_{ij}]$ then $\bar{A} := [\bar{\alpha}_{ij}]$.

Let $\mathcal{A}: U \dashrightarrow V$ be a semilinear mapping. We say that an $m \times n$ matrix $[\mathcal{A}]_{fe}$ is the *matrix of* $\mathcal{A}$ in bases $e_1, \ldots, e_n$ of U and $f_1, \ldots, f_m$ of V if

(4) $$[\mathcal{A}a]_f = \overline{[\mathcal{A}]_{fe}}[a]_e \quad \text{for all } a \in U.$$

Therefore, the columns of $[\mathcal{A}]_{fe}$ are $\overline{[\mathcal{A}e_1]_f}, \ldots, \overline{[\mathcal{A}e_n]_f}$. We write $[\mathcal{A}]_e$ instead of $[\mathcal{A}]_{ee}$ if $U = V$.

If $e'_1, \ldots, e'_n$ and $f'_1, \ldots, f'_m$ are other bases of U and V, then

(5) $$[\mathcal{A}]_{f'e'} = \overline{S^{-1}_{f \to f'}}[\mathcal{A}]_{fe} S_{e \to e'}$$

since by (4)

$$\overline{S^{-1}_{f \to f'}[\mathcal{A}]_{fe} S_{e \to e'}[a]_{e'}} = \overline{S^{-1}_{f \to f'}[\mathcal{A}]_{fe}[a]_e} = \overline{S^{-1}_{f \to f'}[\mathcal{A}a]_f}$$
$$= [\mathcal{A}a]_{f'} = \overline{[\mathcal{A}]_{f'e'}}[a]_{e'}$$

for all $a \in U$.

LEMMA 3.1. *Let* U, V, *and* W *be vector spaces with bases* $e_1, e_2, \ldots,$ $f_1, f_2, \ldots,$ *and* $g_1, g_2, \ldots$.

(a) *The composition of a linear mapping* $\mathcal{A}: U \to V$ *and a semilinear mapping* $\mathcal{B}: V \dashrightarrow W$ *is the semilinear mapping with matrix*

(6) $$[\mathcal{BA}]_{ge} = [\mathcal{B}]_{gf}[\mathcal{A}]_{fe}$$

(b) *The composition of a semilinear mapping* $\mathcal{A}: U \dashrightarrow V$ *and a linear mapping* $\mathcal{B}: V \to W$ *is the semilinear mapping with matrix*

(7) $$[\mathcal{BA}]_{ge} = \overline{[\mathcal{B}]_{gf}}[\mathcal{A}]_{fe}$$

PROOF. The identity (6) follows from observing that $\mathcal{BA}$ is a semilinear mapping and

$$[(\mathcal{BA})a]_g = [\mathcal{B}(\mathcal{A}a)]_g = \overline{[\mathcal{B}]_{gf}}[\mathcal{A}a]_f = \overline{([\mathcal{B}]_{gf}[\mathcal{A}]_{fe})}[a]_e$$

for each $a \in U$. The identity (7) follows from observing that $\mathcal{BA}$ is a semilinear mapping and

$$[(\mathcal{BA})a]_g = [\mathcal{B}(\mathcal{A}a)]_g = [\mathcal{B}]_{gf}[\mathcal{A}a]_f = [\mathcal{B}]_{gf}\overline{[\mathcal{A}]_{fe}[a]_e} = \overline{\left(\overline{[\mathcal{B}]_{gf}}[\mathcal{A}]_{fe}\right)[a]_e}$$

for each $a \in U$. $\square$

Let $\mathcal{A}: V \dashrightarrow V$ be a semilinear mapping; let $[\mathcal{A}]_e$ and $[\mathcal{A}]_{e'}$ be its matrices in bases $e_1, \ldots, e_n$ and $e'_1, \ldots, e'_n$ of V. By (5),

(8) $$[\mathcal{A}]_{e'} = \overline{S^{-1}_{e \to e'}}[\mathcal{A}]_e S_{e \to e'}$$

and so $[\mathcal{A}]_{e'}$ and $[\mathcal{A}]_e$ are consimilar: recall that two matrices A and B are *consimilar* if there exists a nonsingular matrix S such that $\bar{S}^{-1}AS = B$; a canonical form of a square complex matrix under consimilarity is given in [8, Theorem 4.6.12].

Each representation $\mathcal{A}$ of a biquiver G can be given by the set A of matrices A_α of its mappings $\mathcal{A}_\alpha$ in fixed bases of the spaces $\mathcal{A}_1, \ldots, \mathcal{A}_t$. Changing the bases, we can reduce A_α by transformations $S_v^{-1} A_\alpha S_u$ if $\alpha: u \to v$ and $\bar{S}_v^{-1} A_\alpha S_u$ if $\alpha: u \dashrightarrow v$, in which $S_1, \ldots, S_t$ are the transition matrices. This reduces the problem of classifying representations of G up to isomorphism to the problem of classifying the sets A up to these transformations, which leads to the following definition.

DEFINITION 3.2. The *category* $\mathrm{mat}_{\mathbb{C}}\, G$ *of matrix representations* of a biquiver G with vertices $1,\dots,t$ is defined as follows:
- A *matrix representation A of dimension* $d = (d_1,\dots,d_t)$ of G is given by assigning an $d_v \times d_u$ complex matrix A_α to each arrow $\alpha: u \longrightarrow v$ or $u \dashrightarrow v$.
- A *morphism* $F: A \to B$ between matrix representations A and B of dimensions d and d' is given by a family of matrices $F_i \in \mathbb{C}^{d'_i \times d_i}$ ($i = 1,\dots,t$) such that

$$(9) \qquad B_\alpha F_u = \begin{cases} F_v A_\alpha & \text{for every arrow } \alpha: u \longrightarrow v, \\ \overline{F}_v A_\alpha & \text{for every arrow } \alpha: u \dashrightarrow v. \end{cases}$$

Each matrix representation A of dimension $d = (d_1,\dots,d_t)$ can be identified with the representation $\mathcal{A}$ from the Definition 1.1, whose vector spaces have the form
$$\mathcal{A}_v = \mathbb{C} \oplus \cdots \oplus \mathbb{C} \quad (d_v \text{ summands})$$
for all vertices v and whose linear or semilinear mappings $\mathcal{A}_\alpha$ are defined by the matrices A_α.

By (9), two matrix representations A and B of G are isomorphic if there exist nonsingular matrices $S_1,\dots,S_t$ such that

$$(10) \qquad B_\alpha = \begin{cases} S_v^{-1} A_\alpha S_u & \text{for every full arrow } \alpha: u \longrightarrow v, \\ \bar{S}_v^{-1} A_\alpha S_u & \text{for every dashed arrow } \alpha: u \dashrightarrow v. \end{cases}$$

In view of (8),

two matrix representations are isomorphic if and only if they give the same representation $\mathcal{A}$ but in possible different bases.

4. Proof of Theorem 2.1

For each biquiver G and its vertex u, we denote by G^u the biquiver obtained from G by replacing all arrows $u \longrightarrow v$ and $v \longrightarrow u$ for each $v \ne u$ by $u \dashrightarrow v$ and $v \dashrightarrow u$, and vice versa. For example,

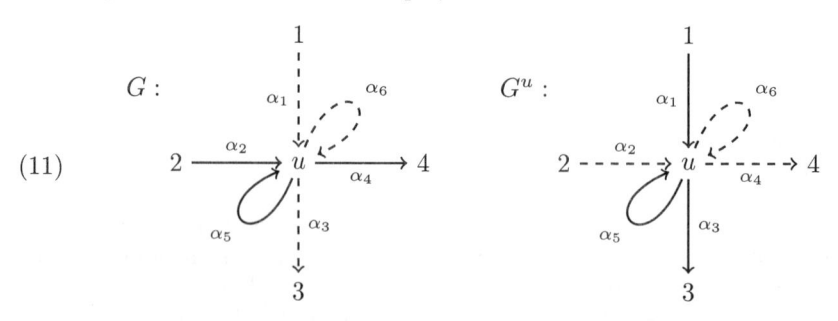

(11)

We say that G^u is obtained from G by *switching* the vertex u (by analogy with the switching a vertex in a signed graph; see the glossary in [16]). For each $A \in \mathrm{mat}_{\mathbb{C}}\, G$, define $A^u \in \mathrm{mat}_{\mathbb{C}}\, G^u$ as follows:

$$(12) \qquad A^u_\alpha := \begin{cases} A_\alpha & \text{if } \alpha \text{ does not start at } u, \\ \bar{A}_\alpha & \text{if } \alpha \text{ starts at } u. \end{cases}$$

LEMMA 4.1. *Let* $A, B \in \mathrm{mat}_\mathbb{C} G$ *and let* u *be any vertex of* G. *Then* $A \simeq B$ *if and only if* $A^u \simeq B^u$.

PROOF. It suffices to prove that

(13)
  if $A \simeq B$ via $S_1, \ldots, S_t$ (see (10)), then $A^u \simeq B^u$
  via $R_1, \ldots, R_t$ in which $R_v := S_v$ if $v \neq u$ and
  $R_u := \bar{S}_u$.

Moreover, it suffices to prove (13) for matrix representations of the biquiver G defined in (11), which contains all possible arrows from the vertex u and to the vertex u.

Let us consider an arbitrary matrix representation A of G and the corresponding matrix representation A^u (see (12)) of G^u:

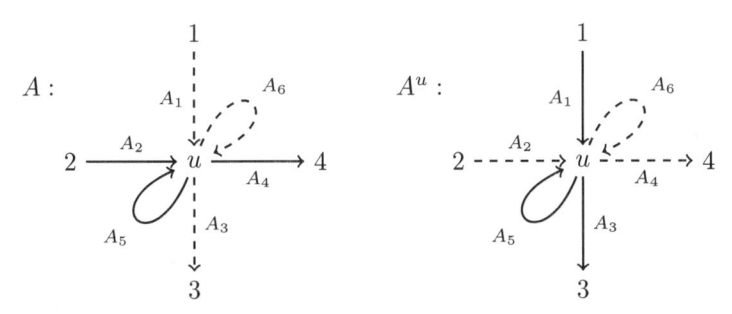

Let B be any matrix representation of G that is isomorphic to A via $S_1, \ldots S_4, S_u$. Then B and the corresponding matrix representation B^u of G^u have the form:

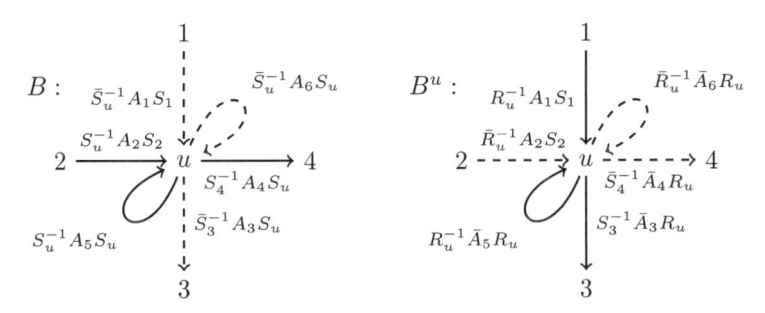

in which R_u is defined by (13).

Therefore, B^u is isomorphic to A^u via $S_1, \ldots S_4, R_u$, which proves (13). □

PROOF OF THEOREM 2.1. Let G be a connected bigraph with t vertices.

(a) Suppose first that G is a tree. Let us prove that by a sequence of switchings we can transform G to the quiver $Q(G)$ obtained from G by replacing each dashed arrow $v \dashrightarrow w$ with the full arrow $v \longrightarrow w$.

Let w be a pendant vertex of G (i.e., a vertex of degree 1). Let α be the arrow for which w is one of its vertices. Denote by $G \backslash \alpha$ the biquiver obtained from G by deleting w and α. Reasoning by induction on the number of vertices, we assume that $G \backslash \alpha$ can be transformed to $Q(G \backslash \alpha)$ by a sequence of switchings. The same sequence of switchings transforms G to some biquiver G' in which only the arrow that is obtained from α can be dashed. If it is dashed, we make it full by switching the vertex w of G' and obtain $Q(G)$. Theorem 2.1 holds for $Q(G)$ by Gabriel's theorem [9]. Lemma 4.1 ensures that the statement (a) of Theorem 2.1 holds for G too.

Suppose now that G is not a tree. Then G contains a cycle C that up to renumbering of vertices of G has the form

(14)
$$C: \quad 1 \xrightarrow{\alpha_1} 2 \xrightarrow{\alpha_2} \cdots \xrightarrow{\alpha_{r-2}} (r-1) \xrightarrow{\alpha_{r-1}} r \overset{\alpha_r}{\frown}$$

If $r > 1$ and the sequence of arrows $\alpha_1, \ldots, \alpha_{r-1}$ contains a dashed arrow, then we take the first dashed arrow α_ℓ and make it full by switching the vertex $\ell + 1$ of G. Repeating this procedure, we make all arrows $\alpha_1, \ldots, \alpha_{r-1}$ full.

For each $n \times n$ matrix M, let us construct the matrix representation $P(M)$ of G by assigning I_n to each of the arrows $\alpha_1, \ldots, \alpha_{r-1}$ (if $r > 1$), M to α_r, and 0_n to the other arrows. It is easy to see that $P(M) \simeq P(N)$ if and only if either α_r is full and M is similar to N, or α_r is dashed and M is consimilar to N (see (8)). The Jordan canonical form and a canonical form under consimilarity [8, Theorem 4.6.12] ensure that G is of infinite type. Since G contains a cycle, it cannot be obtained by directing edges in one of the Dynkin diagrams (3), and so q_G is not positive definite by Gabriel's theorem [9].

(b) If G is of finite type, then G is a tree. By the part (a) of the proof, G can be transformed to the quiver $Q(G)$ by a sequence of switchings. By Lemma 4.1, this sequence of switchings transforms all indecomposable representations of G to all indecomposable representations of $Q(G)$, and nonisomorphic representations are transformed to nonisomorphic representations of the same dimensions. This proves (b) for G since (b) holds for $Q(G)$. □

5. Proof of Theorem 2.2

LEMMA 5.1. *The problem of classifying representations of each of the biquivers*

(15) $\quad G_1: \quad \alpha_1 \overset{\frown}{\dashv} 1 \xrightarrow{\alpha} 2 \qquad G_2: \quad \alpha_1 \overset{\frown}{\dashv} 1 \xleftarrow{\alpha} 2$

contains the problem of classifying representations of the biquiver

(16) $\quad G_3: \quad \alpha_1 \overset{\frown}{\rightharpoonup} 1 \overset{\frown}{\dashv} \alpha_2$

PROOF. The problem of classifying representations of the biquivers (15) is the problem of classifying matrix pairs up to transformations

(17) $\quad (M, N) \mapsto (\bar{S}^{-1}MS, R^{-1}NS),$

(18) $\quad (M, N) \mapsto (\bar{S}^{-1}MS, S^{-1}NR),$

respectively.

Let us consider G_1. Let

(19) $\quad M := \begin{bmatrix} 0 & P \\ I & Q \end{bmatrix}, \quad M' := \begin{bmatrix} 0 & P' \\ I & Q' \end{bmatrix}, \quad N := \begin{bmatrix} 0 & I \end{bmatrix},$

in which all blocks are n-by-n. Let (M, N) be reduced to (M', N) by transformations (17); i.e., there exist nonsingular S and R such that

(20) $\quad MS = \bar{S}M', \quad NS = RN.$

By the second equality in (20), S has the form
$$S = \begin{bmatrix} S_1 & S_2 \\ 0 & R \end{bmatrix}.$$

Equating the 1,1 blocks in the first equality in (20) gives $S_2 = 0$; equating the 2,1 blocks gives $S_1 = \bar{R}$; equating the 1,2 and 2,2 blocks gives
$$PR = RP', \qquad QR = \bar{R}Q'.$$

Therefore, (M, N) and (M', N) define isomorphic representations of G_1 if and only if (P, Q) and (P', Q') define isomorphic representations of (16), and so the problem of classifying representations of G_1 contains the problem of classifying representations of (16).

Let us consider the biquiver G_2. Taking M and M' as in (19), taking $N := \begin{bmatrix} I \\ 0 \end{bmatrix}$, and reasoning as for G_1, we prove that if (M, N) is reduced to (M', N) by transformations (18); i.e., there exist nonsingular S and R such that $MS = \bar{S}M'$ and $NR = SN$, then S is upper block triangular, and so (P, Q) and (P', Q') define isomorphic representations of (16). □

LEMMA 5.2. *The problem of classifying representations of the biquiver G_3 defined in (16) and the problem of classifying representations of the biquiver*

$$G_4: \qquad \alpha_1 \overset{\longrightarrow}{\underset{\longleftarrow}{}} 1 \overset{\longrightarrow}{\underset{\longleftarrow}{}} \alpha_2$$

contain the problem of classifying matrix pairs up to similarity.

PROOF. The problems of classifying representations of the biquivers G_3 and G_4 are the problems of classifying matrix pairs up to transformations

(21) $$(M, N) \mapsto (S^{-1}MS, \bar{S}^{-1}NS),$$
(22) $$(M, N) \mapsto (\bar{S}^{-1}MS, \bar{S}^{-1}NS),$$

respectively.

Let us consider G_3. Let
$$M := \begin{bmatrix} 0 & I \\ 0 & 0 \end{bmatrix}, \qquad N := \begin{bmatrix} P & 0 \\ 0 & Q \end{bmatrix}, \qquad N' := \begin{bmatrix} P' & 0 \\ 0 & Q' \end{bmatrix},$$

in which all blocks are n-by-n. Let (M, N) be reduced to (M, N') by transformations (21); i.e., there exists a nonsingular S such that

(23) $$MS = SM, \qquad NS = \bar{S}N'.$$

By the first equality in (23), S has the form
$$S = \begin{bmatrix} S_1 & S_2 \\ 0 & S_1 \end{bmatrix}.$$

Equating the 1,1 and 2,2 blocks in the second equality in (23) gives
$$\bar{S}_1^{-1}PS_1 = P', \qquad \bar{S}_1^{-1}QS_1 = Q'.$$

Therefore, (M, N) and (M, N') define isomorphic representations of G_3 if and only if (P, Q) and (P', Q') define isomorphic representations of G_4, and so the problem of classifying representations of G_3 contains the problem of classifying representations of G_4.

Let us consider G_4. Let

$$M := \begin{bmatrix} 0 & I & 0 & 0 \\ 0 & 0 & I & 0 \\ 0 & 0 & 0 & I \\ 0 & 0 & 0 & 0 \end{bmatrix}, \quad N := \begin{bmatrix} 0 & 0 & 0 & 0 \\ P & 0 & 0 & 0 \\ 0 & 0 & 0 & 0 \\ 0 & 0 & Q & 0 \end{bmatrix}, \quad N' := \begin{bmatrix} 0 & 0 & 0 & 0 \\ P' & 0 & 0 & 0 \\ 0 & 0 & 0 & 0 \\ 0 & 0 & Q' & 0 \end{bmatrix}$$

in which all blocks are n-by-n. Let (M, N) be reduced to (M, N') by transformations (22); i.e., there exists a nonsingular S such that

(24) $$MS = \bar{S}M, \quad NS = \bar{S}N'.$$

By the first equality in (24), S has the form

$$S = \begin{bmatrix} S_1 & S_2 & S_3 & S_4 \\ 0 & \bar{S}_1 & \bar{S}_2 & \bar{S}_3 \\ 0 & 0 & S_1 & S_2 \\ 0 & 0 & 0 & \bar{S}_1 \end{bmatrix}.$$

Equating the 2,1 and 4,3 blocks in the second equality in (24) gives

$$S_1^{-1} P S_1 = P', \quad S_1^{-1} Q S_1 = Q'.$$

Therefore, (M, N) and (M, N') define isomorphic representations of G_4 if and only if (P, Q) and (P', Q') are similar, and so the problem of classifying representations of G_4 is wild. □

PROOF OF THEOREM 2.2. (a) Suppose first that G is a tree. Reasoning as in the proof of Theorem 2.1, we transform G to the quiver $Q(G)$ by a sequence of switchings. Theorem 2.2 holds for $Q(G)$ by the Donovan–Freislich–Nazarova theorem [4, 11]. Lemma 4.1 ensures that Theorem 2.2 holds for G too.

Suppose now that G is not a tree. Then G contains a cycle C that up to renumbering of vertices of G has the form (14) in which $r \geqslant 1$ and each edge is a full or dashed arrow.

If $G = C$, then G is of tame type; all its representations were classified in [6].

Let us suppose that $G \neq C$ and prove that G is of wild type. The biquiver G contains a biquiver C' obtained by adjoining to C an edge $\alpha : u \to v$ or $v \to u$, in which $u \in \{1, \ldots, r\}$, we suppose that $u = 1$. If C' is of wild type, then G is of wild type too: we can identify all representations of C' with those representations of G, in which the vertices outside of C' are assigned by 0; two representations of C' are isomorphic if and only if the corresponding representations of G are isomorphic. Further we suppose that $G = C'$.

Reasoning as in the proof of Theorem 2.1(a), we can transform the subbiquiver

$$2 \xrightarrow{\alpha_2} 3 \xrightarrow{\alpha_3} \cdots \xrightarrow{\alpha_{r-1}} r \xrightarrow{\alpha_r} 1$$

of G to a quiver by a sequence of vertex switchings. Thus, we can suppose that the arrows $\alpha_2, \ldots, \alpha_r$ of G are full arrows.

Suppose first that v is not a vertex of C. If $\alpha : u - v$ is dashed, we make it full by switching v. If α_1 is a full arrow, then G is a quiver of wild type by the Donovan–Freislich–Nazarova theorem. Thus, we can suppose that α_1 is a dashed arrow. Let $\ell \in \{1, 2\}$ be such that α_1 in G_ℓ defined in (15) has the same direction as α in G. The biquiver G is of wild type since G_ℓ is of wild type and each matrix representation A of G_ℓ can be identified with the matrix representation of G obtained from A by assigning the identity matrix to $\alpha_2,\ldots,\alpha_r$; two representations of G_ℓ are isomorphic if and only if the corresponding representations of G are isomorphic.

Suppose now that v is a vertex of C. If α and α_1 are full arrows, then G is a quiver of wild type by the Donovan–Freislich–Nazarova theorem. Let α or α_1 be a dashed arrow. Denote by G' the biquiver obtained from G by deleting its arrows $\alpha_2,\ldots,\alpha_r$ and its vertices $2,3,\ldots,r-1$, and by identifying the vertices 1 and r. By Lemma 5.2, G' is of wild type. Hence, G is of wild type too since each matrix representation A of G' can be identified with the matrix representation of G obtained from A by assigning the identity matrix to $\alpha_2,\ldots,\alpha_r$; two representations of G' are isomorphic if and only if the corresponding representations of G are isomorphic.

(b) Let G be tame. Then G is a tree or cycle. If G is a tree, then it can be transformed to the quiver $Q(G)$ by a sequence of switchings. By Lemma 4.1, this sequence of switchings transforms all indecomposable representations of G to all indecomposable representations of $Q(G)$, and nonisomorphic representations are transformed to nonisomorphic representations. This proves (b) for G since (b) holds for quivers by the Donovan–Freislich–Nazarova theorem. If G is a cycle, then (b) follows from the classification of its representations given in [6]. $\square$

REMARK 5.3. Theorems 2.1 and 2.2 remain valid if we replace $\mathbb{C}$ by an arbitrary field with involution.

REMARK 5.4. A special class of biquivers, called edge-biparite quivers, are defined and used by Simson [14, 15] for the Coxeter spectral study of edge-bipartite graphs and their matrix morsifications in relation with a description of the structure of Auslander-Reiten quiver of derived categories $\mathcal{D}^b(\mathrm{mod}\, R)$ of finite dimensional algebras R, see [13] for details. We recall that an edge-biparite quiver is a biquiver G such that, given a pair of vertices u, v of G, there is no a pair of arrows $u \longrightarrow v$ and $u \dashrightarrow v$. In [14], a result analogous to Theorems 2.1 and 2.2 is obtained for edge-bipartite graphs. Namely, it is shown there that a loop-free connected edge-bipartite graph Δ has positive definite symmetric Gram matrix G_Δ if and only if Δ is Gram $\mathbb{Z}$-congruent with one of the Dynkin diagrams A_n, D_n, E_6, E_7, E_8. Moreover, the symmetric Gram matrix G_Δ of Δ is positive semidefinite of corank one if and only if Δ is Gram $\mathbb{Z}$-congruent with one of the extended Dynkin diagrams $\tilde{A}_n$, $\tilde{D}_n$, $\tilde{E}_6$, $\tilde{E}_7$, $\tilde{E}_8$.

Acknowledgements

We are grateful to two anonymous referees for helpful and constructive comments and suggestions.

References

[1] Hyman Bass, *Algebraic K-theory*, W. A. Benjamin, Inc., New York-Amsterdam, 1968. MR0249491 (40 #2736)

[2] Genrich R. Belitskii and Vladimir V. Sergeichuk, *Complexity of matrix problems*, Linear Algebra Appl. **361** (2003), 203–222, DOI 10.1016/S0024-3795(02)00391-9. Ninth Conference of the International Linear Algebra Society (Haifa, 2001). MR1955562 (2004b:16014)

[3] I. N. Bernšteĭn, I. M. Gel'fand, and V. A. Ponomarev, *Coxeter functors, and Gabriel's theorem* (Russian), Uspehi Mat. Nauk **28** (1973), no. 2(170), 19–33. MR0393065 (52 #13876)

[4] Peter Donovan and Mary Ruth Freislich, *The representation theory of finite graphs and associated algebras*, Carleton University, Ottawa, Ont., 1973. Carleton Mathematical Lecture Notes, No. 5. MR0357233 (50 #9701)

[5] Debora Duarte de Oliveira, Roger A. Horn, Tatiana Klimchuk, and Vladimir V. Sergeichuk, *Remarks on the classification of a pair of commuting semilinear operators*, Linear Algebra Appl. **436** (2012), no. 9, 3362–3372, DOI 10.1016/j.laa.2011.11.029. MR2900721

[6] Debora Duarte de Oliveira, Vyacheslav Futorny, Tatiana Klimchuk, Dmitry Kovalenko, and Vladimir V. Sergeichuk, *Cycles of linear and semilinear mappings*, Linear Algebra Appl. **438** (2013), no. 8, 3442–3453, DOI 10.1016/j.laa.2012.12.023. MR3023287

[7] Michiel Hazewinkel, Nadiya Gubareni, and V. V. Kirichenko, *Algebras, rings and modules. Vol. 2*, Mathematics and Its Applications (Springer), vol. 586, Springer, Dordrecht, 2007. MR2356157 (2009b:16001)

[8] Roger A. Horn and Charles R. Johnson, *Matrix analysis*, 2nd ed., Cambridge University Press, Cambridge, 2013. MR2978290

[9] Peter Gabriel, *Unzerlegbare Darstellungen. I* (German, with English summary), Manuscripta Math. **6** (1972), 71–103; correction, ibid. 6 (1972), 309. MR0332887 (48 #11212)

[10] P. Gabriel and A. V. Roĭter, *Representations of finite-dimensional algebras*, Algebra, VIII, Encyclopaedia Math. Sci., vol. 73, Springer, Berlin, 1992, pp. 1–177. With a chapter by B. Keller. MR1239447 (94h:16001b)

[11] L. A. Nazarova, *Representations of quivers of infinite type* (Russian), Izv. Akad. Nauk SSSR Ser. Mat. **37** (1973), 752–791. MR0338018 (49 #2785)

[12] Vladimir V. Sergeichuk, *Linearization method in classification problems of linear algebra*, São Paulo J. Math. Sci. **1** (2007), no. 2, 219–240, DOI 10.11606/issn.2316-9028.v1i2p219-240. MR2466889 (2009g:15036)

[13] Daniel Simson, *Mesh geometries of root orbits of integral quadratic forms*, J. Pure Appl. Algebra **215** (2011), no. 1, 13–34, DOI 10.1016/j.jpaa.2010.02.029. MR2678696 (2011g:16029)

[14] Daniel Simson, *A Coxeter-Gram classification of positive simply laced edge-bipartite graphs*, SIAM J. Discrete Math. **27** (2013), no. 2, 827–854, DOI 10.1137/110843721. MR3048204

[15] Daniel Simson, *A framework for Coxeter spectral analysis of edge-bipartite graphs, their rational morsifications and mesh geometries of root orbits*, Fund. Inform. **124** (2013), no. 3, 309–338. MR3100347

[16] T. Zaslavsky, *The home page of signed, gain, and biased graphs*. Available at: http://www.math.binghamton.edu/zaslav/Bsg/

FACULTY OF MECHANICS AND MATHEMATICS, KIEV NATIONAL TARAS SHEVCHENKO UNIVERSITY, KIEV, UKRAINE
E-mail address: klimchuk.tanya@gmail.com

FACULTY OF MECHANICS AND MATHEMATICS, KIEV NATIONAL TARAS SHEVCHENKO UNIVERSITY, KIEV, UKRAINE
E-mail address: kovalenko.d.y@gmail.com

INSTITUTE OF MATHEMATICS, TERESHCHENKIVSKA 3, KIEV, UKRAINE
E-mail address: rybalkina_t@ukr.net

INSTITUTE OF MATHEMATICS, TERESHCHENKIVSKA 3, KIEV, UKRAINE
E-mail address: sergeich@imath.kiev.ua

Invariance of total positivity of a matrix under entry-wise perturbation and completion problems

Mohammad Adm and Jürgen Garloff

ABSTRACT. A totally positive matrix is a matrix having all its minors positive. The largest amount by which the single entries of such a matrix can be perturbed without losing the property of total positivity is given. Also some completion problems for totally positive matrices are investigated.

1. Introduction

In this paper we consider matrices which are totally positive, i.e., all their minors are positive. For the properties of these matrices the reader is referred to the two recent monographs [5] and [11]. In the first part of our paper we are interested in the largest amount by which the single entries of such a matrix can be varied without losing the property of total positivity. This question is answered for a few specified entries in [6], see also [5, Section 9.5]. Similarly, one may ask how much the entries of a totally nonnegative matrix, i.e., a matrix having all its minors nonnegative, can be perturbed without losing the property of total nonnegativity. This problem is solved in [2] for tridiagonal totally nonnegative matrices. A related question is whether all matrices lying between two totally nonnegative matrice with respect to a suitable partial ordering are totally nonnegative, too. The second author conjectured in 1982 [8] that this is true for the nonsingular totally nonnegative matrices and the so-called checkerboard ordering, see also [5, Section 3.2] and [11, Section 3.2]. In [1] we apply the so-called Cauchon algorithm, see, e.g., [3], to settle this conjecture.

In the second part of our paper we solve some completion problems for totally positive matrices. Here we consider the case that in a matrix some entries are specified, while the remaining ones are unspecified and are free to be chosen. All minors contained in the specified part are supposed to be positive. Then the question arises whether values for the unspecified entries can be chosen such that the resulting matrix is totally positive. Solutions of such completion problems can be found in [5, Subsections 9.1-9.4], and [10]. The starting point of our work is the recent paper [9].

2010 *Mathematics Subject Classification.* Primary 15B48.

Key words and phrases. Totally positive matrix, entry-wise perturbation, determinantal inequality, completion problem.

Adm's research is supported by the German Academic Exchange Service (DAAD).

©2016 American Mathematical Society

The organization of our paper is as follows. In the next section we explain our notation and we collect some auxiliary results in Section 3. In Section 4 we present our main results on the perturbation of entries of totally positive matrices and we show that the derived set of determinantal conditions is minimal. In Section 5 we solve a completion problem for some new patterns of unspecified entries, hereby partially settling two conjectures in [9].

2. Notation

We now introduce the notation used in our paper. For κ, n we denote by $Q_{\kappa,n}$ the set of all strictly increasing sequences of κ integers chosen from $\{1, 2, \ldots, n\}$. For $\alpha = \{\alpha_1, \alpha_2, \ldots, \alpha_\kappa\} \in Q_{\kappa,n}$ the *dispersion* of α is $d(\alpha) = \alpha_\kappa - \alpha_1 - \kappa + 1$. If $d(\alpha) = 0$ then the index set α is called *contiguous*. Let A be a real $n \times m$ matrix. For $\alpha = \{\alpha_1, \alpha_2, \ldots, \alpha_\kappa\} \in Q_{\kappa,n}, \beta = \{\beta_1, \beta_2, \ldots, \beta_\mu\} \in Q_{\mu,m}$, we denote by $A[\alpha|\beta]$ the $\kappa \times \mu$ submatrix of A contained in the rows indexed by $\alpha_1, \alpha_2, \ldots, \alpha_\kappa$ and columns indexed by $\beta_1, \beta_2, \ldots, \beta_\mu$. We suppress the braces when we enumerate the indices explicitly. In the case that α or β is obtained by taking the union of two index sets we assume that the resulting index set is ordered increasingly. When $\alpha = \beta$, the principal submatrix $A[\alpha|\alpha]$ is abbreviated to $A[\alpha]$. In the special case where $\alpha = \{1, 2, \ldots, \kappa\}$, we refer to the principal submatrix $A[\alpha]$ as a *leading principal submatrix* (and to det $A[\alpha]$ as a *leading principal minor*). By $A(\alpha|\beta)$ we denote the $(n - \kappa) \times (m - \mu)$ submatrix of A contained in the rows indexed by the elements of $\{1, 2, \ldots, n\} \setminus \{\alpha_1, \alpha_2, \ldots, \alpha_\kappa\}$, and columns indexed by $\{1, 2, \ldots, m\} \setminus \{\beta_1, \beta_2, \ldots, \beta_\mu\}$ (where both sequences are ordered strictly increasingly) with the similar notation $A(\alpha)$ for the complementary principal submatrix.

A minor det $A[\alpha|\beta]$ is called *row-initial* if $\alpha = \{1, 2, \ldots, \kappa\}$ and $\beta \in Q_{\kappa,m}$ is contiguous, it is termed *column-initial* if $\alpha \in Q_{\kappa,n}$ is contiguous while $\beta = \{1, 2, \ldots, \kappa\}$, and *initial* if it is row-initial or column-initial.

The n-by-n matrix whose only nonzero entry is a one in the $(i, j)^{th}$ position is denoted by E_{ij}. We reserve throughout the notation $T_n = (t_{ij})$ for the permutation matrix with $t_{i,n-i+1} = 1$, $i = 1, \ldots, n$. An n-by-m matrix A is called *totally positive* (abbreviated TP henceforth) and *totally nonnegative* (abbreviated TN) if det $A[\alpha|\beta] > 0$ and det $A[\alpha|\beta] \geq 0$, respectively, for all $\alpha, \beta \in Q_{\kappa,n'}$, $\kappa = 1, 2, \ldots, n'$, and $n' := min\{n, m\}$. In passing we note that if an n-by-m matrix A is TP then so are its transpose and $A^\# := T_n A T_m$, see, e.g., [5, Theorem 1.4.1]. We will briefly relate our results to TP_k matrices, $k \leq n'$. A is said to be TP_k if all its minors of order less than or equal to k are positive.

We say that a rectangular array is a *partial matrix* if some of its entries are specified, while the remaining, unspecified, entries are free to be chosen. A partial matrix is *partial TP* if each of its fully specified submatrices is TP. A *completion* of a partial matrix is a choice of values for the unspecified entries, resulting in a matrix that agrees with the a given partial matrix in all its specified positions. A pattern P is TP *completable* if every partial TP matrix with pattern P has a TP completion. TP_k *completion* is defined analogously.

We recall from [9] the following definitions. We associate with a matrix the directions north, east, south, and west. So the entry in position $(1, 1)$ lies north and west. We call a (possibly rectangular) pattern *jagged* if, whenever a position is unspecified, either all positions north and west of it are unspecified or south and east of it are, and we call a (possibly rectangular) pattern *echelon* if, whenever a position is unspecifed, either all positions north and east of it are unspecifed or south and west of it are. Either of these is referred to as *single echelon*, while when both occur, we say *double echelon*. Echelon refers to any of these possibilities.

3. Auxiliary Results

The following proposition shows that it suffices to consider the initial minors if one wants to check a matrix for total positivity.

PROPOSITION 3.1. [**7**, Theorem 4.1], *see also* [**5**, Theorem 3.1.4] *If all initial minors of a matrix A are positive, then A is TP.*

A fundamental tool for proving in the next section inequalities between ratios of determinants is the following proposition.

PROPOSITION 3.2. [**12**, Theorem 4.2] *Let $\alpha, \alpha', \beta, \beta', \gamma, \gamma', \delta, \delta'$ be subsets of $\{1, 2, \ldots, n\}$ with $\alpha \cup \gamma = \{1, 2, \ldots, p\}$ and $\alpha' \cup \gamma' = \{1, 2, \ldots, p'\}$, $q = |\alpha \cap \gamma|$, $q' = |\alpha' \cap \gamma'|$, and $r := \frac{1}{2}(p - q + p' - q')$. Let η be the unique order preserving map*

$$\eta : (\alpha \setminus \gamma) \cup (\gamma \setminus \alpha) \to \{1, 2, \ldots, p - q\},$$

and let η' be the unique order reversing map

$$\eta' : (\alpha' \setminus \gamma') \cup (\gamma' \setminus \alpha') \to \{p - q + 1, \ldots, 2r\}.$$

Define the subsets α'' and β'' of $\{1, 2, \ldots, 2r\}$ by

$$\alpha'' := \eta(\alpha \setminus \gamma) \cup \eta'(\gamma' \setminus \alpha'),$$
$$\beta'' := \eta(\beta \setminus \delta) \cup \eta'(\delta' \setminus \beta').$$

Then the following two statements are equivalent:

(1) *For each square TN matrix A of order at least n the following relation holds:*

$$\det A[\alpha|\alpha'] \det A[\gamma|\gamma'] \leq \det A[\beta|\beta'] \det A[\delta|\delta'].$$

(2) *The relations $\alpha \cup \gamma = \beta \cup \delta$ and $\alpha' \cup \gamma' = \beta' \cup \delta'$ are fulfilled and the sets α'', β'' satisfy the inequality*

(3.1) $$\max\{|\omega \cap \beta''|, |\omega \setminus \beta''|\} \leq \max\{|\omega \cap \alpha''|, |\omega \setminus \alpha''|\}$$

for each subset $\omega \subseteq \{1, 2, \ldots, 2r\}$ of even cardinality.

PROPOSITION 3.3. [**13**], *see also* [**5**, p. 62], [**11**, Theorem 2.6] *The set of the TP n-by-n matrices is dense in the class of TN n-by-n matrices.*

The following propositions are used to solve totally positive completion problems in Section 5.

PROPOSITION 3.4. [**9**, Theorem 5] *Each jagged pattern is TP completable.*

PROPOSITION 3.5. [**5**, Theorem 9.4.4] *Let A be an n-by-m partial TP matrix with only one unspecified entry in the (s,t) position. If $\min\{n,m\} \leq 3$, then A has a TP completion. If $\min\{n,m\} \geq 4$, then any such A has a TP completion if and only if $s + t \leq 4$ or $s + t \geq n + m - 2$.*

We remark that Proposition 3.5 is generalized in [**4**, Theorem 4.5] to the case that the given matrix is partial TP_k, $k \geq 4$.

4. Perturbation of Totally Positive Matrices

In this section we consider the variation of single entries of a TP matrix $A = (a_{ij})$. For simplicity we consider here only the square case ($n = m$). We may restrict the discussion of the off-diagonal entries to the entries which are lying above the main diagonal since the related statements for the entries below the main diagonal follow by consideration of the transposed matrix.

THEOREM 4.1. *Let $A = (a_{ij})$ be a TP matrix and $0 \leq \tau$. Then for $i \leq j$,*

(4.1) $\qquad A \pm \tau E_{ij}$ *is TP if and only if $\tau < \min S$,*

where in each of the following eight cases S is a set of ratios of minors, where the minor in the denominator is obtained from the minor in the numerator by deleting in A additionally row i and column j. If in an index sequence two indices coincide then the respective matrix has to be removed from the listing. In the following cases only the numerator matrices are listed [1]. The cases $(-)$ and $(+)$ refer to the $--$ and $+$-sign in (4.1). In the case that S is empty put $\min S := \infty$.

(1) $i = 2m$, $j = 2k$

$(-) S : \begin{cases} A, A(n-1,n|1,2), \ldots, A(n-2k+3, \ldots, n|1, \ldots, 2k-2), \\ A(1,2|n-1,n), A(1,2,3,4|n-3,n-2,n-1,n), \ldots, \\ A(1, \ldots, 2m-2|n-2m+3, \ldots, n) \end{cases}$

$(+) S : \begin{cases} A(n|1), A(n-2,n-1,n|1,2,3), \ldots, A(n-2k+2, \ldots, n|1, \ldots, 2k-1), \\ A(1|n), A(1,2,3|n-2,n-1,n), \ldots, A(1, \ldots, 2m-1|n-2m+2, \ldots, n) \end{cases}$

(2) $i = 2m$, $j = 2k+1$

$(-) S : \begin{cases} A(n|1), A(n-2,n-1,n|1,2,3), \ldots, A(n-2k+2, \ldots, n|1, \ldots, 2k-1), \\ A(1|n), A(1,2,3|n-2,n-1,n), \ldots, A(1, \ldots, 2m-1|n-2m+2, \ldots, n) \end{cases}$

$(+) S : \begin{cases} A, A(n-1,n|1,2), \ldots, A(n-2k+1, \ldots, n|1, \ldots, 2k), \\ A(1,2|n-1,n), A(1,2,3,4|n-3,n-2,n-1,n), \ldots, \\ A(1, \ldots, 2m-2|n-2m+3, \ldots, n) \end{cases}$

(3) $i = 2m+1$, $j = 2k$

$(-) S : \begin{cases} A(n|1), A(n-2,n-1,n|1,2,3), \ldots, A(n-2k+2, \ldots, n|1, \ldots, 2k-1), \\ A(1|n), A(1,2,3|n-2,n-1,n), \ldots, A(1, \ldots, 2m-1|n-2m+2, \ldots, n) \end{cases}$

$(+) S : \begin{cases} A, A(n-1,n|1,2), \ldots, A(n-2k+3, \ldots, n|1, \ldots, 2k-2), \\ A(1,2|n-1,n), A(1,2,3,4|n-3,n-2,n-1,n), \ldots, \\ A(1, \ldots, 2m|n-2m+1, \ldots, n) \end{cases}$

---

[1] E.g., $A(n-1,n|1,2)$ refers in case 1(-) to the ratio $\frac{\det A(n-1,n|1,2)}{\det A(2m,n-1,n|1,2,2k)}$.

(4) $i = 2m+1$, $j = 2k+1$

$$(-)S: \begin{cases} A, A(n-1,n|1,2), \ldots, A(n-2k+1,\ldots,n|1,\ldots,2k), \\ A(1,2|n-1,n), A(1,2,3,4|n-3,n-2,n-1,n), \ldots, \\ A(1,\ldots,2m|n-2m+1,\ldots,n) \end{cases}$$

$$(+)S: \begin{cases} A(n|1), A(n-2,n-1,n|1,2,3), \ldots, A(n-2k+2,\ldots,n|1,\ldots,2k-1), \\ A(1|n), A(1,2,3|n-2,n-1,n), \ldots, A(1,\ldots,2m-1|n-2m+2,\ldots,n). \end{cases}$$

PROOF. The entries in the positions $(1,1)$ and (n,n) can be increased arbitrarily without loosing the property of total positivity because they enter into the top left and bottom right position, respectively, in every submatrix in which they lie. This corresponds to the fact that in the cases $1(+)$ and $4(+)$ the set S is empty for $i = j = 1, n$. In the remaining cases we present the proof here only for $A(\tau) = A - \tau E_{2m,2k}$ (case $1(-)$); the proof of the other perturbations is similar. If $2m = n$, then $2k = n$, too. The only initial minor containing $a_{nn} - \tau$ is $\det A(\tau)$. By expansion of $\det A(\tau)$ along its bottom row we obtain

$$\det A(\tau) = \det A - \tau \det A(n)$$

from which the condition

$$0 \leq \tau < \frac{\det A}{\det A(n)}$$

follows. We assume now that $2m < n$. For $\alpha \in Q_{\kappa,n}$ we set

$$\phi(\alpha) := \frac{\det A(\alpha|\alpha)}{\det A(\alpha \cup \{2m\}|\alpha \cup \{2k\})}.$$

We further use the intuitive notation

$$\phi(0) := \frac{\det A}{\det A(2m|2k)}.$$

First we show the inequality

(4.2) $$\phi(0) \leq \phi(n).$$

The inequality follows by Proposition 3.2, setting $\alpha := \alpha' := \{1, 2, \ldots, n\}$, $\beta := \beta' := \{1, 2, \ldots, n-1\}$, $\gamma := \beta \setminus \{2m\}$, $\gamma' := \beta' \setminus \{2k\}$, $\delta := \alpha \setminus \{2m\}$, $\delta' := \alpha' \setminus \{2k\}$. Then the assumptions of Proposition 3.2 are fulfilled with $p = p' = n$, $q = q' = n-2$, and therefore $r = 2$, $\alpha'' = \{1, 2\}$, $\beta'' = \{1, 3\}$. For ω the following four sets can be chosen

$$\{1,2\}, \{2,3\}, \{3,4\}, \{1,2,3,4\}.$$

In all four cases the inequality (3.1) is fulfilled. Applying (4.2) to $A(n), A(n-1,n), \ldots, A(2m+1,\ldots,n)$, we obtain the chain of inequalities

(4.3) $$\phi(0) \leq \phi(n) \leq \phi(n-1,n) \leq \cdots \leq \phi(2m+1,\ldots,n).$$

Now we show that all the row-initial minors of $A(\tau)$ are positive; the proof of the positivity of the column-initial minors is similar. Since by expansion of $\det A(\tau)$ along its $2m^{th}$ row

$$\det A(\tau) = \det A - \tau \det A(2m|2k),$$

we obtain the condition $\tau < \phi(0)$. Similarly for $s = 0, 1, \ldots, n - 2m - 1$,

$$\det A(n-s,\ldots,n)(\tau) =$$
$$\det A(n-s,\ldots,n) - \tau \det A(2m, n-s, \ldots, n|2k, n-s, \ldots, n)$$

is positive if $\tau < \phi(n-s,\ldots,n)$. Therefore by (4.3), all leading principal minors of $A(\tau)\,(\beta_1 = 1)$ are positive if $\tau < \phi(0)$.

Now we consider the row-initial minors $\det A[\alpha|\beta](\tau)$, where $\beta = (\beta_1, \beta_2, \ldots, \beta_s)$ with $\beta_1 > 1$. If β_1 is odd these minors are constant or strictly increasing with respect to τ so that they remain positive under the perturbation. If β_1 is even, we apply the proof in the case $\beta_1 = 1$ to the submatrix $A[\alpha|\beta](\tau)$ and obtain the remaining conditions.

By Proposition 3.1 it follows that $A(\tau)$ is TP if τ is taken as the minimum of S in case 1($-$). The necessity follows from the fact that all the initial minors are linear functions in τ and that therefore for $\min S \leq \tau$ there is an initial minor which is nonpositive. $\square$

REMARK 4.2. Cases 1($-$) and 4($-$) give for $i = j = 1, 2$ the bound $\det A/\det A(i)$ on τ; see [6, Theorems 4.2 and 4.3] and [5, Theorems 9.5.4 and 9.5.5] for related statements. In case 3($+$) setting $i = 1$ and $j = 2$ we get the bound $\det A/\det A(1|2)$, see [6, Theorem 4.7] and [5, Theorem 9.5.8].

The next theorem shows that the set S in Theorem 4.1 is minimal.

THEOREM 4.3. *For an arbitrary TP n-by-n matrix A the set S of determinantal ratios listed in Theorem 4.1 cannot be reduced in each of the eight cases.*

PROOF. We present the proof only for the case 1(-); the proof of the other seven cases is similar.

It suffices to show that the following ratios are not comparable if A runs over the set of the n-by-n TP matrices

$$b := \frac{\det A}{\det A(2m \mid 2k)},$$

$$c_\kappa := \frac{\det A(n - 2\kappa + 3, \ldots, n \mid 1, \ldots, 2\kappa - 2)}{\det A(2m, n - 2\kappa + 3, \ldots, n \mid 1, \ldots, 2\kappa - 2, 2k)}, \quad \kappa = 2, \ldots, k,$$

$$d_\mu := \frac{\det A(1, \ldots, 2\mu - 2 \mid n - 2\mu + 3, \ldots, n)}{\det A(1, \ldots, 2\mu - 2, 2m \mid 2k, n - 2\mu + 3, \ldots, n)}, \quad \mu = 2, \ldots, m.$$

We show here only that the ratios c_κ and d_μ are not comparable; the proof of the other cases is similar (and easier). We first prove that the inequality $c_\kappa \leq d_\mu$ does not hold for all TP n-by-n matrices A. To apply Proposition 3.2, we choose

$$\alpha := \{1, \ldots, n - 2\kappa + 2\}, \alpha' := \{2\kappa - 1, \ldots, n\},$$
$$\delta := \{2\mu - 1, \ldots, n\}, \delta' := \{1, \ldots, n - 2\mu + 2\},$$
$$\beta := \alpha \setminus \{2m\}, \beta' = \alpha' \setminus \{2k\},$$
$$\gamma := \delta \setminus \{2m\}, \gamma' = \delta' \setminus \{2k\}.$$

Then we have

$$\alpha \cup \gamma = \beta \cup \delta = \alpha' \cup \gamma' = \beta' \cup \delta'$$

by

$$2\mu - 1 < 2m \leq n - 2\kappa + 2,$$
$$2\kappa - 1 < 2k \leq n - 2\mu + 2,$$

hence $p = p' = n$;

$$\alpha \cap \gamma = \{2\mu - 1, \ldots, n - 2\kappa + 2\} \setminus \{2m\},$$
$$\alpha' \cap \gamma' = \{2\kappa - 1, \ldots, n - 2\mu + 2\} \setminus \{2k\},$$

hence $q = q' = n - 2\kappa - 2\mu + 3$;

$$\eta : \{1, \ldots, 2\mu - 2, 2m\} \cup \{n - 2\kappa + 3, \ldots, n\} \to \{1, \ldots, 2\kappa + 2\mu - 3\},$$
$$\eta' : \{2k, n - 2\mu + 3, \ldots, n\} \cup \{1, \ldots, 2\kappa - 2\} \to \{2\kappa + 2\mu - 2, \ldots, 4\kappa + 4\mu - 6\},$$

$$\begin{aligned}
\alpha'' &= \eta(\{1, \ldots, 2\mu - 2, 2m\}) \cup \eta'(\{1, \ldots, 2\kappa - 2\}) \\
&= \{1, \ldots, 2\mu - 1\} \cup \{2\kappa + 4\mu - 3, \ldots, 4\kappa + 4\mu - 6\}, \\
\beta'' &= \eta(\{1, \ldots, 2\mu - 2\}) \cup \eta'(\{1, \ldots, 2\kappa - 2, 2k\}) \\
&= \{1, \ldots, 2\mu - 2\} \cup \{2\kappa + 4\mu - 4, \ldots, 4\kappa + 4\mu - 6\}.
\end{aligned}$$

Let $w := \{2\kappa + 4\mu - 4, 2\kappa + 4\mu - 3\}$. Then the inequality (3.1) is not fulfilled and by Proposition 3.2 there exists a TN matrix A_1 for which the inequality $c_\kappa > d_\mu$ holds. By interchanging the role of sets $\alpha, \alpha', \gamma, \gamma'$ with the sets $\beta, \beta', \delta, \delta'$, we find by Proposition 3.2 (choosing $w := \{2\mu - 2, 2\mu - 1\}$) a TN matrix A_2 for which the inequality $c_\kappa < d_\mu$ holds. So the ratios c_κ and d_μ are not comparable on the set of the TN matrices. By using Proposition 3.3 we find two TP matrices satisfying the respective inequalities which shows that also on the set of TP matrices the ratios c_κ and d_μ are not comparable. □

EXAMPLE 4.4. Let A be the Pascal matrix of order 4, i.e.,

$$A = \begin{pmatrix} 1 & 1 & 1 & 1 \\ 1 & 2 & 3 & 4 \\ 1 & 3 & 6 & 10 \\ 1 & 4 & 10 & 20 \end{pmatrix}.$$

Then A is TP, see, e.g., [**5**, Example 0.1.6]. In Table 1 we give the largest interval from which τ can be chosen according to Theorem 4.1 such that the matrix $A(\tau) := A + \tau E_{ij}$ is TP, $i, j = 1, \ldots, n$, $i \leq j$. The intervals are given in the (i, j) position of the respective entry. In each case, if τ is chosen as an endpoint of the interval, the respective matrix $A(\tau)$ contains a vanishing minor.

$$(-\tfrac{1}{4}, \infty) \quad (-\tfrac{1}{6}, \tfrac{1}{6}) \quad (-\tfrac{1}{4}, \tfrac{1}{8}) \quad (-\tfrac{1}{3}, \tfrac{1}{3})$$
$$(-\tfrac{1}{14}, \tfrac{1}{4}) \quad (-\tfrac{1}{7}, \tfrac{1}{11}) \quad (-\tfrac{1}{3}, \tfrac{1}{3})$$
$$(-\tfrac{1}{10}, \tfrac{1}{2}) \quad (-1, \tfrac{1}{3})$$
$$(-1, \infty)$$

TABLE 1. The largest perturbation intervals in Example 4.4.

5. *TP* Completion Problems

In this section we consider TP completion problems for some new patterns of the unspecified entries.

We recall from [9] the definition of the patterns P_1, P_1', and P_2. Let A be an n-by-m matrix. We say that A has a P_1 or P_1' pattern if A has just one unspecified entry, viz. in the $(1, m)$ or $(n, 1)$ positions, respectively. The P_2 pattern has just two unspecified entries, viz. in positions $(1, m)$ and $(n, 1)$. Since by Proposition 3.1 a matrix is TP if and only if its initial minors are positive, a partial TP matrix with a P_1 pattern has a TP completion if and only if the upper right entry can be chosen so that the upper right minors with contiguous index sets are all positive. A similar condition holds for a P_1' pattern. We introduce two further patterns. We say that A has a P_3 pattern if $3 \leq m$ and the unspecified entries are a_{ij},

$i = 1, \ldots, l$, $j = 3, \ldots, k$, and $i = r, \ldots, n$, $j = 1, \ldots, m - 2$, with $l \in \{1, \ldots, n-1\}$, $r \in \{l+1, \ldots, n\}$.

A has a P_4 pattern if $2 \leq n$, $4 \leq m$ and the unspecified entries are a_{ij},

$i = 1, \ldots, l$, $j = 1, \ldots, k$, and $i = 1, \ldots, r$, $j = k+3, \ldots, m$, and $i = p, \ldots, n$, $j = 1, \ldots, h$, and $i = t, \ldots, n$, $j = h+3, \ldots, m$, with $1 \leq r \leq l < t \leq p \leq n$, $h, k \in \{1, \ldots, m-3\}$.

Examples 1 and 2 in [9] and 9.1.1 in [5] show that the P_1, P_1', and P_2 patterns are not TP completable if $4 \leq \min\{n, m\}$. This explains why in the sequel the index ranges will often start at 3.

THEOREM 5.1. *Let A be an n-by-m partial TP matrix with the unspecified entries a_{ij}, $i = 1, \ldots, l$, $j = 3, \ldots, k$, where $l \leq n$, $k \leq m$. Then A is TP completable.*

PROOF. Let $B_{lk} := A[l, \ldots, n \mid 1, 2, k, \ldots, m]$. Then by Proposition 3.5, B_{lk} is TP completable. We enter the value for the unspecified entry a_{lk} into the matrix A and call the resulting matrix A_{lk}. If $l > 1$ let $B_{l-1,k} := A_{lk}[l-1, \ldots, n \mid 1, 2, k, \ldots, m]$. Then by Proposition 3.5, $B_{l-1,k}$ is TP completable and similarly as for the entry a_{lk} we obtain the n-by-m partial TP matrix $A_{l-1,k}$ which has one unspecified entry less than A_{lk}. Now we continue in this manner until we find values for all the unspecified entries in column k resulting in the partial TP matrix A_{1k}. If $k > 3$ repeat the above process with the partial TP matrix A_{1k} to find values for the unspecified entries in the columns $k-1, \ldots, 3$. At the end of this process we get the matrix A_{13} which is TP. □

COROLLARY 5.2. *Let A be an n-by-m partial TP matrix with the unspecified entries a_{ij}, $i = r, \ldots, n$, $j = k, \ldots, m-2$, where $r \leq n$, $k \leq m-2$. Then A is TP completable.*

PROOF. The matrix $A^{\#}$ is a partial TP matrix with the same pattern of unspecified entries as the one considered in Theorem 5.1, whence $A^{\#}$ is TP completable. Then A is TP completable, too. □

By application of Theorem 5.1 and Corollary 5.2 to A, $A^{\#}$, or A^T it follows that a partial TP matrix A whose pattern is a single echelon pattern is TP completable if and only if it contains no P_1 or P_1' as a subpattern. This settles [9, Conjecture 1] in a special case.

THEOREM 5.3. *Let $3 \leq m$ and A be an n-by-m partial TP matrix with the unspecified entries a_{1j}, $j = 3, \ldots, m$, and a_{i1}, $i = l, \ldots, n$, with $l \leq 4$. Then A is TP completable.*

PROOF. If $l = 1$ or 2 then it is easy to find values for the unspecified entries in the positions $(1, 1)$ and $(2, 1)$, respectively, so let $l \in \{3, 4\}$. Let $B_1 := A[2, \ldots, n \mid 1, \ldots, m]$. Then by taking the transpose of B_1 and using Theorem 5.1 (with m

replaced by n), B_1 is TP completable. We enter the values for the unspecified entries a_{i1}, $i = l, \ldots, n$, into the matrix A and call the resulting matrix A_1. The matrix A_1 is a partial TP matrix with the unspecified entries in the first row. By Theorem 5.1 A_1 is TP completable, and so A is TP completable. □

THEOREM 5.4. *Let $3 \leq m$, $4 \leq n$, and A be an n-by-m partial TP matrix with the unspecified entries a_{ij}, $i = 1, \ldots, l$, $j = 3, \ldots, m$, and $i = l+3, \ldots, n$, $j = 1, \ldots, h$, where $h \leq m$. Then A is TP completable.*

PROOF. Let $B_h := [l+1, \ldots, n \mid h, \ldots, m]$. Then B_h is a partial TP matrix and by Theorem 5.1 (taking the transpose), B_h is TP completable. We enter the values for the unspecified entries of B_h into the matrix A and call the resulting matrix A_h. Repeat the last step to find values for the unspecified entries in the lower part of the columns $h-1, \ldots, 2$. At the end of this process we get the partial TP matrix A_2 having the unspecified entries in the upper right corner and in the first column below the position $(1, l+3)$. To find values for the unspecified entries in the first column, we proceed analogously to the proof of Theorem 5.3. Let $C := A_2[l+1, \ldots, n \mid 1, \ldots, m]$. Then by Theorem 5.1 C is TP completable. Since only the initial minors of A need to be positive, C can be completed independently of the entries of $A[1, \ldots, l \mid 1, 2]$. We enter the values for the unspecified entries of the first column into the matrix A_2 and call the resulting matrix A_1; then A_1 is a partial TP matrix with the unspecified entries in the upper right corner. We apply Theorem 5.1 on A_1 and it follows that A_1 is TP completable. Therefore A is TP completable. □

The following theorem combines the patterns of Theorem 5.1 and Corollary 5.2 and is related to Proposition 3.4.

THEOREM 5.5. *If $A = (a_{ij})$ is an n-by-m partial TP matrix with a P_3 pattern, then A is TP completable.*

PROOF. Let A be an n-by-m partial TP matrix with a P_3 pattern. We distinguish two cases.
Case (1): $k = m$. Let $B_1 := A[l+1, \ldots, n \mid 1, m-1, m]$. Then by applying Proposition 3.5 successively to $C_\rho := B_1[l+1, \ldots, \rho \mid 1, m-1, m]$ we find values for the unspecified entries $a_{\rho 1}$, $\rho = r, \ldots, n$. Therefore B_1 is TP completable. We enter the values for the unspecified entries of B_1 into the matrix A and call the resulting matrix A_1. To find values for the unspecified entries in the second column of A_1 in and below position $(r, 2)$ we similarly apply Proposition 3.5 to the submatrix $B_2 := A_1[l+1, \ldots, n \mid 1, 2, m-1, m]$. This completion can be accomplished independently of the entries of $A[1, \ldots, l \mid 1, 2]$, see the proof of Theorem 5.4. We enter the values for the unspecified entries of B_2 into the matrix A and call the resulting matrix A_2. Let $B_3 := A_2[l+1, \ldots, n \mid 1, 2, \ldots, m]$. Then B_3 is partial TP and by Corollary 5.2 TP completable. We enter the values for the unspecified entries of B_3 into the matrix A_2 and call the resulting matrix A_3. By Theorem 5.1 A_3 is TP completable and hence A is TP completable.
Case (2): $k < m$. We consider first the case $k = m-1$. If $l+1 = r$, then we can choose a positive number for the unspecified entry $a_{l, m-1}$ such that the matrix remains partial TP. If $l+1 < r$, let $B_{l, m-1} := A[l, \ldots, r-1 \mid 1, 2, m-1, m]$ and $B'_{l, m-1} := A[l, \ldots, n \mid m-1, m]$. Then by Proposition 3.5 both submatrices are TP completable. Moreover, we can choose in both matrices a common value for the

unspecified entry $a_{l,m-1}$ because the only nontrivial initial minor of the submatrix $B'_{l,m-1}$ containing this entry, viz. $\det B'_{l,m-1}[l, l+1|m-1, m]$, is also an initial minor of the submatrix $B_{l,m-1}$.

We enter the value for the unspecified entry $a_{l,m-1}$ into the matrix A and call the resulting matrix $A_{l,m-1}$. Repeating this process we find values for the unspecified entries in column $m-1$ and finally obtain the partial TP matrix $A_{1,m-1}$. Let $C := A_{1,m-1}[1, \ldots, r-1 \mid 1, \ldots, m]$. By Theorem 5.1, C is TP completable. We enter the values for the unspecified entries in $A_{1,m-1}$ and call the resulting matrix A', which is a partial TP matrix. By Corollary 5.2, A' is TP completable and we can conclude that A is also TP completable.

If $k < m-1$ we follow the proof in the case $k = m-1$ but we may end the proof already with the definition of the matrix C. □

THEOREM 5.6. *If $A = (a_{ij})$ is an n-by-m partial TP matrix with a P_4 pattern, then A is TP completable.*

PROOF. Let A be an n-by-m partial TP matrix with a P_4 pattern. Without loss of generality we may assume that $l = r$ and $p = t$. Otherwise let $B_0 := A[r+1, \ldots, p-1|1, \ldots, m]$. Then B_0 has a jagged pattern and can be TP completed by Proposition 3.4 (independently of the entries of $A[1, \ldots, r|k+1, k+2]$ and $A[p, \ldots, n|h+1, h+2]$). In what follows we therefore assume that $l = r$ and $p = t$. Let $B_1 := A[1, \ldots, p-1|k+1, \ldots, m]$; then B_1 is TP completable by Theorem 5.1. We enter the values for the unspecified entries of B_1 into the matrix $A[1, \ldots, p-1|1, \ldots, m]$ and call the resulting matrix B_2. Since B_2 is a partial TP matrix with a jagged pattern it is TP completable by Proposition 3.4. We enter the values for the unspecified entries of B_2 into the matrix A and call the resulting matrix A_1. Proceeding with $A_1^{\#}$ we obtain similarly the TP completion of A_1 and in this way of A, too. □

THEOREM 5.7. *If $A = (a_{ij})$ is an n-by-m partial TP matrix with the unspecified entries a_{ij} $i = 1, \ldots, l$, $j = 1, \ldots, k$, and $i = 1, \ldots, r$, $j = k+3, \ldots, m$, and $i = t, \ldots, n$, $j = 1, \ldots, m-2$, with $r \leq l < t$, then A is TP completable.*

PROOF. Without loss of generality we may assume that $r = l$, see the proof of Theorem 5.6. Let $C_1 := A[r+1, \ldots, n \mid 1, \ldots, k, m-1, m]$. Then C_1 is TP completable by Corollary 5.2. We enter the values for the unspecified entries of C_1 into the matrix A and call the resulting matrix A_1. Let $C_2 := A_1[r+1, \ldots, n \mid 1, \ldots, k+1, m-1, m]$. Then by an argument similar to that used in the proof of Theorem 5.1, by starting from the unspecified entry $a_{t,k+1}$, and proceeding downwards, we can find values for the unspecified entries in C_2; so C_2 is TP completable. We enter the values for the unspecified entries of C_2 into the matrix A_1 and call the resulting matrix A_2. Similarly, we can find values for the unspecified entries in the column $k+2$, then we enter these values into the matrix A_2 and call the resulting matrix A_3. Let $D := A_3[r+1, \ldots, n \mid 1, \ldots, m]$; then D is a partial TP matrix with the same type of pattern as the one treated in Corollary 5.2, thus D is TP completable. We enter the values for the unspecified entries in D into the matrix A_3 and call the resulting matrix A_4, where A_4 is a partial TP matrix. Let $D_1 := A_4[1, \ldots, n|k+1, \ldots, m]$; then D_1 is TP completable by Theorem 5.1 (with $k = m$). We enter the values for the unspecified entries of D_1 into A_4 and call the resulting matrix A_5 which is partial TP. Since A_5 has a jagged pattern, it is completable by Proposition 3.4, whence A is TP completable. □

THEOREM 5.8. *If $A = (a_{ij})$ is an n-by-m partial TP matrix with the unspecified entries a_{ij} $i = 1, \ldots, l$, $j = 1, \ldots, k$, and $i = 1, \ldots, r$, $j = k+3, \ldots, m$, and $i = t, \ldots, n$, $j = h, \ldots, m$, with $r \leq l < t$, and $h < k$, then A is TP completable.*

PROOF. We may assume without loss of generality that $r = l$. Otherwise we proceed as follows: Let $B_1 := A[r+1, \ldots, t-1 | h, \ldots, m]$. Then B_1 is partial TP with a jagged pattern, thus TP completable. If $h > 1$ we have to take into account the entries of $A[t, \ldots, n | 1, \ldots, h-1]$ when we want to extend the completion to the left. We proceed element-wise by taking successively the entries $a_{l,h-1}$, $a_{l-1,h-1}$, ..., $a_{r+1,h-1}$, $a_{l,h-2}$, ..., $a_{r+1,h-2}$, ..., $a_{r+1,1}$. For a fixed entry we consider the submatrices which have the chosen entry as the only unspecified entry, viz in position $(1,1)$. For each such submatrix we can find a positive number such that the matrix is TP. Then we take the maximum of all these positive numbers (for the chosen entry).

The matrix $C_1 := A[r+1, \ldots, n \mid 1, \ldots, k+1]$ is a partial TP matrix with a jagged pattern and by Proposition 3.4 C_1 is TP completable. We enter the values for the unspecified entries of C_1 into the matrix A and call the resulting matrix A_1. Let $C_2 := A_1[r+1, \ldots, n \mid 1, \ldots, k+2]$, and $C_2' := A_1[1, \ldots, n \mid k+1, k+2]$. Then both submatrices are partial TP matrices with jagged patterns, and so by the argument used in the proof of Theorem 5.5, Case (2), common values for the unspecified entries can be found. We enter the values for the unspecified entries of C_2 into the matrix A_1 and call the resulting matrix A_2. Let $C_3 := A_2[r+1, \ldots, n \mid 1, \ldots, m]$. Then C_3 is a partial TP matrix with a jagged pattern, thus C_3 is TP completable. We enter the values for the unspecified entries of C_3 into the matrix A_2 and call the resulting matrix A_3. Since A_3 is a partial TP matrix with the same type of pattern as the one considered in the proof of Theorem 5.6, A_3 is TP completable. Hence A is TP completable. □

Theorems 5.7 and 5.8 represent certain instances of a jagged echelon pattern with no P_1, P_1', or P_2 as a subpattern. Thus both theorems and the following remark settle [9, Conjecture 2] in some special cases leaving this conjecture unresolved only in the case of a double echelon pattern.

REMARK 5.9. The following patterns can be proven to be TP completable by using similar methods as in the proofs of the Theorems 5.6, 5.7, and 5.8.
The entries a_{ij} are unspecified for
(1) $i = 1, \ldots, l$, $j = 1, \ldots, k$, and $i = r, \ldots, n$, $j = 1, \ldots, m-2$, with $l < r$;
(2) $i = 1, \ldots, l$, $j = 3, \ldots, m$, and $i = r, \ldots, n$, $j = k, \ldots, m$, with $l < r$;
(3) $i = 1, \ldots, l$, $j = 3, \ldots, m$, and $i = r, \ldots, n$, $j = 1, \ldots, k$, and $i = t, \ldots, n$, $j = k+3, \ldots, m$, with $l < t < r$;
(4) $i = 1, \ldots, l$, $j = 1, \ldots, k$, and $i = r, \ldots, n$, $j = 1, \ldots, h$, and $i = t, \ldots, n$, $j = h+3, \ldots, m$, with $l < t < r$ and $h < k$.

REMARK 5.10. By using [4, Theorem 5.4] instead of Proposition 3.5 all the results of Section 5 carry over to the case that the given matrix A is partial TP_k, $k \geq 4$.

References

[1] Mohammad Adm and Jürgen Garloff, *Intervals of totally nonnegative matrices*, Linear Algebra Appl. **439** (2013), no. 12, 3796–3806, DOI 10.1016/j.laa.2013.10.021. MR3133459

[2] Mohammad Adm and Jürgen Garloff, *Invariance of total nonnegativity of a tridiagonal matrix under element-wise perturbation*, Oper. Matrices **8** (2014), no. 1, 129–137, DOI 10.7153/oam-08-06. MR3202931

[3] Mohammad Adm and Jürgen Garloff, *Improved tests and characterizations of totally nonnegative matrices*, Electron. J. Linear Algebra **27** (2014), 588–610. MR3266168

[4] V. Akin, Charles R. Johnson, and Shahla Nasserasr, TP_K *completion of partial matrices with one unspecified entry*, Electron. J. Linear Algebra **27** (2014), 426–443. MR3240024

[5] Shaun M. Fallat and Charles R. Johnson, *Totally nonnegative matrices*, Princeton Series in Applied Mathematics, Princeton University Press, Princeton, NJ, 2011. MR2791531 (2012d:15001)

[6] Shaun M. Fallat, Charles R. Johnson, and Ronald L. Smith, *The general totally positive matrix completion problem with few unspecified entries*, Electron. J. Linear Algebra **7** (2000), 1–20. MR1737244 (2000k:15044)

[7] Mariano Gasca and Juan Manuel Peña, *Total positivity and Neville elimination*, Linear Algebra Appl. **165** (1992), 25–44, DOI 10.1016/0024-3795(92)90226-Z. MR1149743 (93d:15031)

[8] Jürgen Garloff, *Criteria for sign regularity of sets of matrices*, Linear Algebra Appl. **44** (1982), 153–160, DOI 10.1016/0024-3795(82)90010-6. MR657704 (84f:15024)

[9] Charles R. Johnson and Zhen Wei, *Asymmetric TP and TN completion problems*, Linear Algebra Appl. **438** (2013), no. 5, 2127–2135, DOI 10.1016/j.laa.2012.11.008. MR3005280

[10] Cristina Jordán and Juan R. Torregrosa, *The totally positive completion problem*, Linear Algebra Appl. **393** (2004), 259–274, DOI 10.1016/j.laa.2004.03.015. MR2098618 (2005h:15065)

[11] Allan Pinkus, *Totally positive matrices*, Cambridge Tracts in Mathematics, vol. 181, Cambridge University Press, Cambridge, 2010. MR2584277 (2010k:15065)

[12] Mark Skandera, *Inequalities in products of minors of totally nonnegative matrices*, J. Algebraic Combin. **20** (2004), no. 2, 195–211, DOI 10.1023/B:JACO.0000047282.21753.ae. MR2104676 (2005k:15046)

[13] Anne M. Whitney, *A reduction theorem for totally positive matrices* (English, with Hebrew summary), J. Analyse Math. **2** (1952), 88–92. MR0053173 (14,732c)

DEPARTMENT OF MATHEMATICS AND STATISTICS, UNIVERSITY OF KONSTANZ, D-78464 KONSTANZ, GERMANY

E-mail address: mjamathe@yahoo.com

FACULTY OF COMPUTER SCIENCE, UNIVERSITY OF APPLIED SCIENCES / HTWG KONSTANZ, D-78405 KONSTANZ, GERMANY

E-mail address: garloff@htwg-konstanz.de

Subsystems and regular quotients of C-systems

Vladimir Voevodsky

ABSTRACT. C-systems were introduced by J. Cartmell under the name "contextual categories". In this note we study sub-objects and quotient-objects of C-systems. In the case of the sub-objects we consider all sub-objects while in the case of the quotient-objects only *regular* quotients that in particular have the property that the corresponding projection morphism is surjective both on objects and on morphisms.

It is one of several short papers based on the material of the "Notes on Type Systems" by the same author.

1. Introduction. C-systems were introduced by John Cartmell ([2], [3, p.237]) and studied further by Thomas Streicher (see [7, Def. 1.2, p.47]). Both authors used the name "contextual categories" for these structures. We feel it to be important to use the word "category" only for constructions which are invariant under equivalences of categories. For the essentially algebraic structure with two sorts "morphisms" and "objects" and operations "source", "target", "identity" and "composition" we suggest to use the word pre-category. Since the additional structures introduced by Cartmell are not invariant under equivalences we can not say that they are structures on categories but only that they are structures on pre-categories. Correspondingly, Cartmell objects should be called "contextual pre-categories". We suggest to use the name C-systems instead[1].

Our first result, Proposition 2.4, shows that C-systems can be defined in two equivalent ways: one, as was originally done by Cartmell, using the condition that certain squares are pull-back and another using an additional operation $f \mapsto s_f$ which is almost everywhere defined and satisfies simple algebraic conditions.

This description is useful for the study of quotients and homomorphisms of C-systems.

To any C-system CC we associate a set $\widetilde{Ob}(CC)$ and eight partially defined operations on the pair of sets $(Ob(CC), \widetilde{Ob}(CC))$.

In Proposition 4.3 we construct a bijection between C-subsystems of a given C-system CC and pairs of subsets $(C, \widetilde{C})$ in $(Ob(CC), \widetilde{Ob}(CC))$ which are closed under the eight operations. This provides, through the results established in [9],

---

2010 *Mathematics Subject Classification.* Primary 03B15, 03B22, 03F50, 03G25.
Work on this paper was supported by NSF grant 1100938.

[1] The distinction between categories and pre-categories becomes precise in the univalent foundations where not all collections of objects are constructed from sets. See [1] for a detailed discussion.

an algebraic justification for what is known as the "structural" or "basic" rules of the dependent type theory (see [**5**, p.585]). More precisely, the description of subsystems constructed in the present paper provides a justification for the subset of the "structural" rules that concern the behavior of the type and term judgements.

The algebraic justification for the rules that concern the type equality and the term equality judgements is achieved in Proposition 5.4 where we construct a bijection between *regular congruence relations* on CC and pairs of equivalence relations on $(Ob(CC), \widetilde{Ob}(CC))$ which are compatible with the eight operations and satisfy some additional properties.

Besides their role in the mathematical theory of the syntactic structures that arise in dependent type theory these two results strongly suggest that the theory of C-systems is equivalent to the theory with the sorts $(Ob, \widetilde{Ob})$ and the eight operations which we consider together with some relations among these operations.

The essentially algebraic version of this other theory is called the theory of B-systems and will be considered in the sequel [**10**].

This is one of the short papers based on the material of [**8**] by the same author. I would like to thank the Institute Henri Poincare in Paris and the organizers of the "Proofs" trimester for their hospitality during the preparation of this paper. The work on this paper was facilitated by discussions with Richard Garner and Egbert Rijke.

2. C-systems. By a pre-category C we mean a pair of sets $Mor(C)$ and $Ob(C)$ with four maps

$$\partial_0, \partial_1 : Mor(C) \to Ob(C)$$

$$Id : Ob(C) \to Mor(C)$$

and

$$\circ : Mor(C)_{\partial_1} \times_{\partial_0} Mor(C) \to Mor(C)$$

which satisfy the well known conditions of unity and associativity (note that we write composition of morphisms in the form $f \circ g$ or fg where $f : X \to Y$ and $g : Y \to Z$). These objects would be usually called categories but we reserve the name "category" for those uses of these objects that are invariant under the equivalences.

DEFINITION 2.1. *A C0-system is a pre-category CC with additional structure of the form*

(1) *a function $l : Ob(CC) \to \mathbf{N}$,*
(2) *an object pt,*
(3) *a map $ft : Ob(CC) \to Ob(CC)$,*
(4) *for each $X \in Ob(CC)$ a morphism $p_X : X \to ft(X)$,*
(5) *for each $X \in Ob(CC)$ such that $l(X) > 0$ and each morphism $f : Y \to ft(X)$ an object f^*X and a morphism $q(f, X) : f^*X \to X$,*

which satisfies the following conditions:

(1) $l^{-1}(0) = \{pt\}$
(2) *for X such that $l(X) > 0$ one has $l(ft(X)) = l(X) - 1$*
(3) $ft(pt) = pt$
(4) *pt is a final object,*

(5) for $X \in Ob(CC)$ such that $l(X) > 0$ and $f : Y \to ft(X)$ one has $l(f^*(X)) > 0$, $ft(f^*X) = Y$ and the square

(1)
$$\begin{array}{ccc} f^*X & \xrightarrow{q(f,X)} & X \\ p_{f^*X} \downarrow & & \downarrow p_X \\ Y & \xrightarrow{f} & ft(X) \end{array}$$

commutes,

(6) for $X \in Ob(CC)$ such that $l(X) > 0$ one has $id^*_{ft(X)}(X) = X$ and $q(id_{ft(X)}, X) = id_X$,

(7) for $X \in Ob(CC)$ such that $l(X) > 0$, $g : Z \to Y$ and $f : Y \to ft(X)$ one has $(gf)^*(X) = g^*(f^*(X))$ and $q(gf, X) = q(g, f^*X)q(f, X)$.

REMARK 2.2. In this definition pt stands for "point" as a common notation for a final object of a category. The name "ft" stands for "father" which is the name given to this map in [7, Def. 1.1].

For $f : Y \to X$ in CC we let $ft(f) : Y \to ft(X)$ denote the composition $f \circ p_X$.

DEFINITION 2.3. A C-system is a C0-system together with an operation $f \mapsto s_f$ defined for all $f : Y \to X$ such that $l(X) > 0$ and such that

(1) $s_f : Y \to (ft(f))^*(X)$,
(2) $s_f \circ p_{(ft(f))^*(X)} = Id_Y$,
(3) $f = s_f \circ q(ft(f), X)$,
(4) if $X = g^*(U)$ where $g : ft(X) \to ft(U)$ then $s_f = s_{f \circ q(g,U)}$.

PROPOSITION 2.4. Let CC be a C0-system. Then the following are equivalent:

(1) the canonical squares (1) of CC are pull-back squares,
(2) there is given a structure of a C-system on CC.

Proof: Let us show first that if we are given an operation $f \mapsto s_f$ satisfying the conditions of Definition 2.3 then the canonical squares of CC are pull-back squares.

Let $l(X) > 0$ and $f : Y \to ft(X)$. We want to show that for any Z the map

$$(g : Z \to f^*(X)) \mapsto (ft(g), g \circ q(f, X))$$

is injective and that for any $g_1 : Z \to Y$, $g_2 : Z \to X$ such that $g_1 \circ f = ft(g_2)$ there exists a unique $g : Z \to Y$ such that $ft(g) = g_1$ and $g \circ q(f, X) = g_2$.

Let $g, g' : Z \to f^*(X)$ be such that $ft(g) = ft(g')$ and $g \circ q(f, X) = g' \circ q(f, X)$. Then

$$g = s_g \circ q(ft(g), f^*(X)) = s_{g \circ q(f,X)} \circ q(ft(g), f^*(X))$$
$$= s_{g' \circ q(f,X)} \circ q(ft(g'), f^*(X)) = s_{g'} \circ q(ft(g'), f^*(X)) = g'.$$

If we are given g_1, g_2 as above let $g = s_{g_2} \circ q(g_1, f^*(X))$. Then:

$$ft(g) = s_{g_2} \circ ft(q(g_1, f^*(X))) = s_{g_2} \circ p_{g_1^*(f_*(X))} \circ g_1 = g_1$$

$$g \circ q(f, X) = s_{g_2} \circ q(g_1, f^*(X)) \circ q(f, X) = s_{g_2} \circ q(g_1 \circ f, X) = s_{g_2} \circ q(ft(g_2), X) = g_2.$$

If on the other hand the canonical squares of CC are pull-back then we can define the operation s_f in the obvious way and moreover such an operation is unique because of the uniqueness part of the definition of pull-back. This implies the assertion of the proposition.

REMARK 2.5. As was pointed out by one of the referees, operation s_f was considered for contextual categories by Cartmell who denoted it by $f \mapsto \text{'}f\text{'}$, see [**2**, 2.19].

REMARK 2.6. Let

$$Ob_n(CC) = \{X \in Ob(CC) \,|\, l(X) = n\}$$

$$Mor_{n,m}(CC) = \{f : Mor(CC) | \partial_0(f) \in Ob_n \text{ and } \partial_1(f) \in Ob_m\}.$$

One can reformulate the definitions of C0-systems and C-systems using $Ob_n(CC)$ and $Mor_{n,m}(CC)$ as the underlying sets together with the obvious analogs of maps and conditions of the definition given above. In this reformulation there will be no use of the function l and of the condition $l(X) > 0$.

This shows that C0-systems and C-systems can be considered as models of essentially algebraic theories with sorts Ob_n, and $Mor_{n,m}$ and in particular all the results of [**6**] are applicable to C-systems.

REMARK 2.7. Note that as defined C0-systems and C-systems can not be described, in general, by generators and relations. For example, what is a C0-system generated by $X \in Ob$? There is no such universal object because we do not know what $l(X)$ is.

This problem is, of course, eliminated by using the definition with two infinite families of sorts Ob_n and $Mor_{n,m}$.

REMARK 2.8. The notion of a homomorphism of C0-systems and C-systems and the associated definitions of the categories of C0-systems and C-systems are obtained by the specialization of the corresponding general notions for models of essentially algebraic theories. Equivalently homomorphisms are defined as homomorphisms of pre-categories that commute with the length functions and the operations. The category of C-systems is a full subcategory of the category of C0-systems. Since they are categories of models of essentially algebraic theories they have all limits and colimits. According to the results and observations in [**2**] the category of C-systems is equivalent to a suitably defined category of the GATs (Generalized Algebraic Theories). The category of GATs is studied in [**4**].

Presentation of C-systems in terms of GATs uses constructions that are substantially non-finitary - a C-system given by finite sets of generators and relations can rarely be represented by a generalized algebraic theory with finitely many generating objects.

The C-systems that correspond to finitely presented GATs may play a special role in the theory of C-systems but what such a role might be remains to be discovered.

REMARK 2.9. Note that the additional structure on a pre-category that defines a C0-system is not an additional essentially algebraic structure and can not be made to be such by modification of definitions. Indeed, the pre-category underlying the product of two C0-systems (defined as the categorical product in the category of C0-systems and their homomorphisms) is not the product of the underlying pre-categories but a sub-pre-category in this product which consists of pairs of objects (X, Y) such that $l(X) = l(Y)$.

3. The set $\widetilde{Ob}$ of a C-system. For a C-system CC denote by $\widetilde{Ob}(CC)$ the subset of $Mor(CC)$ which consists of elements s of the form $s: ft(X) \to X$ where $l(X) > 0$ and such that $s \circ p_X = Id_{ft(X)}$. In other words, $\widetilde{Ob}$ is the set of sections of the canonical projections p_X for X such that $l(X) > 0$.

Note that $f \mapsto s_f$ is an operation from $\{f: Y \to X | l(X) > 0\}$ to $\widetilde{Ob}$.

For $X \in Ob(CC)$ and $i \geq 0$ such that $l(X) \geq i$ denote by $p_{X,i}$ the composition of the canonical projections $X \to ft(X) \to \ldots \to ft^i(X)$ such that $p_{X,0} = Id_X$ and for $l(X) > 0$, $p_{X,1} = p_X$. If $l(X) < i$ we will consider $p_{X,i}$ to be undefined. *All of the considerations involving $p_{X,i}$'s below are modulo the qualification that $p_{X,i}$ is defined, i.e., that $l(X) \geq i$.*

For X such that $l(X) \geq i$ and $f: Y \to ft^i(X)$ denote by $f^*(X, i)$ the objects and by $q(f, X, i): f^*(X, i) \to X$ the morphisms defined inductively by the rule

$$f^*(X, 0) = Y \qquad q(f, X, 0) = f,$$

$$f^*(X, i+1) = q(f, ft(X), i)^*(X) \qquad q(f, X, i+1) = q(q(f, ft(X), i), X).$$

If $l(X) < i$, then $q(f, X, i)$ is undefined since $q(-, X)$ is undefined for $X = pt$ and again, as in the case of $p_{X,i}$, *all of the considerations involving $q(f, X, i)$ are modulo the qualification that $l(X) \geq i$.*

For $i \geq 1$, $(s: ft(X) \to X) \in \widetilde{Ob}$ such that $l(X) \geq i$, and $f: Y \to ft^i(X)$ let

$$f^*(s, i): f^*(ft(X), i-1) \to f^*(X, i)$$

be the pull-back of the section $s: ft(X) \to X$ along the morphism $q(f, ft(X), i-1)$ i.e. the only morphism such that

$$f^*(s, i) \circ p_{f^*(X,i)} = Id_{f^*(ft(X), i-1)}$$

$$f^*(s, i) \circ q(f, X, i) = q(f, ft(X), i-1) \circ s$$

We again use the agreement that always when $f^*(s, i)$ is used the condition $l(X) \geq i$ is part of the assumptions.

Consider the following operations on the pair of sets $Ob = Ob(CC)$ and $\widetilde{Ob} = \widetilde{Ob}(CC)$:

(1) $pt \in Ob$,
(2) $ft: Ob \to Ob$,
(3) $\partial: \widetilde{Ob} \to Ob$ of the form $(s: ft(X) \to X) \mapsto X$,
(4) T which is defined on pairs $(Y, X) \in Ob \times Ob$ such that $l(Y) > 0$ and there exists (a necessarily unique) $i \geq 1$ with $ft(Y) = ft^i(X)$ and for such pairs $T(Y, X) = p_Y^*(X, i)$,
(5) $\widetilde{T}$ which is defined on pairs $(Y, (r: ft(X) \to X)) \in Ob \times \widetilde{Ob}$ such that $l(Y) > 0$ and there exists (a necessarily unique) $i \geq 1$ such that $ft(Y) = ft^i(X)$ and for such pairs $\widetilde{T}(Y, r) = p_Y^*(r, i)$,
(6) S which is defined on pairs $((s: ft(Y) \to Y), X) \in \widetilde{Ob} \times Ob$ such that there exists (a necessarily unique) $i \geq 1$ such that $Y = ft^i(X)$ and for such pairs $S(s, X) = s^*(X, i)$,
(7) $\widetilde{S}$ which is defined on pairs $((s: ft(Y) \to Y), (r: ft(X) \to X)) \in \widetilde{Ob} \times \widetilde{Ob}$ such that there exists (a necessarily unique) $i \geq 1$ such that $Y = ft^i(X)$ and for such pairs $\widetilde{S}(s, r) = s^*(r, i)$,
(8) δ which is defined on elements $X \in Ob$ such that $l(X) > 0$ and for such elements $\delta(X) \in \widetilde{Ob}$ is $s_{Id_X}: X \to p_X^*(X)$.

4. C-subsystems. A C-subsystem CC' of a C-system CC is a sub-pre-category of the underlying pre-category which is closed, in the obvious sense under the operations which define the C-system on CC.

A C-subsystem is itself a C-system with respect to the induced structure.

LEMMA 4.1. *Let CC be a C-system and CC', CC'' be two C-subsystems such that $Ob(CC') = Ob(CC'')$ (as subsets of $Ob(CC)$) and $\widetilde{Ob}(CC') = \widetilde{Ob}(CC'')$ (as subsets of $\widetilde{Ob}(CC)$). Then $CC' = CC''$.*

Proof: Let $f : Y \to X$ be a morphism in CC'. We want to show that it belongs to CC''. Proceed by induction on $m = l(X)$. For $m = 0$ the assertion is obvious. Suppose that $m > 0$. Since CC' is a C-subsystem we have a commutative diagram

(2)
$$\begin{array}{ccc} Y & & \\ {\scriptstyle s_f}\downarrow & & \\ (f \circ p_X)^*X & \xrightarrow{q(f \circ p_X, X)} & X \\ \downarrow & & \downarrow p_X \\ Y & \xrightarrow{f \circ p_X} & ft(X) \end{array}$$

in CC' such that $f = s_f\, q(p_X f, X)$. By the inductive assumption $f \circ p_X$ is in CC'' and since the square is the canonical pull-back square we conclude that $q(p_X f, X)$ is in CC''. On the other hand $s_f \in CC''$ since $\widetilde{Ob}(CC') = \widetilde{Ob}(CC'')$. Therefore $f \in CC''$.

REMARK 4.2. In Lemma 4.1, it is sufficient to assume that $\widetilde{Ob}(CC') = \widetilde{Ob}(CC'')$. The condition $Ob(CC') = Ob(CC'')$ is then also satisfied. Indeed, let $X \in Ob(CC')$ and $l(X) > 0$. Then $p_X^* X$ is the product $X \times_{ft(X)} X$ in CC. Consider the diagonal section $\delta_X : X \to p_X^* X$ of $p_{p_X^*(X)}$. Since CC' is assumed to be a C-subsystem we conclude that $\delta_X \in \widetilde{Ob}(CC') = \widetilde{Ob}(CC'')$ and therefore $X \in Ob(CC'')$. It is however more convenient to think of C-subsystems in terms of subsets of both Ob and $\widetilde{Ob}$.

PROPOSITION 4.3. *A pair $(B, \widetilde{B})$ where $B \subset Ob(CC)$ and $\widetilde{B} \subset \widetilde{Ob}(CC)$ corresponds to a C-subsystem of CC if and only if the following conditions hold:*

(1) $pt \in B$,
(2) if $X \in B$ then $ft(X) \in B$,
(3) if $s \in \widetilde{B}$ then $\partial(s) \in B$,
(4) if $Y \in B$ and $r \in \widetilde{B}$ then $\widetilde{T}(Y,r) \in \widetilde{B}$,
(5) if $s \in \widetilde{B}$ and $r \in \widetilde{B}$ then $\widetilde{S}(s,r) \in \widetilde{B}$,
(6) if $X \in B$ then $\delta(X) \in \widetilde{B}$.

Conditions (4) and (5) are illustrated by the following diagrams:

$$
\begin{array}{ccc}
p_Y^*(ft(X), i-1) & \xrightarrow{q(p_Y, ft(X), i-1)} & ft(X) \\
\downarrow q(p_Y, ft(X), i-1)^*(r) & & \downarrow r \\
p_Y^*(X, i) & \xrightarrow{q(p_Y, X, i)} & X \\
\downarrow & & \downarrow p_X \\
p_Y^*(ft(X), i-1) & \xrightarrow{q(p_Y, ft(X), i-1)} & ft(X) \\
\downarrow & & \downarrow \\
\vdots & & \vdots \\
\downarrow & & \downarrow \\
Y & \xrightarrow{p_Y} & ft^i(X)
\end{array}
\qquad
\begin{array}{ccc}
s^*(ft(X), i-1) & \xrightarrow{q(s, ft(X), i-1)} & ft(X) \\
\downarrow q(s, ft(X), i-1)^*(r) & & \downarrow r \\
s^*(X, i) & \xrightarrow{q(s, X, i)} & X \\
\downarrow & & \downarrow p_X \\
s^*(ft(X), i-1) & \xrightarrow{q(s, ft(X), i-1)} & ft(X) \\
\downarrow & & \downarrow \\
\vdots & & \vdots \\
\downarrow & & \downarrow \\
ft^{i+1}(X) & \xrightarrow{s} & ft^i(X)
\end{array}
$$

Proof: The "only if" part of the proposition is straightforward. Let us prove that for any $(B, \widetilde{B})$ satisfying the conditions of the proposition there exists a C-subsystem CC' of CC such that $B = Ob(CC')$ and $\widetilde{B} = \widetilde{Ob}(CC')$.

Define a candidate subcategory CC' setting $Ob(CC') = B$ and defining the set $Mor(CC')$ of morphisms of CC' inductively by the conditions:

(1) $Y \to pt$ is in $Mor(CC')$ if and only if $Y \in B$,
(2) $f : Y \to X$ is in $Mor(CC')$ if and only if $X \in B$, $ft(f) \in Mor(CC')$ and $s_f \in \widetilde{B}$.

(Note that for $(f : Y \to X) \in Mor(CC')$ one has $Y \in B$ since $s_f : Y \to (ft(f))^*(X)$).

Let us show that if the conditions of the proposition are satisfied then $(Ob(CC'), Mor(CC'))$ form a C-subsystem of CC.

The subset $Ob(CC')$ contains pt and is closed under ft map by the first two conditions. The following lemma shows that $Mor(CC')$ contains identities and the compositions of the canonical projections.

LEMMA 4.4. *Under the assumptions of the proposition, if $X \in B$ and $i \geq 0$ then $p_{X,i} : X \to ft^i(X)$ is in $Mor(CC')$.*

Proof: Let $n = l(X)$ and proceed by decreasing induction on i starting with n. The morphism $p_{X,n}$ is of the form $X \to pt$ and therefore it belongs to $Mor(CC')$ by the first constructor of $Mor(CC')$. By induction it remains to show that if $X \in B$ and $p_{X,i} \in Mor(CC')$ then $p_{X,i-1} \in Mor(CC')$. We have $ft(p_{X,i-1}) = p_{X,i}$ and

$$s_{p_{X,i-1}} = (p_{X,i-1}, 1)^*(\delta(ft^{i-1}(X)))$$

We have $\delta(ft^{i-1}(X)) \in \widetilde{B}$ by conditions (2) and (6). The pull-back $(p_{X,i-1}, 1)^*$ can be expressed as the composition of operations $\widetilde{T}(ft^j(X), -)$, $j = i-1, \ldots, 1$ and therefore $s_{p_{X,i-1}}$ is in $\widetilde{B}$ by repeated application of condition (4).

LEMMA 4.5. *Under the assumptions of the proposition, let $(r : ft(X) \to X) \in \widetilde{B}$, $i \geq 1$, and $(f : Y \to ft^i(X)) \in Mor(CC')$. Then $f^*(r, i) : ft(f^*(X, i)) \to f^*(X, i)$ is in $\widetilde{B}$.*

Proof: Proceed by increasing induction on the length of $ft^i(X)$. Suppose first that $ft^i(X) = pt$. Then $f = p_{Y,n}$ for some n and the statement of the lemma follows from repeated application of condition (4). Suppose that the lemma is proved for all morphisms to objects of length $j-1$ and let the length of $ft^i(X)$ be j. Consider the canonical decomposition $f = s_f q_f$. From it we have $f^*(r, i) = s_f^*(q_f^*(r, i), i)$. Since q_f is the canonical pull-back of $ft(f)$ we further have $q_f^*(r, i) = (ft(f))^*(r, i+1)$ and therefore
$$f^*(r, i) = s_f^*(ft(f)^*(r, i+1), i)$$
By induction $(ft(f))^*(r, i+1) \in \widetilde{B}$ and therefore $f^*(r, i) \in \widetilde{B}$ by condition (5).

LEMMA 4.6. *Under the assumptions of the proposition, let $g : Z \to Y$ and $f : Y \to X$ be in $Mor(CC')$. Then $gf \in Mor(CC')$.*

Proof: If $X = pt$ the the statement is obvious. Assume that it is proved for all f whose codomain is of length $< j$ and let X be of length j. We have $ft(gf) = g\,ft(f)$ and therefore $ft(gf) \in Mor(CC')$ by the inductive assumption. It remains to show that $s_{gf} \in \widetilde{B}$. We have the following diagram whose squares are canonical pull-back squares

$$\begin{array}{ccccc} X_{gf} & \longrightarrow & X_f & \longrightarrow & X \\ \downarrow & & \downarrow & & \downarrow p_X \\ Z & \xrightarrow{g} & Y & \xrightarrow{ft(f)} & ft(X) \end{array}$$

which shows that $s_{gf} = g^*(s_f, 1)$. Therefore, $s_{gf} \in Mor(CC')$ by Lemma 4.5.

LEMMA 4.7. *Under the assumptions of the proposition, let $X \in B$ and let $f : Y \to ft(X)$ be in $Mor(CC')$, then $f^*(X) \in B$ and $q(f, X) \in Mor(CC')$.*

Proof: Consider the diagram

$$\begin{array}{ccc} f^*(X) & \xrightarrow{q(f,X)} & X \\ s_{q(f,X)} \downarrow & & \downarrow s_{Id_X} \\ q(f,X)^*(p_X^*(X)) & \longrightarrow p_X^*(X) \longrightarrow & X \\ \downarrow & \downarrow & \downarrow \\ f^*(X) & \xrightarrow{q(f,X)} X & \longrightarrow ft(X) \\ p_{f^*(X)} \downarrow & \downarrow p_X & \\ Y & \xrightarrow{f} ft(X) & \end{array}$$

where the squares are canonical. By condition (6) we have $s_{Id_X} = \delta(X) \in \widetilde{B}$. Therefore, by Lemma 4.5, we have
$$s_{q(f,X)} = f^*(\delta(X), 2) \in \widetilde{B}.$$
By condition (3), $\partial(s_{q(f,X)}) \in B$ and therefore
$$f^*(X) = ft(\partial(s_{q(f,X)})) \in B.$$

by condition (2). Together with the previous lemmas this shows that
$$ft(q(f,X)) = p_{f^*(X)}f \in Mor(CC')$$
and therefore $q(f,X) \in Mor(CC')$.

LEMMA 4.8. *Under the assumptions of Lemma 4.7, the square*

$$\begin{array}{ccc} f^*(X) & \xrightarrow{q(f,X)} & X \\ {\scriptstyle p_{f^*(X)}}\downarrow & & \downarrow{\scriptstyle p_X} \\ Y & \xrightarrow{f} & ft(X) \end{array}$$

is a pull-back square in CC'.

Proof: We need to show that for a morphism $g : Z \to f^*(X)$ such that $gp_{f^*(X)}$ and $gq(f,X)$ are in $Mor(CC')$ one has $g \in Mor(CC')$. We have $ft(g) = gp_{f^*(X)}$, therefore by definition of $Mor(CC')$ it remains to check that $s_g \in \widetilde{B}$. The diagram of canonical pull-back squares

$$\begin{array}{ccccc} (f^*(X))_g & \longrightarrow & f^*(X) & \xrightarrow{q(f,X)} & X \\ \downarrow & & \downarrow & & \downarrow \\ Z & \xrightarrow{ft(g)} & Y & \xrightarrow{f} & ft(X) \end{array}$$

shows that $s_g = s_{gq(f,X)}$ and therefore $s_g \in Mor(CC')$.

To finish the proof of the proposition it remains to show that $Ob(CC') = B$ and $\widetilde{Ob}(CC') = \widetilde{B}$. The first assertion is tautological. The second one follows immediately from the fact that for $(s : ft(X) \to X) \in \widetilde{Ob}(CC)$ one has $ft(s) = Id_{ft(X)}$ and $s_s = s$.

5. Regular congruence relations on C-systems. The following definition of a regular congruence relation is an abstraction to the contextual categories of the structure that arises from the "definitional" equalities between types and terms of a type in dependent type theory. This connection is studied further in [9].

DEFINITION 5.1. *Let CC be a C-system. A regular congruence relation on CC is a pair of equivalence relations $\sim_{Ob}, \sim_{Mor}$ on $Ob(CC)$ and $Mor(CC)$ respectively such that:*

(1) $\sim_{Ob}$ and $\sim_{Mor}$ are compatible with $\partial_0, \partial_1, id, ft, (X \mapsto p_X), ((f,g) \mapsto fg), ((X,f) \mapsto f^*(X)), (X,f) \mapsto q(f,X)$ and $f \mapsto s_f$,
(2) $X \sim_{Ob} Y$ implies $l(X) = l(Y)$,
(3) *for any $X, F \in Ob(CC)$, $l(X) > 0$ such that $ft(X) \sim_{Ob} F$ there exists X_F such that $X \sim_{Ob} X_F$ and $ft(X_F) = F$,*
(4) *for any $f : X \to Y$ and X', Y' such that $X' \sim_{Ob} X$ and $Y' \sim_{Ob} Y$ there exists $f' : X' \to Y'$ such that $f' \sim_{Mor} f$,*

LEMMA 5.2. *If $R = (\sim_{Ob}, \sim_{Mor})$ is a regular congruence relation on CC then there exists a unique C-system CC/R on the pair of sets $(Ob(CC)/\sim_{Ob}, Mor(CC)/\sim_{Mor})$ such that the obvious function from CC is a homomorphism of C-systems.*

Proof: Since operations such as composition, $(X, f) \mapsto f^*(X)$ and $(X, f) \mapsto q(f, X)$ are not everywhere defined the condition that $\sim_{Ob}$ and $\sim_{Mor}$ are compatible with operations does not imply that the operations can be descended to the quotient sets. However when we add conditions (3) and (4) of Definition 5.1 we see that the functions from the quotients of the domains of definitions of operations to the domains where quotient operations should be defined are surjective and therefore the quotient operations are defined and satisfy all the relations which the original operations satisfied.

LEMMA 5.3. *Let $R = (\sim_{Ob}, \sim_{Mor})$ be a regular congruence relation on CC and let $\sim_{\widetilde{Ob}}$ be the restriction of $\sim_{Mor}$ to $\widetilde{Ob}$. Then one has:*

$$\widetilde{Ob}(CC/R) = \widetilde{Ob}(CC)/\sim_{\widetilde{Ob}}$$

Proof: It is sufficient to verify that for $X \in Ob(CC)$ and $t : ft(X) \to X$ such that $l(X) > 0$ and $ft(t) \sim_{Mor} Id_{ft(X)}$ there exists $(s : ft(X) \to X) \in \widetilde{Ob}(CC)$ such that $t \sim_{Mor} s$.

We have $t = s_t \circ q(ft(t), X)$. Since $ft(t) \sim_{Mor} Id_{ft(X)}$ we have $t \sim_{Mor} s_t$.

PROPOSITION 5.4. *The function which maps a regular congruence relation $(\sim_{Ob}, \sim_{Mor})$ to the pair of equivalence relations $(\sim_{Ob}, \sim_{\widetilde{Ob}})$ on $Ob(CC)$ and $\widetilde{Ob}(CC)$, where $\sim_{\widetilde{Ob}}$ is obtained by the restriction of $\sim_{Mor}$, is a bijection to the set of pairs of relations $(\sim, \simeq)$ satisfying the following conditions:*

(1) compatibilities with operations ft, ∂, T, $\widetilde{T}$, S, $\widetilde{S}$ and δ,
(2) $X \sim Y$ implies $l(X) = l(Y)$,
(3) for any $X, F \in Ob(CC)$, $l(X) > 0$ such that $ft(X) \sim F$ there exists X_F such that $X \sim X_F$ and $ft(X_F) = F$,
(4) for any $(s : ft(X) \to X) \in \widetilde{Ob}$ and $X' \sim X$ there exists $(s' : ft(X') \to X') \in \widetilde{Ob}$ such that $s' \simeq s$.

Proof: Let us show first that the pair defined by a regular congruence relation satisfies the conditions (1)-(4). The compatibilities with operations follow from our definitions of these operations in terms of the C-system structure and the assertion of Lemma 5.2 that the projection to the quotient by a regular congruence relation is a homomorphism of C-systems.

Conditions (2) and (3) follow directly from the definition of a regular congruence relation. Condition (4) follows easily from condition (4) of Definition 5.1 and Lemma 5.3.

Let now $(\sim_{Ob}, \sim_1)$ and $(\sim_{Ob}, \sim_2)$ be two regular congruence relations such that the restrictions of $\sim_1$ and $\sim_2$ to $\widetilde{Ob}(CC)$ coincide. Let us show that $f \sim_1 f'$ implies that $f \sim_2 f'$. Let $f \sim_1 f'$. By induction we may assume that $ft(f) \sim_2 ft(f')$. Then $q(ft(f), \partial_1(f)) \sim_2 q(ft(f'), \partial_1(f'))$ and $s_f \sim_2 s_{f'}$. Therefore

$$f = s_f \circ q(ft(f), \partial_1(f)) \sim_2 s_{f'} \circ q(ft(f'), \partial_1(f')) = f'$$

This proves injectivity.

To prove surjectivity let $(\sim, \simeq)$ be a pair of equivalence relations satisfying conditions (1)-(4). Let us show that it can be extended to a regular congruence relation on CC.

Define $\sim_{Mor}$ on $Mor_{*,m}$ by induction on m as follows. For $m = 0$ we say that $(X_1 \to pt) \sim_{Mor} (X_2 \to pt)$ iff $X_1 \sim X_2$.

For $(f_1 : X_1 \to Y_1)$, $(f_2 : X_2 \to Y_2)$ where $l(Y_1) = l(Y_2) = m + 1$ we let $f_1 \sim_{Mor} f_2$ iff $ft(f_1) \sim_{Mor} ft(f_2)$ and $s_{f_1} \simeq s_{f_2}$.

Let us show that if $X_1 \sim X_2$ and $i \le n = l(X_1) = l(X_2)$ then $p_{X_1,i} \sim_{Mor} p_{X_2,i}$. We show it by decreasing induction i. For $i = n$ it immediately follows from our definition. Let $i < n$. By induction we may assume that
$$ft(p_{X_1,i}) = p_{X_1,i+1} \sim_{Mor} p_{X_2,i+1} = ft(p_{X_2,i})$$
On the other hand since $i < l(X)$ one has
$$s_{p_{X,i}} = \widetilde{T}(X, \widetilde{T}(ft(X), \ldots, \widetilde{T}(ft^{i-1}(X), \delta(ft^i(X))) \ldots))$$
which implies that $s_{p_{X_1,i}} \simeq s_{p_{X_2,i}}$ and therefore $p_{X_1,i} \sim_{Mor} p_{X_2,i}$.

In particular, if $X_1 \sim X_2$ then $Id_{X_1} = p_{X_1,0} \sim_{Mor} p_{X_2,0} = Id_{X_2}$.

This also shows that the restriction of $\sim_{Mor}$ to $\widetilde{Ob}$ coincides with $\simeq$. Indeed, for $(s : ft(X) \to X) \in \widetilde{Ob}$ one has $s_s = s$ and $ft(s) = Id_{ft(X)}$. Therefore
$$(s_1 \sim_{Mor} s_2) = (Id_{ft(X_1)} \sim_{Mor} Id_{ft(X_2)}) \wedge (s_1 \simeq s_2) \Leftrightarrow (s_1 \simeq s_2).$$
The rest of the required properties of $\sim_{Mor}$ are verified similarly.

REMARK 5.5. It is straightforward to see that the projection from a C-system on which a regular congruence relation is defined to the C-system that is defined by this congruence relation according to Lemma 5.2 is an epimorphism in the category of C-systems and their homomorphisms. Categorical characterization of such epimorphisms remains at the moment unknown.

References

[1] Benedikt Ahrens, Chris Kapulkin, and Michael Shulman. Univalent categories and Rezk completion. http://arxiv.org/abs/1303.0584, 2011.

[2] John Cartmell. Generalised algebraic theories and contextual categories. Ph.D. Thesis, Oxford University, 1978. https://uf-ias-2012.wikispaces.com/Semantics+of+type+theory.

[3] John Cartmell, Generalised algebraic theories and contextual categories, Ann. Pure Appl. Logic 32 (1986), no. 3, 209–243, DOI 10.1016/0168-0072(86)90053-9. MR865990 (88j:03046)

[4] Richard Garner, Combinatorial structure of type dependency, J. Pure Appl. Algebra 219 (2015), no. 6, 1885–1914, DOI 10.1016/j.jpaa.2014.07.015. MR3299712

[5] Bart Jacobs, Categorical logic and type theory, Studies in Logic and the Foundations of Mathematics, vol. 141, North-Holland Publishing Co., Amsterdam, 1999. MR1674451 (2001b:03077)

[6] E. Palmgren and S. J. Vickers, Partial horn logic and Cartesian categories, Ann. Pure Appl. Logic 145 (2007), no. 3, 314–353, DOI 10.1016/j.apal.2006.10.001. MR2286417 (2007k:03189)

[7] Thomas Streicher, Semantics of type theory, Progress in Theoretical Computer Science, Birkhäuser Boston, Inc., Boston, MA, 1991. Correctness, completeness and independence results; With a foreword by Martin Wirsing. MR1134134 (93d:68042)

[8] Vladimir Voevodsky. Notes on type systems. https://github.com/vladimirias/old_notes_on_type_systems, 2009-2012.

[9] Vladimir Voevodsky. C-system of a module over a monad on sets. http://arxiv.org/abs/1407.3394, 2014.

[10] Vladimir Voevodsky. B-systems. http://arxiv.org/abs/1410.5389, October 2014.

SCHOOL OF MATHEMATICS, INSTITUTE FOR ADVANCED STUDY, PRINCETON, NEW JERSEY
E-mail address: vladimir@ias.edu

Landweber-type operator and its properties

Andrzej Cegielski

ABSTRACT. Our aim is to present several properties of a Landweber operator as well as a Landweber-type one. These operators are widely used in methods for solving the split feasibility problem and the split common fixed point problem. The presented properties can be used in proofs of convergence of related algorithms.

1. Preliminaries

In this section we recall some notions and facts which we will use in the further part of the paper. Let $\mathcal{H}$ be a real Hilbert space equipped with an inner product $\langle \cdot, \cdot \rangle$ and with the corresponding norm $\|\cdot\|$. We say that an operator $S : \mathcal{H} \to \mathcal{H}$ is *nonexpansive* (NE) if for all $x, y \in \mathcal{H}$ it holds $\|Sx - Sy\| \leq \|x - y\|$. We say that S is α-*averaged*, where $\alpha \in (0, 1)$, if $S = (1 - \alpha)\operatorname{Id} + \alpha N$ for a nonexpansive operator N. We say that S is *firmly nonexpansive* (FNE) if for all $x, y \in \mathcal{H}$ it holds

$$\langle Sx - Sy, x - y \rangle \geq \|Sx - Sy\|^2. \tag{1}$$

By Id we denote the *identity* operator. If follows from (1) that S is FNE if and only if $\operatorname{Id} - S$ is FNE. Denote by $S_\lambda := \operatorname{Id} + \lambda(S - \operatorname{Id})$ the λ-*relaxation* of S, where $\lambda \geq 0$. S is FNE if and only if S_λ is NE for all $\lambda \in [0, 2]$. Moreover, S is the λ-relaxation of an FNE operator if and only if S is $\frac{\lambda}{2}$-averaged, $\lambda \in (0, 2)$. Let $C \subseteq \mathcal{H}$ be a nonempty closed convex subset. Then for any $x \in \mathcal{H}$ there is a unique $y \in C$ satisfying $\|y - x\| \leq \|z - x\|$ for all $z \in C$. This point is denoted by $P_C x$ and is called the *metric projection* of x onto C. The metric projection P_C is an FNE operator. By $\operatorname{Fix} S := \{z \in \mathcal{H} : Sx = x\}$ we denote the subset of *fixed points* of S. We say that an operator $S : \mathcal{H} \to \mathcal{H}$ having a fixed point is α-*strongly quasi-nonexpansive* (α-SQNE), where $\alpha \geq 0$, if for all $x \in \mathcal{H}$ and all $z \in \operatorname{Fix} S$ it holds

$$\|Sx - z\|^2 \leq \|x - z\|^2 - \alpha \|Sx - x\|^2. \tag{2}$$

If $\alpha > 0$, then we call an α-SQNE operator *strongly quasi-nonexpansive* (SQNE). A 0-SQNE operator is called *quasi-nonexpansive* (QNE). The λ-relaxation of an FNE operator having a fixed point is $\frac{2-\lambda}{\lambda}$-SQNE, $\lambda \in (0, 2]$. We say that S is a *cutter* if

$$\langle x - Sx, z - Sx \rangle \leq 0 \tag{3}$$

2010 *Mathematics Subject Classification.* Primary 47J25, 47N10, 65J15, 90C25.

for all $x \in \mathcal{H}$ and all $z \in \operatorname{Fix} S$. An operator $U : \mathcal{H} \to \mathcal{H}$ is α-SQNE, if and only if

(4) $$\lambda \langle Ux - x, z - x \rangle \geq \|Ux - x\|^2$$

for all $x \in \mathcal{H}$ and all $z \in \operatorname{Fix} U$, where $\lambda = \frac{2}{\alpha+1}$. A FNE operator having a fixed point is a cutter. An operator S is a cutter if and only if S_λ is $\frac{2-\lambda}{\lambda}$-SQNE, $\lambda \in (0,2]$. An extended collection of properties of FNE as well as SQNE operators can be found, e.g., in [**3**, Chapter 2].

2. Landweber-type operator

Let $\mathcal{H}_1, \mathcal{H}_2$ be real Hilbert spaces, $A : \mathcal{H}_1 \to \mathcal{H}_2$ be a bounded linear operator with $\|A\| > 0$, $C \subseteq \mathcal{H}_1$ and $Q \subseteq \mathcal{H}_2$ be nonempty closed convex subsets.

DEFINITION 1. ([**2**], [**4**]) An operator $V : \mathcal{H}_1 \to \mathcal{H}_1$ defined by

(5) $$V := \operatorname{Id} + \frac{1}{\|A\|^2} A^*(P_Q - \operatorname{Id})A$$

is called a *Landweber operator*. An operator $U : \mathcal{H}_1 \to \mathcal{H}_1$ defined by

(6) $$U := P_C(\operatorname{Id} + \frac{1}{\|A\|^2} A^*(P_Q - \operatorname{Id})A)$$

is called a *projected Landweber operator*. An operator $R_\lambda : \mathcal{H}_1 \to \mathcal{H}_1$

(7) $$R_\lambda := P_C(\operatorname{Id} + \frac{\lambda}{\|A\|^2} A^*(P_Q - \operatorname{Id})A),$$

where $\lambda \in (0,2)$, is called a *projected relaxation* of a Landweber operator V.

A Landweber operator was applied by Landweber [**10**] in a method for approximating least-squares solution of a first kind integral equation. Censor and Elfving [**7**] introduced the following problem:

(8) $$\text{find } x^* \in C \text{ with } Ax^* \in Q$$

and called it the *split feasibility problem* (SFP). Byrne [**2**] applied a projected relaxation of a Landweber operator for solving the SFP. Now we introduce a more general operator than the defined by (5).

DEFINITION 2. Let $T : \mathcal{H}_2 \to \mathcal{H}_2$ be quasi-nonexpansive. An operator $V : \mathcal{H}_1 \to \mathcal{H}_1$ defined by

(9) $$V := \operatorname{Id} + \frac{1}{\|A\|^2} A^*(T - \operatorname{Id})A$$

is called a *Landweber-type operator* (related to T).

Because P_Q is quasi-nonexpansive, (5) is a special case of (9). The operators defined by (5), (7) and (9) were applied by many authors for solving the split feasibility problem, the multiple split feasibility problem and the split common fixed point problem (see [**2, 4, 8, 11, 12, 14, 16**] and the references therein).

3. Properties of a Landweber-type operator

Let $A : \mathcal{H}_1 \to \mathcal{H}_2$ be a bounded linear operator with $\|A\| > 0$ and $Q \subseteq \mathcal{H}_2$ be closed convex. By $\operatorname{im} A$ we denote the image of A. The Landweber operator V is closely related to the proximity function $f : \mathcal{H}_1 \to \mathbb{R}_+$ defined by the following equality

$$f(x) = \frac{1}{2} \|P_Q(Ax) - Ax\|^2. \tag{10}$$

This relation is expressed in the following result, which proof can be found in [**2**, Proposition 2.1] or [**3**, Lemma 4.6.2].

PROPOSITION 3. *The proximity function* $f : \mathcal{H}_1 \to \mathbb{R}$ *defined by* (10) *is a differentiable convex function and* $Df = A^*(Ax - P_Q(Ax))$. *Moreover,* $\operatorname{Fix} V = \operatorname{Argmin}_{x \in \mathcal{H}_1} f$, *where* $V := \operatorname{Id} + \frac{1}{\|A\|^2} A^*(P_Q - \operatorname{Id})A$ *denotes the Landweber operator.*

A sequence $\{U^k x\}_{k=0}^{\infty}$, where $x \in \mathcal{H}$ is called an *orbit* of the operator $U : \mathcal{H} \to \mathcal{H}$. Byrne [**2**, Theorem 2.1] proved that any orbit of the a projected relaxation R_λ of a Landweber operator converges to a solution of the SFP in the case $\mathcal{H}_1$ and $\mathcal{H}_2$ are Euclidean spaces, $\lambda \in (0, 2)$. Xu [**15**] observed that any orbit of R_λ converges weakly in the infinite dimensional case.

Below we give an important property of a Landweber-type operator.

PROPOSITION 4. *If* $T : \mathcal{H}_2 \to \mathcal{H}_2$ *is firmly nonexpansive, then a Landweber-type operator defined by* (9) *is firmly nonexpansive.*

Proof. (cf. [**3**, Theorem 4.6.3], where the case $T = P_Q$ was proved) Recall that U is FNE if and only if $\operatorname{Id} - U$ is FNE. Suppose now that T is FNE, i.e.,

$$\langle (u - Tu) - (v - Tv), u - v \rangle \geq \|(u - Tu) - (v - Tv)\|^2 \tag{11}$$

for all $u, v \in \mathcal{H}_2$. Let $G := \operatorname{Id} - V = \frac{1}{\|A\|^2} A^*(\operatorname{Id} - T)A$. We prove that G is FNE. If we take $u := Ax$ and $v := Ay$ for $x, y \in \mathcal{H}_1$ in inequality (11) and apply the inequality $\|A^* z\| \leq \|A^*\| \cdot \|z\|$ and the equality $\|A^*\| = \|A\|$, then we obtain

$$\begin{aligned}
\langle G(x) - G(y), x - y \rangle &= \frac{1}{\|A\|^2} \langle A^*(Ax - TAx) - A^*(Ay - TAy), x - y \rangle \\
&= \frac{1}{\|A\|^2} \langle (Ax - TAx) - (Ay - TAy), Ax - Ay \rangle \\
&\geq \frac{1}{\|A\|^2} \|(Ax - TAx) - (Ay - TAy)\|^2 \\
&= \frac{1}{\|A\|^4} \|A^*\|^2 \|(Ax - TAx) - (Ay - TAy)\|^2 \\
&\geq \frac{1}{\|A\|^4} \|A^*(Ax - TAx) - A^*(Ay - TAy)\|^2 \\
&= \left\| \frac{1}{\|A\|^2} A^*(Ax - TAx) - \frac{1}{\|A\|^2} A^*(Ay - TAy) \right\|^2 \\
&= \|G(x) - G(y)\|^2,
\end{aligned}$$

i.e., G is FNE. This yields that a Landweber-type operator $V = \operatorname{Id} - G$ is FNE. ∎

COROLLARY 5. *If $T : \mathcal{H}_2 \to \mathcal{H}_2$ is nonexpansive, then a Landweber-type operator V defined by (9) is nonexpansive.*

Proof. The Corollary follows easily from Proposition 4 by an application the fact that S is FNE if and only if $2S - \text{Id}$ is NE. ∎

DEFINITION 6. We say that an operator $S : \mathcal{H} \to \mathcal{H}$ is *asymptotically regular* (AR) if
$$\lim_{k \to \infty} \|S^{k+1}x - S^k x\| = 0$$
for all $x \in \mathcal{H}$.

Asymptotically regular nonexpansive operators play an important role in fixed point iterations. An example of an asymptotically regular operator is an SQNE one [**3**, Theorem 3.4.3]. Opial [**13**, Theorem 1] proved that any orbit of an AR and NE operator S with $\text{Fix}\, S \neq \emptyset$ converges weakly to a fixed point of S. It turns out that the same result one can obtain in a more general case. First we recall a notion of the demi-closedness principle.

DEFINITION 7. We say that an operator $U : \mathcal{H} \to \mathcal{H}$ satisfies the *demi-closedness* (DC) *principle* if

(12) $\qquad (x^k \rightharpoonup x \text{ and } \|Ux^k - x^k\| \to 0) \Longrightarrow x \in \text{Fix}\, U.$

If (12) holds then we also say that $U - \text{Id}$ is *demi-closed* at 0.

Opial proved that a nonexpansive operator satisfies the DC principle [**13**, Lemma 2]. Basing on this fact Opial proved his famous result [**13**, Theorem 1] that any orbit of a nonexpansive and asymptotically regular operator T having a fixed point converges weakly to $x^* \in \text{Fix}\, T$. Note that the demi-closedness principle also holds for a subgradient projection P_f, for a continuous convex function $f : \mathcal{H} \to \mathbb{R}$ with $S(f, 0) := \{x \in \mathcal{H} : f(x) \leq 0\} \neq \emptyset$, which is Lipschitz continuous on bounded subsets (see [**3**, Theorem 4.2.7]).

Now we present further properties of a Landweber-type operator. We apply ideas of [**3**, Section 2.4] to a Landweber-type operator V. Recall that an operator $S_\tau : \mathcal{H}_1 \to \mathcal{H}_1$ defined by $S_\tau x := x + \tau(x)(Sx - x)$ is called a *generalized relaxation* of $S : \mathcal{H}_1 \to \mathcal{H}_1$, where $\tau : \mathcal{H}_1 \to (0, +\infty)$ is called a *step-size function*. If $\tau(x) \geq 1$ for all $x \in \mathcal{H}_1$, then S_τ is called an *extrapolation* of S.

Let $V := \text{Id} + \frac{1}{\|A\|^2} A^*(T - \text{Id})A$ be a Landweber-type operator. Denote

(13) $\qquad U := \text{Id} + A^*(T - \text{Id})A.$

Then, obviously,

(14) $\qquad Ux - x = \|A\|^2 (Vx - x).$

Let a step-size function $\sigma : \mathcal{H}_1 \to (0, +\infty)$ be defined by

(15) $\qquad \sigma(x) := \begin{cases} \frac{\|TAx - Ax\|^2}{\|A^*(TAx - Ax)\|^2}, & \text{if } Ax \notin \text{Fix}\, T, \\ 1, & \text{otherwise.} \end{cases}$

An operator U_σ defined by $U_\sigma x := x + \sigma(x) A^*(TAx - Ax)$ is a generalized relaxation of U. Denoting

(16) $\qquad \tau(x) = \begin{cases} \|A\|^2 \sigma(x), & \text{if } Ax \notin \text{Fix}\, T, \\ 1, & \text{otherwise.} \end{cases}$

we obtain

(17) $$V_\tau x = \begin{cases} x + \frac{\|TAx - Ax\|^2}{\|A^*(TAx - Ax)\|^2} A^*(TAx - Ax), & \text{if } Ax \notin \text{Fix } T, \\ x, & \text{otherwise.} \end{cases}$$

By $\|A^*u\| \leq \|A^*\| \cdot \|u\| = \|A\| \cdot \|u\|$, we have $\sigma(x) \geq \|A\|^{-2}$ and $\tau(x) \geq 1$. Therefore, V_τ is an extrapolation of the Landweber-type operator V. Note, however, that we do not need to know the norm of A in order to evaluate $V_\tau x$. In the theorem below we present important properties of V_τ.

THEOREM 8. *Let $A : \mathcal{H}_1 \to \mathcal{H}_2$ be a bounded linear operator with $\|A\| > 0$, $T : \mathcal{H}_2 \to \mathcal{H}_2$ be an α-SQNE operator with $\text{im } A \cap \text{Fix } T \neq \emptyset$, where $\alpha \geq 0$. Further, let V_τ be the extrapolation of a Landweber-type operator, defined by (17). Then:*
  (i) $\text{Fix } V_\tau = \text{Fix } V = A^{-1}(\text{Fix } T)$;
  (ii) *V_τ is α-SQNE;*
  (iii) *If $\alpha > 0$ then V_τ is asymptotically regular;*
  (iv) *If T satisfies the demi-closedness principle then V_τ also satisfies the demi-closedness principle.*

Proof. By (4), T is α-SQNE, where $\alpha \geq 0$, if and only if

(18) $$\lambda \langle Tu - u, y - u \rangle \geq \|Tu - u\|^2$$

for all $u \in \mathcal{H}_2$ and all $y \in \text{Fix } T$, where $\lambda = \frac{2}{\alpha+1} \in (0, 2]$. Note that $z \in A^{-1}(\text{Fix } T)$ if and only if $Az \in \text{Fix } T$.

(i) The equality $\text{Fix } V_\tau = \text{Fix } V$ is clear, because $\sigma(x) > 0$ for all $x \in \mathcal{H}_1$. Now we prove the second equality.

$\supseteq$ Let $Az \in \text{Fix } T$. Then $Vz = z + (\frac{1}{\|A\|^2} A^*(TAz - Az) = z$.

$\subseteq$ Let $z \in \text{Fix } V$. Then, of course, $A^*(TAz - Az) = 0$. Let $w \in \mathcal{H}_1$ be such that $Aw \in \text{Fix } T$. By (18), we have

$$\|TAz - Az\|^2 \leq \lambda \langle TAz - Az, Aw - Az \rangle = \lambda \langle A^*(TAz - Az), w - z \rangle = 0,$$

i.e., $Az \in \text{Fix } T$ which is equivalent to $z \in A^{-1}(\text{Fix } T)$.

(ii) Let $z \in \text{Fix } V_\sigma = \text{Fix } V$. By (i), $Az \in \text{Fix } T$. (14) and (18) yield

$$\lambda \langle Vx - x, z - x \rangle$$
$$= \frac{\lambda}{\|A\|^2} \langle Ux - x, z - x \rangle = \frac{\lambda}{\|A\|^2} \langle A^*(TAx - Ax), z - x \rangle$$
$$= \frac{\lambda}{\|A\|^2} \langle TAx - Ax, Az - Ax \rangle \geq \frac{1}{\|A\|^2} \|TAx - Ax\|^2$$
$$= \frac{\sigma(x)}{\|A\|^2} \|A^*(TAx - Ax)\|^2 = \frac{\sigma(x)}{\|A\|^2} \|Ux - x\|^2$$
$$= \tau(x) \|Vx - x\|^2,$$

where $\lambda = \frac{2}{\alpha+1} \in (0, 2]$, $U(x)$, $\sigma(x)$ and $\tau(x)$ are defined by (13) and (15)–(16), $x \in \mathcal{H}_1$. This yields $\lambda \langle Vx - x, z - x \rangle \geq \tau(x) \|Vx - x\|^2$. Multiplying both sides by $\tau(x)$ we obtain $\lambda \langle V_\tau x - x, z - x \rangle \geq \|V_\tau x - x\|^2$. By (4), the operator V_τ is α-SQNE.

(iii) Follows from (ii) and from the fact that any SQNE operator is AR (see [**3**, Theorem 3.4.3]).

(iv) Suppose, that T satisfies the DC principle, i.e., $y^k \rightharpoonup y$ together with $\|Ty^k - y^k\| \to 0$ implies $y \in \operatorname{Fix} T$. We prove that V_τ also satisfies the demi-closedness principle. Let $x^k \rightharpoonup x$ and $\|V_\tau x^k - x^k\| \to 0$. Then, obviously, $\|Vx^k - x^k\| \to 0$, because $\tau(x^k) \geq 1$. Clearly, for any $u \in \mathcal{H}_2$,
$$\lim_k \langle Ax^k - Ax, u \rangle = \lim_k \langle x^k - x, A^*u \rangle = 0,$$
i.e., $Ax^k \rightharpoonup Ax$. Choose an arbitrary $z \in \operatorname{Fix} V_\tau$. By (i), $Az \in \operatorname{Fix} T$. By (18), the Cauchy–Schwarz inequality and the boundedness of x^k, we have
$$\begin{aligned}
\|TAx^k - Ax^k\|^2 &\leq \lambda \langle TAx^k - Ax^k, Az - Ax^k \rangle \\
&= \lambda \|A\|^2 \langle \frac{1}{\|A\|^2} A^*(TAx^k - Ax^k), z - x^k \rangle \\
&\leq \lambda \|A\|^2 \|Vx^k - x^k\| \cdot \|z - x^k\| \to 0 \text{ as } k \to \infty.
\end{aligned}$$
Consequently, $\lim_k \|T(Ax^k) - Ax^k\| = 0$. This, together with $Ax^k \rightharpoonup Ax$ and with the assumption that T satisfies the demi-closedness principle, gives $Ax \in \operatorname{Fix} T$, i.e., $x \in A^{-1}(\operatorname{Fix} T) = \operatorname{Fix} V_\tau$ and the proof is completed. ∎

By an application of the inequality $\tau(x) \geq 1$, $x \in \mathcal{H}_1$, one can prove that Theorem 8 implies the following result which is a slight generalization of a result due to Wang and Xu [**14**, Lemma 3.1 and Theorem 3.3].

COROLLARY 9. *Let $A : \mathcal{H}_1 \to \mathcal{H}_2$ be a bounded linear operator with $\|A\| > 0$, $T : \mathcal{H}_2 \to \mathcal{H}_2$ be an α-SQNE operator with $\operatorname{im} A \cap \operatorname{Fix} T \neq \emptyset$, where $\alpha \geq 0$. Further, let $V := \operatorname{Id} + \frac{1}{\|A\|^2} A^*(T - \operatorname{Id})A$. Then:*

(i) *$\operatorname{Fix} V = A^{-1}(\operatorname{Fix} T)$ and*
(ii) *V is α-SQNE.*

If, moreover, T satisfies the DC principle, then V also satisfies the DC principle.

A proof of Corollary 9 can be also found in [**4**, Lemma 4.1]. One can obtain similar results for a composition of an extrapolation of a Landweber-type operator and an SQNE operator.

COROLLARY 10. *Let $T : \mathcal{H}_2 \to \mathcal{H}_2$ α-SQNE, $U : \mathcal{H}_1 \to \mathcal{H}_1$ be β-SQNE with $\operatorname{Fix} U \cap A^{-1}(\operatorname{Fix} T) \neq \emptyset$, where $\alpha, \beta > 0$. Further, let $R := UV_\tau$, where V_τ is defined by (17). Then*

(i) *$\operatorname{Fix} R = \operatorname{Fix} U \cap A^{-1}(\operatorname{Fix} T)$;*
(ii) *R is γ-SQNE, where $\gamma = (\frac{1}{\alpha} + \frac{1}{\beta})^{-1}$;*
(iii) *R is asymptotically regular.*

If, moreover, T and U satisfy the DC principle, then R also satisfies the DC principle.

Proof. By Theorem 8(i)-(ii) and by the assumption $\operatorname{Fix} U \cap A^{-1}(\operatorname{Fix} T) \neq \emptyset$, V_τ and U are SQNE operators having a common fixed point. Therefore, (i) follows from [**1**, Proposition 2.10(i)]. Part (ii) follows now from Theorem 8(ii), from the assumption that U is β-SQNE and from [**3**, Theorem 2.1.48(ii)]. Part (iii) follows from (i), (ii) and from [**3**, Theorem 3.4.3]. Suppose now, that T and U satisfy the DC principle. By Theorem 8, V_τ satisfies the DC principle. Therefore, [**4**, Theorem 4.2] yields, that R also satisfies the DC principle. ∎

The results presented in Theorem 8 and Corollary 10 can be applied to a proof of weak convergence of sequences generated by the iteration defined below. Let $T : \mathcal{H}_2 \to \mathcal{H}_2$ and $U : \mathcal{H}_1 \to \mathcal{H}_1$ be QNE operators with $\operatorname{Fix} U \cap A^{-1}(\operatorname{Fix} T) \neq \emptyset$, $T_k := T_{\lambda_k}$ and $U_k := U_{\mu_k}$ be their relaxations, where the relaxation parameters $\lambda_k, \mu_k \in [\varepsilon, 1 - \varepsilon]$ for some small $\varepsilon > 0$. Further, let

$$(19) \quad V_k(x) := \begin{cases} x + \dfrac{\|T_k Ax - Ax\|^2}{\|A^*(T_k Ax - Ax)\|^2} A^*(T_k Ax - Ax), & \text{if } Ax \notin \operatorname{Fix} T, \\ x, & \text{otherwise,} \end{cases}$$

be an extrapolation of a Landweber-type operator related to T_k. Consider the following iterative process

$$(20) \quad x^{k+1} = U_k V_k x^k,$$

where $x^0 \in \mathcal{H}_1$.

COROLLARY 11. *Let $T : \mathcal{H}_2 \to \mathcal{H}_2$ and $U : \mathcal{H}_1 \to \mathcal{H}_1$ be QNE operators with $\operatorname{Fix} U \cap A^{-1}(\operatorname{Fix} T) \neq \emptyset$ and satisfying the demi-closedness principle. Then for any starting point $x^0 \in \mathcal{H}_1$ the sequence $\{x^k\}_{k=0}^\infty$ generated by (20) converges weakly to a point $x^* \in \operatorname{Fix} U \cap A^{-1}(\operatorname{Fix} T)$.*

Proof. It is easily seen that U_k is α_k-SQNE and T_k is β_k-SQNE, where $\alpha_k = \frac{1-\lambda_k}{\lambda_k}$, $\beta_k = \frac{1-\mu_k}{\mu_k}$ and $\alpha_k, \beta_k \in [\varepsilon, \frac{1}{\varepsilon}]$. By Corollary 10(i)-(ii), $\operatorname{Fix}(U_k V_k) = \operatorname{Fix} U \cap A^{-1}(\operatorname{Fix} T)$ and $U_k V_k$ is γ_k-SQNE with $\gamma_k = (\frac{1}{\alpha_k} + \frac{1}{\beta_k})^{-1} \in [\frac{\varepsilon}{2}, \frac{1}{2\varepsilon}]$. Therefore, the corollary follows from a more general result [**4**, Theorem 5.1]. ∎

REMARK 12. A special case of iteration (20) was recently presented by López et al. [**11**], where T_k and U_k are subgradient projections related to subdifferentiable convex functions c and q, respectively, with $C := \{x \in \mathcal{H}_1 : c(x) \leq 0\}$ and $Q := \{y \in \mathcal{H}_2 : q(y) \leq 0\}$. López et al. [**11**, Theorem 4.3] proved the weak convergence of x^k to a solution of the SFP (8). Very recently, Cui and Wang [**9**] studied the split fixed point problem: find $x \in \operatorname{Fix} V$ with $Ax \in \operatorname{Fix} T$, where $V : \mathcal{H}_1 \to \mathcal{H}_1$ and $T : \mathcal{H}_2 \to \mathcal{H}_2$ are β-demicontractive operators. Recall that an operator $S : \mathcal{H} \to \mathcal{H}$ having a fixed point is called β-*contractive*, where $\beta \in [0, 1)$, if

$$\|Sx - z\|^2 \leq \|x - z\|^2 + \beta \|Sx - x\|^2,$$

for all $x \in \mathcal{H}_1$ and $z \in \operatorname{Fix} S$ (cf. [**12**]). Cui and Wang applied a generalized relaxation of the operator S defined by (13) for solving this problem. Roughly spoken, Cui and Wang [**9**, Theorem 3.3] observed that any orbit of the operator $V_\lambda U_\rho$, where the step-size $\rho(x) = \frac{1-\beta}{2}\sigma(x)$ and the relaxation parameter $\lambda \in (0, 1 - \beta)$, converges weakly to a solution of the problem. Their result is interesting, however, an application of demicontractive operators seems to be artificial, because $T_{\frac{1-\beta}{2}}$ is a cutter, consequently U_ρ is a cutter, and V_λ is strongly quasi-nonexpansive.

Before we present the next property of a Landweber-type operator, we recall a notion of an approximately shrinking operator.

DEFINITION 13. ([**6**, Definition 3.1]) *Let $U : \mathcal{H} \to \mathcal{H}$ be quasi-nonexpansive. We say that U is approximately shrinking (AS) if for any bounded subset $D \subseteq \mathcal{H}$ and for any $\eta > 0$ there is $\gamma > 0$ such that for any $x \in D$ it holds*

$$(21) \quad \|Ux - x\| < \gamma \Longrightarrow d(x, \operatorname{Fix} U) < \eta.$$

Important examples of AS operators are the metric projection P_C onto a closed convex subset $C \subseteq \mathcal{H}$ and the subgradient projection P_f for a convex function $f : \mathbb{R}^n \to \mathbb{R}$ [5, Lemma 24].

THEOREM 14. *Let $T : \mathbb{R}^m \to \mathbb{R}^m$ be a QNE operator with $\operatorname{im} A \cap \operatorname{Fix} T \neq \emptyset$. Further, let $V := \operatorname{Id} + \frac{1}{\|A\|^2} A^*(T - \operatorname{Id})A$ be a Landweber-type operator and V_τ be its extrapolation, where τ is given by (16). If T is approximately shrinking then V and V_τ are also approximately shrinking.*

The theorem will be proved elsewhere. It turns out that the properties presented in Theorems 8, 14 and in Corollary 10 can be applied to a proof of the strong convergence of sequences generated by a hybrid steepest descent method with an application of a Landweber-type operators to a solution of a variational inequality (see [5] and [6] for the details).

4. Examples

EXAMPLE 15. Let $a \in \mathcal{H}$ with $\|a\| > 0$, $A : \mathcal{H} \to \mathbb{R}$ be defined by $Ax = \langle a, x \rangle$. Then $\|A\| = \|a\|$. If $Q := (-\infty, \beta]$, where $\beta \in \mathbb{R}$, then the Landweber operator V and its extrapolation V_τ with a step-size function τ defined by (16) coincide and

$$Vx = V_\tau x = P_C x = x - \frac{(\langle a, x \rangle - \beta)_+}{\|a\|^2} a,$$

where $C = H(a, \beta)_- := \{u \in \mathcal{H} : \langle a, u \rangle \leq \beta\}$ and $\alpha_+ := \max\{\alpha, 0\}$.

EXAMPLE 16. Let $A : \mathcal{H}_1 \to \mathcal{H}_2$ be a bounded linear operator with $\|A\| > 0$. If $Q = \{b\}$, where $b \in \mathcal{H}_2$, then the SFP is to find a solution of a linear system $Ax = b$. The Landweber operator V related to this problem and its extrapolation V_τ with a step-size function τ defined by (16) have the forms

$$Vx = x - \frac{1}{\|A\|^2} A^*(Ax - b)$$

and

$$V_\tau x = x - \frac{\|Ax - b\|^2}{\|A^*(Ax - b)\|^2} A^*(Ax - b).$$

By Theorem 8 and Corollary 9, V and V_τ satisfy the DC principle.

EXAMPLE 17. Let A be an $m \times n$ real matrix representing a linear operator and $b \in \mathbb{R}^m$. Suppose without loss of generality that the rows a_i of A, $i = 1, 2, \ldots, m$, are nonzero vectors. If $Q := \{u \in \mathbb{R}^m : u \leq b\} = b - \mathbb{R}^m_+$, then the SFP is to find a solution of a system of linear inequalities $Ax \leq b$. The Landweber operator V related to this problem and its extrapolation V_τ with a step-size function τ defined by (16) have the forms

$$Vx = x - \frac{1}{\lambda_{\max}(A^T A)} A^T (Ax - b)_+$$

and

$$V_\tau x = x - \frac{\|(Ax - b)_+\|^2}{\|A^*(Ax - b)_+\|^2} A^T (Ax - b)_+,$$

where A^T denotes the transpose of the matrix A and $\lambda_{\max}(A^T A)$ denotes the maximal eigenvalue of $A^T A$. By Theorem 8 and Corollary 9, V and V_τ satisfy the DC principle. Moreover, by Theorem 14, V and V_τ are approximately shrinking.

EXAMPLE 18. Let $f : \mathcal{H}_2 \to \mathbb{R}$ be a continuous convex function with $S(f,0) := \{y \in \mathcal{H}_2 : f(y) \leq 0\} \neq \emptyset$. Define a subgradient projection $P_f : \mathcal{H}_2 \to \mathcal{H}_2$, related to f by

$$T(y) = \begin{cases} y - \frac{f(y)}{\|g_f(y)\|^2} g_f(y), & \text{if } f(y) > 0, \\ y, & \text{otherwise,} \end{cases}$$

where $g_f(y)$ denotes a subgradient of f at $y \in \mathcal{H}_2$. Suppose that f is Lipschitz continuous on bounded subsets. Let $A : \mathcal{H}_1 \to \mathcal{H}_2$ be a bounded linear operator with $\|A\| > 0$, $\lambda \in (0,2]$ and $T := P_{f,\lambda}$. Then T is α-SQNE with $\alpha = \frac{2-\lambda}{\lambda}$. A Landweber-type operator V related to P_f and its extrapolation V_τ with a step-size function τ defined by (16) have the forms

$$Vx = \begin{cases} x - \frac{\lambda f(Ax)}{(\|A\| \cdot \|g_f(Ax)\|)^2} A^* g_f(Ax), & \text{if } f(Ax) > 0, \\ x, & \text{otherwise} \end{cases}$$

and

$$V_\tau x = \begin{cases} x - \frac{\lambda f(Ax)}{\|A^* g_f(Ax)\|^2} A^* g_f(Ax), & \text{if } f(Ax) > 0, \\ x, & \text{otherwise.} \end{cases}$$

Because P_f satisfies the DC principle (see [3, Theorem 4.2.7]), Theorem 8 and Corollary 9 yield, that V and V_τ satisfy the DC principle. Moreover, if $\mathcal{H}_2$ is finite-dimensional, then P_f is approximately shrinking (see [5, Lemma 24]). Therefore, Theorem 14 yields that in this case V and V_τ are also approximately shrinking.

References

[1] H. H. Bauschke and J. M. Borwein, *On projection algorithms for solving convex feasibility problems*, SIAM Rev. **38** (1996), no. 3, 367–426, DOI 10.1137/S0036144593251710. MR1409591 (98f:90045)

[2] C. Byrne, *Iterative oblique projection onto convex sets and the split feasibility problem*, Inverse Problems **18** (2002), no. 2, 441–453, DOI 10.1088/0266-5611/18/2/310. MR1910248 (2003g:65059)

[3] A. Cegielski, *Iterative methods for fixed point problems in Hilbert spaces*, Lecture Notes in Mathematics, vol. 2057, Springer, Heidelberg, 2012. MR2931682

[4] A. Cegielski, *General Method for Solving the Split Common Fixed Point Problem*, J. Optim. Theory Appl. **165** (2015), no. 2, 385–404, DOI 10.1007/s10957-014-0662-z. MR3328201

[5] A. Cegielski and R. Zalas, *Methods for variational inequality problem over the intersection of fixed point sets of quasi-nonexpansive operators*, Numer. Funct. Anal. Optim. **34** (2013), no. 3, 255–283, DOI 10.1080/01630563.2012.716807. MR3011888

[6] A. Cegielski and R. Zalas, *Properties of a class of approximately shrinking operators and their applications*, Fixed Point Theory **15** (2014), no. 2, 399–426. MR3235660

[7] Y. Censor and T. Elfving, *A multiprojection algorithm using Bregman projections in a product space*, Numer. Algorithms **8** (1994), no. 2-4, 221–239, DOI 10.1007/BF02142692. MR1309222 (95h:90093)

[8] Y. Censor and A. Segal, *The split common fixed point problem for directed operators*, J. Convex Anal. **16** (2009), no. 2, 587–600. MR2559961 (2010i:47115)

[9] H. Cui and F. Wang, *Iterative methods for the split common fixed point problem in Hilbert spaces*, Fixed Point Theory Appl., posted on 2014, 2014:78, 8, DOI 10.1186/1687-1812-2014-78. MR3248626

[10] L. Landweber, *An iteration formula for Fredholm integral equations of the first kind*, Amer. J. Math. **73** (1951), 615–624. MR0043348 (13,247c)

[11] G. López, V. Martín-Márquez, F. Wang, and H.-K. Xu, *Solving the split feasibility problem without prior knowledge of matrix norms*, Inverse Problems **28** (2012), no. 8, 085004, 18, DOI 10.1088/0266-5611/28/8/085004. MR2948743

[12] A. Moudafi, *The split common fixed-point problem for demicontractive mappings*, Inverse Problems **26** (2010), no. 5, 055007, 6, DOI 10.1088/0266-5611/26/5/055007. MR2647149 (2011f:47097)

[13] Z. Opial, *Weak convergence of the sequence of successive approximations for nonexpansive mappings*, Bull. Amer. Math. Soc. **73** (1967), 591–597. MR0211301 (35 #2183)

[14] F. Wang and H.-K. Xu, *Cyclic algorithms for split feasibility problems in Hilbert spaces*, Nonlinear Anal. **74** (2011), no. 12, 4105–4111, DOI 10.1016/j.na.2011.03.044. MR2802990

[15] H.-K. Xu, *A variable Krasnosel'skiĭ-Mann algorithm and the multiple-set split feasibility problem*, Inverse Problems **22** (2006), no. 6, 2021–2034, DOI 10.1088/0266-5611/22/6/007. MR2277527 (2007j:47135)

[16] H.-K. Xu, *Iterative methods for the split feasibility problem in infinite-dimensional Hilbert spaces*, Inverse Problems **26** (2010), no. 10, 105018, 17, DOI 10.1088/0266-5611/26/10/105018. MR2719779 (2011i:65094)

FACULTY OF MATHEMATICS, COMPUTER SCIENCE AND ECONOMETRICS, UNIVERSITY OF ZIELONA GÓRA, UL. SZAFRANA 4A, 65-516 ZIELONA GÓRA, POLAND – AND – DEPARTMENT OF MATHEMATICS, KUWAIT UNIVERSITY, P.O. BOX 5969, SAFAT 13060, KUWAIT

E-mail address: a.cegielski@wmie.uz.zgora.pl

Recovering the conductances on grids: A theoretical justification

C. Araúz, Á. Carmona, A. M. Encinas, and M. Mitjana

ABSTRACT. In this work, we present an overview of the work developed by the authors in the context of inverse problems on finite networks and moreover, we display the steps needed to recover the conductances in a 3–dimensional grid. This study performs an extension of the pioneer studies by E. B. Curtis and J. A. Morrow, and sets the theoretical basis for solving inverse problems on networks. We present just a glance of what we call overdetermined partial boundary value problems, in which any data are not prescribed on a part of the boundary, whereas in another part of the boundary both the values of the function and of its normal derivative are given. The resolvent kernels associated with these problems are described and they are the fundamental tool to perform an algorithm for the recovery of the conductance of a 3–dimensional grid. We strongly believe that the columns of the overdetermined partial Poisson kernel are the discrete counterpart of the so–called CGO solutions (complex geometrical optic solutions) that, in their turn, are the key to solve inverse continuous problems on planar domains.

1. Introduction

The first applications of inverse boundary value problems are found in geophysical electrical prospection and electrical impedance tomography. The objective in physical electrical prospection is to deduce internal terrain properties from surface electrical measurements, which is of great interest in the engineering field. Electrical impedance tomography is a medical imaging technique whose aim is to obtain visual information of the body densities from some electrodes placed on the skin of the patient.

The corresponding mathematical problem is whether it is possible to determine the conductivity of a body from boundary measurements and global equilibrium conditions. That is, an *inverse boundary value problem* consists in the recovery of the internal structure or conductivity information of a body using only external data and general conditions on the body. In general, inverse problems are exponentially ill–posed, since they are highly sensitive to changes in the boundary data. For this problem is poorly arranged, at times the target is only a partial reconstruction of

---

2010 *Mathematics Subject Classification.* Primary 35R30; Secondary 34B45.

Key words and phrases. Inverse problem, Schrödinger operator, partial overdetermined boundary value problems, Schur complement.

Supported by the Spanish Research Council under projects MTM2014-60450-R and MTM2011-28800-C02-02.

the conductivity data or the addition of morphological conditions so as to perform a full internal recovery. However, in this work we deal with a situation where the recovery of the conductance is feasible: grid networks.

De Verdière *et alt.* [**D1, D2**] and Curtis *et alt.* [**C1, C2**] studied the inverse problem for circular planar electrical networks when the Dirichlet–to–Neumann map associated with the combinatorial Laplacian is supposed to be known. Specifically, they considered networks embedded in a disk without crossings and with boundary vertices located on the boundary of the disk and proved that the edge conductances of a critical circular planar electrical network can be recovered uniquely from the response matrix; that is, the matrix associated with the Dirichlet–to–Neumann map. Moreover, they proved that the response matrices realizable by circular planar networks are all singular diagonally dominant M–matrices having all circular minors nonnegative.

Some of the present authors have developed an adequate framework to the study of discrete inverse problems through a series of three papers [**A1, A2, A3**]. These works represent an extension of the ones above mentioned and define a global setting for the study of discrete inverse problems. In [**A2**], we established the theoretical foundations for the study of overdetermined partial boundary value problems associated with a Schrödinger operator on a network, which constitute the appropriate framework for the recovery of the conductances of a network. In it, we gave a necessary and sufficient condition for the existence and uniqueness of solution, as well as a recovery algorithm for the conductances of spider networks when the data is the Dirichlet–to–Robin map associated with a Schrödinger operator.

The data for an inverse problem on a network is the Dirichlet–to–Robin map since it contains the boundary information. Therefore, we raised in [**A1**] the analysis of the properties of the Dirichlet–to–Robin map, proving in particular the alternating property for general networks. The consideration of Schrödinger operators has allowed us to consider as response matrices a wide class of matrices, not necessarily singular nor weakly diagonally dominant. Therefore, our results represent a generalization of those obtained in [**C2, C3**]. Once we have characterized those matrices that are the response matrices of certain networks, we raise the problem of constructing an algorithm to recover the conductances. With this end, we provided a new necessary and sufficient condition for the existence and uniqueness of solution of overdetermined partial boundary value problems and described its resolvent kernels; named *overdetermined partial Green, Poisson and Robin kernels*, see [**A3**].

In the present study we want to reflect the power of the mentioned works in an unified approach. For this, we summarize the main results of the mentioned papers and we present some new results in the setting of the recovery of the conductances. The first one establishes a boundary spike formula that generalizes the one obtained in [**C2**], since we consider arbitrary networks and we work with positive semi–definite Schrödinger operators. Then, in Section 4 we present a new algorithm for the recovery of the conductances in a 3–dimensional grid with arbitrary conductances when the data is the Dirichlet–to–Robin map. This algorithm makes use of the mentioned boundary spike formula and of the characteristics relative to the zero zone of the overdetermined partial Poisson kernel. A draw of the algorithm for the case of 2–dimensional grids was given in [**A4**]. In an unpublished paper by

Oberlin [**O**], we can also find a sketch of an algorithm for 2–dimensional grids where the hypothesis of diagonal dominance of the response matrix is required.

2. Preliminaires

Let $\Gamma = (V, c)$ be a finite network; that is, a finite connected graph without loops nor multiple edges, with vertex set V. If E denotes the set of edges of the network, each $(x, y) \in E$ has been assigned a *conductance* $c(x, y)$, where $c : V \times V \longrightarrow [0, +\infty)$. Moreover, $c(x, y) = c(y, x)$ and $c(x, y) = 0$ if $(x, y) \notin E$. Then, $x, y \in V$ are *adjacent*, $x \sim y$, iff $c(x, y) > 0$. We denote by $N_r(x)$, the set of vertices of V that are at distance less than or equal to r from x.

The set of real value functions on a subset $F \subseteq V$, denoted by $\mathcal{C}(F)$, and the set of non–negative functions on F, $\mathcal{C}^+(F)$, are naturally identified with $\mathbb{R}^{|F|}$ and the nonnegative cone of $\mathbb{R}^{|F|}$, respectively. Moreover, if F is a non empty subset of V, its characteristic function is denoted by $\mathbf{1}_F$. When $F = \{x\}$, its characteristic function is denoted by ε_x. If $u \in \mathcal{C}(V)$, we define the *support* of u as $\mathsf{supp}(u) = \{x \in V : u(x) \neq 0\}$. Clearly, $\mathcal{C}(F)$ is identified with the set of functions in $\mathcal{C}(V)$ that vanish outside F.

If we consider a proper subset $F \subset V$, then its *boundary* $\delta(F)$ is given by the vertices of $V \setminus F$ that are adjacent to at least one vertex of F. The vertices of $\delta(F)$ are called *boundary vertices* and when a boundary vertex $x \in \delta(F)$ has a unique neighbour in F we call the edge joining them a *boundary spike*. It is easy to prove that $\bar{F} = F \cup \delta(F)$ is connected when F is. Of course networks do not have boundaries by themselves, but starting from a network we can define a *network with boundary* as $\Gamma = (\bar{F}, c_F)$ where F is a proper subset and $c_F = c \cdot \mathbf{1}_{(\bar{F} \times \bar{F}) \setminus (\delta(F) \times \delta(F))}$. From now on we will work with networks with boundary. Moreover, for the sake of simplicity we denote $c = c_F$. In addition, we assume that $|F| = n$ and $|\delta(F)| = m$.

On $\mathcal{C}(\bar{F})$ we consider the standard inner product defined as $\langle u, v \rangle = \sum_{x \in \bar{F}} u(x) v(x)$. Any function $\omega \in \mathcal{C}^+(\bar{F})$ such that $\mathsf{supp}(\omega) = \bar{F}$ and $\sum_{x \in \bar{F}} \omega^2(x) = 1$ is called *weight* on $\bar{F}$. The set of weights is denoted by $\Omega(\bar{F})$. We call *(generalized) degree* of $x \in V$, the value $\kappa(x) = \sum_{y \in V} c(x, y)$.

Clearly, functions and operators can be identified with vectors and matrices, after giving a label on the vertex set. Along the paper we use the convention that operators and their associated matrices, and functions and their associated vectors, are denoted with the same letter, operators in calligraphic font and matrices and vectors in sans serif font. In general, given a matrix M and A, B sets of vertices, M$(A; B)$ denotes the matrix obtained from M with rows indexed by the vertices of A and columns indexed by the vertices of B. Also, given a vector v and a set of vertices A, v(A) denotes the entries of v indexed by the vertices of A.

The *combinatorial Laplacian* of Γ is the linear operator $\mathcal{L} : \mathcal{C}(\bar{F}) \longrightarrow \mathcal{C}(\bar{F})$ that assigns to each $u \in \mathcal{C}(\bar{F})$ the function defined for all $x \in \bar{F}$ as

$$\mathcal{L}(u)(x) = \sum_{y \in \bar{F}} c(x, y) \big(u(x) - u(y) \big).$$

Given $q \in \mathcal{C}(\bar{F})$, the *Schrödinger operator* on Γ with *potential* q is the linear operator $\mathcal{L}_q : \mathcal{C}(\bar{F}) \longrightarrow \mathcal{C}(\bar{F})$ that assigns to each $u \in \mathcal{C}(\bar{F})$ the function $\mathcal{L}_q(u) = \mathcal{L}(u) + qu$. It is well–known that any Schrödinger operator is self–adjoint. Moreover, if L denotes the matrix associated with $\mathcal{L}_q$, then L is an irreducible and symmetric M–matrix.

We define the *normal derivative of* $u \in \mathcal{C}(\bar{F})$ *on* F as the function in $\mathcal{C}(\delta(F))$ given by

$$\left(\frac{\partial u}{\partial \mathsf{n}_F}\right)(x) = \sum_{y \in F} c(x,y)\bigl(u(x) - u(y)\bigr), \quad \text{for any } x \in \delta(F).$$

For any weight $\sigma \in \Omega(\bar{F})$, the so–called *potential associated with* σ is the function in $\mathcal{C}(\bar{F})$ defined as $q_\sigma = -\sigma^{-1}\mathcal{L}(\sigma)$ on F, $q_\sigma = -\sigma^{-1}\dfrac{\partial \sigma}{\partial \mathsf{n}_F}$ on $\delta(F)$. It is worth to note that the definition of q_σ is a discrete analogous of the Liouville transform, see [**S**]. In [**A1**], we proved that the energy is positive semi–definite on $\mathcal{C}(\bar{F})$ if there exist $\lambda \geq 0$ and $\sigma \in \Omega(\bar{F})$ such that $q = q_\sigma + \lambda\chi_{\delta(F)}$. In this case, it is positive definite iff $\lambda > 0$. So, through this paper, we will suppose that the above condition $q = q_\sigma + \lambda\chi_{\delta(F)}$ holds with $\sigma \in \Omega(\bar{F})$ and $\lambda \geq 0$. Therefore, for any $f \in \mathcal{C}(F)$ and $g \in \mathcal{C}(\delta(F))$ the following Dirichlet problem

$$\mathcal{L}_q(u) = f \quad \text{on} \quad F \quad \text{and} \quad u = g \quad \text{on} \quad \delta(F),$$

has a unique solution. The existence and uniqueness of solution implies that the operator $\mathcal{L}_q$ is invertible on F and its inverse is called the *Green operator* for F and it is denoted by $\mathcal{G}_q$. Observe that $\mathcal{G}_q$ is a self–adjoint operator whose associated matrix will be denoted by G.

When $f = 0$, we denote the unique solution of the corresponding Dirichlet problem as u_g. Then, the map $\Lambda_q \colon \mathcal{C}(\delta(F)) \longrightarrow \mathcal{C}(\delta(F))$ that assigns to any function $g \in \mathcal{C}(\delta(F))$ the function $\Lambda_q(g) = \dfrac{\partial u_g}{\partial \mathsf{n}_F} + qg$ is called *Dirichlet–to–Robin map*.

In [**A1**], the authors proved that the Dirichlet–to–Robin map is a self–adjoint, positive semi–definite operator. Moreover, λ is the lowest eigenvalue of Λ_q and its associated eigenfunctions are multiple of σ. In addition, the matrix, Λ, associated with Λ_q is an irreducible and symmetric M–matrix, usually called the *response matrix* of the network and it is the Schur complement of $\mathsf{L}(F;F)$ in L (see [**H**]); that is,

$$\Lambda = \mathsf{L}/\mathsf{L}(F;F) = \mathsf{D}(\delta(F);\delta(F)) - \mathsf{C}(\delta(F);F) \cdot \mathsf{G} \cdot \mathsf{C}(\delta(F);F)^\top,$$

where D is the diagonal matrix whose diagonal entries are given by $\kappa + q$ and $\mathsf{C} = \bigl(c(x,y)\bigr)_{x,y \in \bar{F}}$. This map is the response matrix of a Schrödinger type matrix, and therefore it can be assumed to be known, since it provides boundary reactions to boundary actions, the type of data that we can measure.

Moreover, we proved that the edge conductances of a critical circular planar electrical network can be recovered from the Dirichlet–to–Robin map associated with a positive semi–definite Schrödinger operator. In addition, we proved that the response matrices realizable by circular planar networks are all M-matrices having all circular minors nonnegative. Therefore, these results are a generalization of those proved in [**D2, C2**].

Unlike the planar network case, very few results have been obtained for general networks. In [**L**], the authors carried out a study of the inverse Dirichlet–to–Neumann problem for networks embedded in a cylinder. For the class of purely cylindrical networks, they obtained a characterization of the response matrices and proved that the network can be recovered from the Dirichlet–to–Neumann map.

3. Overdetermined Partial boundary value problems

In this section we summarize some of the results obtained in [**A2**, **A3**] that allows us to perform the algorithm for the recovery of the conductances in a grid. We fix a proper and connected subset $F \subset V$ and $A, B \subset \delta(F)$ non–empty subsets such that $A \cap B = \emptyset$. Moreover we denote by R the set $R = \delta(F) \setminus (A \cup B)$, so $\delta(F) = A \cup B \cup R$ is a partition of $\delta(F)$. We remark that R can be an empty set. We consider a new type of non self–adjoint boundary value problems in which the values of the functions and their normal derivatives are known at the same part of the boundary, which represents an overdetermined problem, and there exists another part of the boundary where no data is known.

For any $f \in \mathcal{C}(F)$, $g \in \mathcal{C}(A \cup R)$ and $h \in \mathcal{C}(A)$, the *overdetermined partial Dirichlet–Neumann boundary value problem* on F with data f, g, h consists in finding a function $u \in \mathcal{C}(\bar{F})$ such that

$$(3.1) \qquad \mathcal{L}_q(u) = f \text{ on } F, \qquad \frac{\partial u}{\partial n_F} = h \text{ on } A \quad \text{and} \quad u = g \text{ on } A \cup R.$$

In [**A2**] the authors proved the existence and uniqueness of solution of this problem for any data $f \in \mathcal{C}(F)$, $g \in \mathcal{C}(A \cup R)$, $h \in \mathcal{C}(A)$ iff $|A| = |B|$ and $\Lambda(A; B)$ is invertible, or equivalently iff $|A| = |B|$ and $\mathsf{L}(A \cup F; F \cup B)$ is invertible, see [**A3**].

From now on, we will suppose that $\mathsf{L}(A \cup F; F \cup B)$ is invertible. In this case, the following result holds.

PROPOSITION 3.1. [**A3**, Corollary 2.10] *If $u \in \mathcal{C}(\bar{F})$ is the unique solution of the overdetermined partial boundary value problem (3.1), then*

$$\mathsf{u}(B) = \Lambda(A; B)^{-1} \cdot \Big(\mathsf{C}(A; F) \cdot \mathsf{G} \cdot \mathsf{f} - \Lambda(A; A \cup R) \cdot \mathsf{g} + \mathsf{h}\Big),$$
$$\mathsf{u}(F) = \mathsf{G} \cdot \Big(\mathsf{f} + \mathsf{C}(F; B) \cdot \mathsf{u}_B + \mathsf{C}(F; A \cup R) \cdot \mathsf{g}\Big)$$

and, clearly, $\mathsf{u}(A \cup R) = \mathsf{g}$.

We call *overdetermined partial Green operator*, $\widetilde{\mathcal{G}}_q : \mathcal{C}(F) \longrightarrow \mathcal{C}(F \cup B)$, the operator that assigns to any $f \in \mathcal{C}(F)$ the unique solution of problem

$$(3.2) \qquad \mathcal{L}_q(u) = f \text{ on } F, \qquad \frac{\partial u}{\partial n_F} = 0 \text{ on } A \quad \text{and} \quad u = 0 \text{ on } A \cup R.$$

We call *overdetermined partial Poisson operator*, $\widetilde{\mathcal{P}}_q : \mathcal{C}(A \cup R) \longrightarrow \mathcal{C}(\bar{F})$, the operator that assigns to any $g \in \mathcal{C}(A \cup R)$ the unique solution of problem

$$(3.3) \qquad \mathcal{L}_q(u) = 0 \text{ on } F, \qquad \frac{\partial u}{\partial n_F} = 0 \text{ on } A \quad \text{and} \quad u = g \text{ on } A \cup R.$$

The matrix associated with the overdetermined partial Poisson operator plays an important role in the following section, it will be denoted by $\widetilde{\mathsf{P}}$. In fact, the functions $\widetilde{\mathcal{P}}_q(\varepsilon_x) = \widetilde{\mathsf{P}}(\cdot, x)$ for each $x \in A \cup R$ are the key tool for the recovery algorithm and hence they can be seen as the discrete counterpart of the so-called *complex geometrical optic solutions*, see [**U**]. In [**A3**], we have obtained explicit expressions for the overdetermined partial resolvent kernels for a generalized cylinder.

On the other hand, a consequence of the invertibility of $\mathsf{L}(A \cup F; F \cup B)$ is the following extension of the boundary spike formula obtained in [**C2**] for circular planar networks and for the combinatorial Laplacian. This so–called extension covers all networks, not only circular planar ones, and all Schrödinger operators, not only Laplacian operators. Moreover, this formula is a powerful tool for recovery algorithms since when a certain diagonal value of the overdetermined partial Green matrix $\widetilde{\mathsf{G}}$ is null it allows to recover the conductance of a certain boundary spike.

PROPOSITION 3.2 (Boundary spike formula). *If $x \in R$ has a unique neighbour $y \in F$, then*

$$\Lambda(x;x) - \Lambda(x;B) \cdot \Lambda(A;B)^{-1} \cdot \Lambda(A;x) = \lambda + \frac{\omega(y)}{\omega(x)} c(x,y) - \widetilde{\mathsf{G}}(y,y) c(x,y)^2.$$

Proof. As the matrices $\mathsf{L}(F;F)$ and $\mathsf{L}(A \cup F; F \cup B)$ are invertible, applying the properties of the Schur complement we get that

$$\mathsf{L}(x;x) - \mathsf{L}(x; F \cup B) \cdot \mathsf{L}(A \cup F; F \cup B)^{-1} \cdot \mathsf{L}(A \cup F; x)$$

$$= \mathsf{L}(\{x\} \cup A \cup F; \{x\} \cup F \cup B) \big/ \mathsf{L}(A \cup F; F \cup B)$$

$$= \mathsf{L}(\{x\} \cup A \cup F; \{x\} \cup F \cup B) \big/ \mathsf{L}(F;F) \, \Big/ \, \mathsf{L}(A \cup F; F \cup B) \big/ \mathsf{L}(F;F).$$

The last equality can be written as

$$\mathsf{L}(x;x) - \mathsf{L}(x; F \cup B) \cdot \mathsf{L}(A \cup F; F \cup B)^{-1} \cdot \mathsf{L}(A \cup F; x)$$

(3.4)
$$= \Lambda(\{x\} \cup A; \{x\} \cup B) \big/ \Lambda(A;B)$$

$$= \Lambda(x;x) - \Lambda(x;B) \cdot \Lambda(A;B)^{-1} \cdot \Lambda(A;x).$$

On the other hand, it is clear that

(3.5) $\quad \mathsf{L}(x,x) = \mathcal{L}_q(\varepsilon_x)(x) = c(x,y) - \dfrac{1}{\omega(x)} \dfrac{\partial \omega}{\partial \mathsf{n}_F}(x) + \lambda = \dfrac{\omega(y)}{\omega(x)} c(x,y) + \lambda$

and

$$\mathsf{L}(x; B \cup F) \cdot \mathsf{L}(A \cup F; B \cup F)^{-1} \cdot \mathsf{L}(A \cup F; x)$$

(3.6)
$$= \mathsf{C}(x;y) \cdot \left[\mathsf{L}(A \cup F; B \cup F) \right]^{-1}(y;y) \cdot \mathsf{C}(y;x)$$

$$= c(x,y)^2 \left[\mathsf{L}(A \cup F; B \cup F) \right]^{-1}(y;y)$$

because the unique neighbour of x is y. The result follows when we properly join Equations (3.4), (3.5), (3.6) and using the equality

$$\left[\mathsf{L}(A \cup F; B \cup F) \right]^{-1}(y;y) = \widetilde{\mathsf{G}}(y,y)$$

proved in [**A3**, Proposition 2.12]. □

4. Recovering the conductance on grids

Our purpose is to use the results of the previous section in order to recover all the conductances on 3–dimensional grids. To the best of our knowledge, this is the first work that shows a recovery algorithm for non–planar networks. First, we define this family of networks for $n = 3$ and then we prove all the steps for the recovery algorithm. The 2–dimensional case was presented in the IX JMDA conference held in Tarragona, Spain, and whose extended abstract, without proofs, can be found in [**A4**]. A similar algorithm for spider networks was presented in [**A1**].

A 3–dimensional grid is the discretization of any cuboid in $\mathbb{R}^3$. Let us take three integers $\ell_i \in \mathbb{N}$, $i = 1, 2, 3$. We define the *three dimensional grid with boundary* as the network $\Gamma = (V, c)$ with vertex set

$$V = \{x_{ijk} : i = 0, \ldots, \ell_1 + 1, j = 0, \ldots, \ell_2 + 1, k = 0, \ldots, \ell_3 + 1\}$$

and conductivity function c given for all $i = 1, \ldots, \ell_1$, $j = 1, \ldots, \ell_2$ and $k = 1, \ldots, \ell_3$, by

$$c(x_{ijk}, x_{rst}) > 0 \text{ when } \begin{cases} r = i \pm 1, & s = j \text{ and } t = k, \\ r = i, & s = j \pm 1 \text{ and } t = k, \\ r = i, & s = j \text{ and } t = k \pm 1, \end{cases}$$

$c(x_{ij0}, x_{ij1}) > 0$, $c(x_{ij\ell_3}, x_{ij\ell_3+1}) > 0$, $c(x_{0jk}, x_{1jk}) > 0$, $c(x_{\ell_1 jk}, x_{\ell_1+1 jk}) > 0$, $c(x_{i0k}, x_{i1k}) > 0$ and $c(x_{i\ell_2 k}, x_{i\ell_2+1 k}) > 0$ and $c(x, y) = 0$ otherwise.

We define for $j = 0, \ldots, \ell_2 + 1$, the following sets of vertices

$$A_j = \{x_{ijk} : i = 1, \ldots, \ell_1, k = 1, \ldots, \ell_3\},$$

and for $j = 1, \ldots, \ell_2$, the sets

$$R_j = \{x_{ijk} : i = 0, \ell_1 + 1, k = 1, \ldots, \ell_3\} \cup \{x_{ijk} : i = 1, \ldots, \ell_1, k = 0, \ell_3 + 1\}.$$

If we consider the set $F = \bigcup_{j=1}^{\ell_2} A_j$, then $\delta(F) = A \cup R \cup B$, where

$$A = A_0, \quad R = \bigcup_{i=1}^{\ell_2} R_i \text{ and } B = A_{\ell_2+1}.$$

Figure 1 displays a three dimensional grid with $\ell_1 = \ell_3 = 2$ and $\ell_2 = 3$, and the boundary sets A, B and $R_1, \ldots, R_{\ell_2}$.

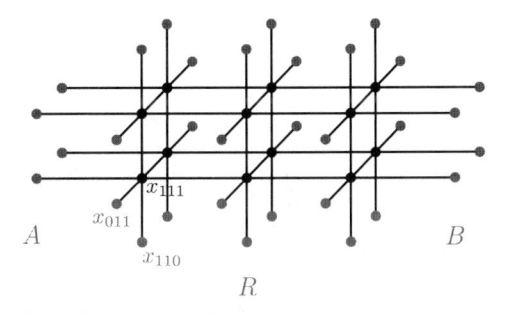

FIGURE 1. An example of 3 dimensional grid with $\ell_1 = \ell_3 = 2$ and $\ell_2 = 3$.

For fixed $\sigma \in \Omega(\bar{F})$ and $\lambda \geq 0$, such that $q = q_\sigma + \lambda\chi_{\delta(F)}$, it is satisfied that

$$L(A \cup F; F \cup B) = \begin{pmatrix} -\mathsf{C}_{01} & 0 & 0 & \cdots & 0 & 0 & 0 \\ \mathsf{Q}_1 & -\mathsf{C}_{12} & 0 & \cdots & 0 & 0 & 0 \\ -\mathsf{C}_{12}^\top & \mathsf{Q}_2 & -\mathsf{C}_{23} & \cdots & 0 & 0 & 0 \\ \vdots & \ddots & \ddots & \ddots & \vdots & \vdots & \vdots \\ 0 & \cdots & \cdots & \cdots & -\mathsf{C}_{\ell_1-2\ell_1-1} & 0 & 0 \\ 0 & \cdots & \cdots & \cdots & \mathsf{Q}_{\ell-1} & -\mathsf{C}_{\ell_1-1\ell_1} & 0 \\ 0 & \cdots & \cdots & \cdots & -\mathsf{C}_{\ell_1-1\ell_1}^\top & \mathsf{Q}_{\ell_1} & -\mathsf{C}_{\ell_1\ell_1+1} \end{pmatrix}$$

where $\mathsf{C}_{ii+1} = \mathsf{C}(A_i; A_{i+1})$ for any $i = 0, \ldots, \ell_1$ and $\mathsf{Q}_i = \mathsf{D}(A_i; A_i) - \mathsf{C}(A_i; A_i)$. We recall that D is the diagonal matrix whose diagonal entries are given by $\kappa + q$.

PROPOSITION 4.1. *For any $f \in \mathcal{C}(F)$, $g \in \mathcal{C}(A \cup R)$ and $h \in \mathcal{C}(A)$, the overdetermined partial Dirichlet–Neumann boundary value problem on F with data f, g, h on a 3-dimensional grid*

$$\mathcal{L}_q(u) = f \quad \text{on} \quad F, \quad \frac{\partial u}{\partial \mathsf{n}_\mathsf{F}} = h \quad \text{on} \quad A \quad \text{and} \quad u = g \quad \text{on} \quad A \cup R,$$

has a unique solution.

Proof. Note that the matrix $L(A \cup F; F \cup B)$ is invertible, since C_{ii+1} is a diagonal matrix whose diagonal entries are positive. Then, the result follows by applying Theorem 2.1 in [**A3**]. □

In the following results we analyze some properties of the overdetermined partial Green and Poisson kernels for a grid that will be useful for the recovery algorithm. Notice that a pole $y \in R \cup A$, is of the form $y = x_{ijk}$ for $i = 1, \ldots, \ell_1$, $j = 1, \ldots, \ell_2$ and $k = 0, \ell_3+1$ when $y \in R_j$ or $y = x_{i0k}$ for $i = 1, \ldots, \ell_1$, $k = 1, \ldots, \ell_3$ when $y \in A$. So, from now on we denote by $u_{ijk} = \widetilde{\mathsf{P}}(\cdot, x_{ijk})$.

In order to prove the properties of the overdetermined partial resolvent kernels, we will need the following result that was proved in [**A3**].

LEMMA 4.1. [**A3**, Corollary 2.11] *It is satisfied that*

$$\widetilde{\mathsf{G}}(A_1; F) = 0, \quad \widetilde{\mathsf{P}}(A_1; R) = 0 \quad \text{and} \quad \widetilde{\mathsf{P}}(A_1; A) = \mathsf{I}.$$

In the following result we analyze the zero–set of the function u_{ijk} as well as the sign pattern along certain paths starting from $A \cup R$. This behavior is in accordance with the alternating property proved in [**A1**].

PROPOSITION 4.2. *The following properties hold:*

(i) *If $k = 0, \ell_3 + 1$, then for any $i = 1, \ldots, \ell_1$, $j = 1, \ldots, \ell_2$,*

$$u_{ijk}(x) = 0 \quad \text{if} \quad x \in \bigcup_{s=0}^{j} A_s \cup \bigcup_{r=j+1}^{\ell_2+1} (A_r \setminus N_{r-j}(x_{irk})).$$

Moreover, for any $h = 0, \ldots, \ell_2 - j$, it is satisfied that

$$c(x_{ij+h|k-h-1|}, x_{ij+h+1|k-h-1|})u_{ijk}(x_{ij+h+1|k-h-1|})$$
$$= -c(x_{ij+h|k-h-1|}, x_{ij+h|k-h|})u_{ijk}(x_{ij+h|k-h|}),$$

which implies $(-1)^h u_{ijk}(x_{ij+h|k-h|}) > 0$.

(ii) If $j = 0$, then for $i = 1, \ldots, \ell_1$, $k = 1, \ldots, \ell_3$

$$u_{i0k}(x) = 0 \quad \text{if} \quad x \in \bigcup_{r=1}^{\ell_2+1} \left(A_r \setminus N_{r-1}(x_{irk})\right)$$

and $u_{i0k}(x_{i1k}) = 1$. Moreover, for any $h = 1, \ldots, \ell_3 - 1$, it is satisfied

$$c(x_{ihh+1}, x_{ih+1h+1})u_{i01}(x_{ih+1h+1}) = -c(x_{ihh+1}, x_{ihh})u_{i01}(x_{ihh}),$$

which implies $(-1)^h u_{i01}(x_{ih+1h+1}) > 0$.

Proof. (i) Since $x_{ijk} \in R_j$, then

$$\mathcal{L}_q(u_{ijk}) = 0 \quad \text{on} \quad F, \quad u_{ijk} = \varepsilon_{x_{ijk}} \quad \text{on} \quad A \cup R \quad \text{and} \quad \frac{\partial u_{ijk}}{\partial \mathsf{n}_F} = 0 \quad \text{on} \quad A.$$

The proof is by induction on s. The case $s = 0$, is trivial from the definition of u_{ijk} and the case $s = 1$, follows from Lemma 4.1. Suppose now that the result is true for any $s \leq j - 1$. Let $x_{ms+1t} \in A_{s+1}$ and then, by hypothesis of induction,

$$0 = \mathcal{L}_q(u_{ijk})(x_{mst}) = -c(x_{mst}, x_{ms+1t})u_{ijk}(x_{ms+1t}),$$

and hence $u_{ijk} = 0$ on A_{s+1}.

Suppose now that $s = j + 1$, then for any $x_{mj+1t} \in A_s \setminus N_1(x_{ij+1k})$,

$$0 = \mathcal{L}_q(u_{ijk})(x_{mjt}) = -c(x_{mjt}, x_{mj+1t})u_{ijk}(x_{mj+1t}),$$

and hence $u_{ijk}(x_{mj+1t}) = 0$. Suppose now that the result is true for any $r < \ell_2-j+1$ and let $r+1$, then for any $x_{mj+r+1t} \in A_{j+r+1} \setminus N_{r+1}(x_{ij+r+1k})$

$$0 = \mathcal{L}_q(u_{ijk})(x_{mj+rt}) = -c(x_{mj+rt}, x_{mj+r+1t})u_{ijk}(x_{mj+r+1t})$$

and hence $u_{ijk}(x_{mj+r+1t}) = 0$.

Moreover, for any $h \in \{0, \ldots, \ell_2 - j\}$, it holds that

$$u_{ijk}(x_{ij+h|k-h-1|}) = u_{ijk}(x_{ij+h-1|k-h-1|}) = u_{ijk}(x_{i-1j+h|k-h-1|})$$
$$= u_{ijk}(x_{i+1j+h|k-h-1|}) = u_{ijk}(x_{ij+h|k-h-2|}) = 0,$$

and hence

$$c(x_{ij+h|k-h-1|}, x_{ij+h+1|k-h-1|})u_{ijk}(x_{ij+h+1|k-h-1|})$$
$$= -c(x_{ij+h|k-h-1|}, x_{ij+h|k-h|})u_{ijk}(x_{ij+h|k-h|})$$

since $\mathcal{L}_q(u_{ijk})(x_{ij+h|k-h-1|}) = 0$.

As $u_{ijk}(x_{ijk}) = 1$, the above identity implies that $(-1)^h u_{ijk}(x_{ij+h|k-h|}) > 0$.

(ii) Suppose now that $j = 0$. Then,

$$\mathcal{L}_q(u_{i0k}) = 0 \quad \text{on} \quad F, \quad u_{i0k} = \varepsilon_{x_{i0k}} \quad \text{on} \quad A \cup R \quad \text{and} \quad \frac{\partial u_{i0k}}{\partial \mathsf{n}_F} = 0 \quad \text{on} \quad A.$$

The proof is by induction on r. Suppose that $r = 1$, then for any $x_{m1t} \in A_1 \setminus \{x_{i1k}\}$

$$0 = \frac{\partial u_{i0k}}{\partial \mathsf{n}_F}(x_{m0t}) = -c(x_{m0t}, x_{m1t})u_{i0k}(x_{m1t}).$$

Therefore, $u_{i0k} = 0$ on $A_1 \setminus \{x_{i1k}\}$. Moreover,

$$0 = \frac{\partial u_{i0k}}{\partial \mathsf{n}_F}(x_{i0k}) = c(x_{i0k}, x_{i1k})\bigl(1 - u_{i0k}(x_{i1k})\bigr)$$

and hence $u_{i0k}(x_{i1k}) = 1$. The rest of the proof is analogue to the case $x_{ijk} \in R_j$.

Moreover, for any $h = 1, \ldots, \ell_3 - 1$, it holds that

$$u_{i01}(x_{ihh+1}) = u_{i01}(x_{ihh+2}) = u_{i01}(x_{ih-1h+1}) = u_{i01}(x_{i+1hh+1}) = u_{i01}(x_{i-1hh+1}) = 0,$$

and hence

$$c(x_{ihh+1}, x_{ih+1h+1})u_{i01}(x_{ih+1h+1}) = -c(x_{ihh+1}, x_{ihh})u_{i01}(x_{ihh})$$

since $\mathcal{L}_q(u_{i01})(x_{ihh+1}) = 0$.

As $u_{i01}(x_{i11}) = 1$, the above identity implies that $(-1)^h u_{i01}(x_{ih+1h+1}) > 0$. □

PROPOSITION 4.3. *For a 3–dimensional grid it is satisfied that*

$$\widetilde{\mathsf{G}}(A_i, A_j) = 0, \text{ for all } j = 1, \ldots, \ell_2 \text{ and } 1 \le i \le j.$$

Proof. The proof is analogous to the one of Proposition 4.2 but considering the overdetermined partial boundary value problem

$$\mathcal{L}_q(v) = \varepsilon_{x_{ijk}} \text{ on } F, \quad v = 0 \text{ on } A \cup R \quad \text{and} \quad \frac{\partial v}{\partial n_F} = 0 \text{ on } A$$

instead; that is, $v = \widetilde{\mathcal{G}}_q(\varepsilon_{x_{ijk}})$ on $\bar{F}$. □

4.1. Recovery algorithm. Without loss of generality we assume that $\ell_3 \le \ell_2$ and we consider the *partial layers* of vertices

$$D_0 = \{x_{ij0} \in \bar{F} : i = 1, \ldots, \ell_1, j = 1, \ldots, \ell_2\},$$
$$D_k = \{x_{ijk} \in \bar{F} : i = 1, \ldots, \ell_1, \ j = k, \ldots, \ell_2\}, \text{ for } k = 1, \ldots, \ell_3 + 1.$$

In particular, $D_{\ell_3+1}, D_0 \subset R$. Note that $D_{\ell_3+1} = \emptyset$ when $\ell_3 = \ell_2$.

The recovery of conductances on a 3–dimensional grid from its response matrix is an iterative process, for we are not able to give explicit formulae for all the conductances at the same time but we can give a recovery algorithm instead. Hence, we describe the algorithm in steps, each of them requiring the information obtained in the previous one.

To start with, let Λ be an irreducible and symmetric M–matrix of order $2(\ell_1\ell_2 + \ell_1\ell_3 + \ell_2\ell_3)$ satisfying that is a response matrix and which is supposed to be known. Let $\lambda \ge 0$ be the lowest eigenvalue of Λ and $\sigma \in \Omega(\delta(F))$ the eigenvector associated with λ. In addition, we choose $\omega \in \Omega(\bar{F})$ such that $\omega = k\sigma$ on $\delta(F)$ with $0 < k < 1$.

Step 1. Let us fix the indices $\mathfrak{r} \in \{1, \ldots, \ell_1\}$ and $\mathfrak{s} \in \{1, \ldots, \ell_2\}$ and consider $u_{\mathfrak{rs}0}$ and $u_{\mathfrak{r}01}$. We already know that $u_{\mathfrak{rs}0} = 0$ on $A \cup (R \smallsetminus \{x_{\mathfrak{rs}0}\})$, $u_{\mathfrak{rs}0}(x_{\mathfrak{rs}0}) = 1$, $u_{\mathfrak{r}01} = 0$ on $R \cup (A \smallsetminus \{x_{\mathfrak{r}01}\})$ and $u_{\mathfrak{r}01}(x_{\mathfrak{r}01}) = 1$; see Figure 2 (A) and (B). Moreover, the values of $u_{\mathfrak{rs}0}$ and of $u_{\mathfrak{r}01}$ on B are given by

$$\mathsf{u}_{\mathfrak{rs}0}(B) = -\Lambda(A;B)^{-1} \cdot \Lambda(A;x_{\mathfrak{rs}0}) \quad \text{and} \quad \mathsf{u}_{\mathfrak{r}01}(B) = -\Lambda(A;B)^{-1} \cdot \Lambda(A;x_{\mathfrak{r}01})$$

because of Proposition 3.1.

Notice that this means that all the values of $u_{\mathfrak{rs}0}$ and of $u_{\mathfrak{r}01}$ on B are known, for the Dirichlet–to–Robin map is known. In Figure 2 (C) and (D) we show all the information obtained at the end of this step.

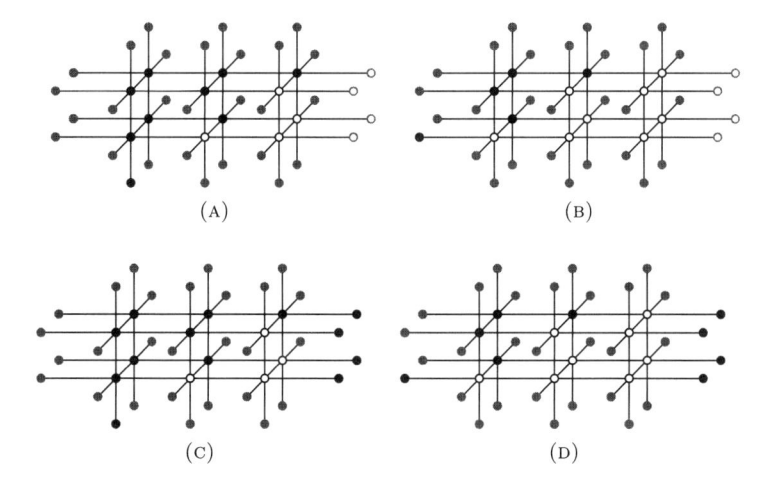

FIGURE 2. The bold red items are the poles and the ones known at the end of Step 1 for the case $\mathfrak{r} = \mathfrak{s} = 1$.

Step 2. From Proposition 4.3, $\widetilde{\mathsf{G}}(y;y) = 0$ for any $y \in F$ and hence applying the boundary spike formula given in Proposition 3.2 we can recover the conductance of the boundary edges.

COROLLARY 4.1. *The conductances of the edges joining the vertices of D_0 with the vertices of D_1 are given by*

$$c(x_{ij0}, x_{ij1}) = \frac{\omega(x_{ij0})}{\omega(x_{ij1})} \left(\Lambda(x_{ij0}; x_{ij0}) - \Lambda(x_{ij0}; B) \cdot \Lambda(A; B)^{-1} \cdot \Lambda(A; x_{ij0}) - \lambda \right)$$

for all $i = 1, \ldots, \ell_1$ and $j = 1, \ldots, \ell_2$.

By rotating the grid conveniently; that is, by considering another labelings of the vertices of the grid that fit the description at the begining of this section, we obtain the conductances of all the boundary edges of the grid. In Figure 3 we show all the information obtained at the end of this step.

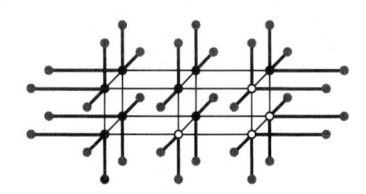

FIGURE 3. The bold red items are the ones known at the end of Step 2 for the case $\mathfrak{r} = \mathfrak{s} = 1$.

Step 3. In this step we recover all the values of $u_{\mathfrak{r}\mathfrak{s}0}$ and of $u_{\mathfrak{p}0\mathfrak{t}}$ on D_1.

LEMMA 4.2. *Given $\mathfrak{r} = 1, \ldots, \ell_1$ and $\mathfrak{s} = 1 \ldots, \ell_3$, the values of $u_{\mathfrak{r}\mathfrak{s}0}$ and of $u_{\mathfrak{r}01}$ on D_1 are given for all $i = 1, \ldots, \ell_1$ and $j = 1, \ldots, \ell_2$ by*

$$u_{\mathfrak{r}01}(x_{ij1}) = \frac{1}{c(x_{ij0}, x_{ij1})} \left(-\Lambda(x_{ij0}; x_{\mathfrak{r}01}) + \Lambda(x_{ij0}; B) \cdot \Lambda(A; B)^{-1} \cdot \Lambda(A; x_{\mathfrak{r}01}) \right)$$

whereas when $i \neq \mathfrak{r}$ or $j \neq \mathfrak{s}$

$$u_{\mathfrak{rs}0}(x_{ij1}) = \frac{1}{c(x_{ij0}, x_{ij1})}\Big(-\Lambda(x_{ij0}; x_{\mathfrak{rs}0}) + \Lambda(x_{ij0}; B) \cdot \Lambda(A; B)^{-1} \cdot \Lambda(A; x_{\mathfrak{rs}0})\Big)$$

and $u_{\mathfrak{rs}0}(x_{\mathfrak{rs}1}) = 0$.

Proof. Consider the function $u_{\mathfrak{rs}0}$. Observe that $u_{\mathfrak{rs}0}(x_{\mathfrak{rs}1}) = 0$ from Proposition 4.2. Suppose now that $i \neq \mathfrak{r}$ or $j \neq \mathfrak{s}$ then, we can rewrite Problem (3.3) as the Dirichlet problem

$$\mathcal{L}_q(u_{\mathfrak{rs}0}) = 0 \text{ on } F \quad \text{and} \quad u_{\mathfrak{rs}0} = \varepsilon_{x_{\mathfrak{rs}0}} + u_{\mathfrak{rs}0_{|B}} \text{ on } \delta(F)$$

with the additional condition $\dfrac{\partial u_{\mathfrak{rs}0}}{\partial \mathsf{n}_F} = 0$ on A. Therefore, by the definition of the Dirichlet–to–Robin map, for all $x_{ij0} \in \delta(F)$ it is satisfied that

$$\Lambda(x_{ij0}; x_{\mathfrak{rs}0}) + \Lambda(x_{ij0}; B) \cdot \mathsf{u}_{\mathfrak{rs}0}(B) = \Lambda_q\Big(\varepsilon_{x_{\mathfrak{rs}0}} + u_{\mathfrak{rs}0_{|B}}\Big)(x_{ij0})$$

$$= \frac{\partial u_{\mathfrak{rs}0}}{\partial \mathsf{n}_F}(x_{ij0}) + q(x_{ij0})u_{\mathfrak{rs}0}(x_{ij0})$$

$$= \Big(\lambda + \frac{\omega(x_{ij1})}{\omega(x_{ij0})}c(x_{ij0}, x_{ij1})\Big)u_{\mathfrak{rs}0}(x_{ij0}) - c(x_{ij0}, x_{ij1})u_{\mathfrak{rs}0}(x_{ij1})$$

$$= -c(x_{ij0}, x_{ij1})u_{\mathfrak{rs}0}(x_{ij1}).$$

Observe that all the terms of this equality, except the value $u_{\mathfrak{rs}0}(x_{ij1})$, are already known. Therefore, we get the result. We proceed analogously to obtain the values of $u_{\mathfrak{r}01}$. □

In Figure 4 (A) and (B) we show all the data gathered from the 3–dimensional grid at the end of this step.

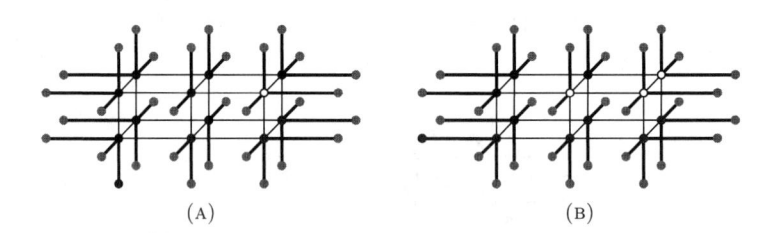

FIGURE 4. The bold red items are the ones known at the end of Step 3 for the case $\mathfrak{r} = \mathfrak{s} = 1$.

Step 4. Here we find the conductances of all the edges with both ends in D_1 and such that the second indices of their ends are different. However, we state a more general result which is a direct consequence of Proposition 4.2 (i) when $k = 0$.

PROPOSITION 4.4. *Let $h \in \{0, \ldots, \ell_3 - 1\}$. For every $\mathfrak{r} = 1, \ldots, \ell_1$ and $\mathfrak{s} = 1, \ldots, \ell_2$, let us suppose that we know the values of $u_{\mathfrak{rs}0}$ on D_h and D_{h+1}. Also, we suppose that the conductances of all the edges joining vertices from D_h and D_{h+1}*

are known. Then, for every $\mathfrak{r} = 1, \ldots, \ell_1$ and $\mathfrak{s} = 1, \ldots, \ell_2 - h - 1$ the conductances $c(x_{\mathfrak{rs}+hh+1}, x_{\mathfrak{rs}+h+1h+1})$ are given by

$$c(x_{\mathfrak{rs}+hh+1}, x_{\mathfrak{rs}+h+1h+1}) = -\frac{u_{\mathfrak{rs}0}(x_{\mathfrak{rs}+hh})}{u_{\mathfrak{rs}0}(x_{\mathfrak{rs}+h+1h+1})} c(x_{\mathfrak{rs}+h}, x_{\mathfrak{rs}+hh+1}).$$

When $h = 0$, Proposition 4.4 shows that $c(x_{\mathfrak{rs}1}, x_{\mathfrak{rs}+11})$ is known for all $\mathfrak{r} = 1, \ldots, \ell_1$ and $\mathfrak{s} = 1, \ldots, \ell_2 - 1$. See Figure 5 in order to see all the known information at the end of this step.

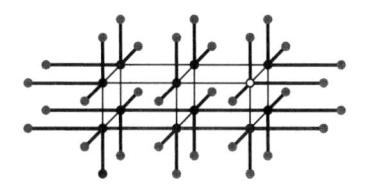

FIGURE 5. The bold red items are the ones known at the end of Step 4 for the case $\mathfrak{r} = \mathfrak{s} = 1$.

Step 5. In this step we give the conductances of all the edges with both ends in D_1 that are still unknown. Furthermore, we state a more general result.

PROPOSITION 4.5. *Let $h \in \{0, \ldots, \ell_3 - 1\}$. For every $\mathfrak{r} = 1, \ldots, \ell_1$ and $\mathfrak{s} = 1, \ldots, \ell_2$, let us suppose that we know the values of $u_{\mathfrak{rs}0}$ and of $u_{\mathfrak{r}01}$ on D_h and D_{h+1}. Also, let us assume that we know the conductances of all the edges joining vertices from D_h and D_{h+1}, and the ones of the edges with both ends in D_{h+1} and such that the ends have different second component. Then, for every $\mathfrak{r} = 1, \ldots, \ell_1$ and $\mathfrak{s} = 1, \ldots, \ell_2 - h - 1$ the conductances $c(x_{\mathfrak{r}+1\mathfrak{s}+h+1h+1}, x_{\mathfrak{rs}+h+1h+1})$ are given by*

$$c(x_{\mathfrak{rs}+h+1h+1}, x_{\mathfrak{r}+1\mathfrak{s}+h+1h+1}) = -c(x_{\mathfrak{r}+1\mathfrak{s}+h+1h+1}, x_{\mathfrak{r}+1\mathfrak{s}+h+2h+1}) \frac{u_{\mathfrak{rs}0}(x_{\mathfrak{r}+1\mathfrak{s}+h+2h+1})}{u_{\mathfrak{rs}0}(x_{\mathfrak{rs}+h+1h+1})}$$

$$- c(x_{\mathfrak{r}+1\mathfrak{s}+h+1h}, x_{\mathfrak{r}+1\mathfrak{s}+h+1h+1}) \frac{u_{\mathfrak{rs}0}(x_{\mathfrak{r}+1\mathfrak{s}+h+1h})}{u_{\mathfrak{rs}0}(x_{\mathfrak{rs}+h+1h+1})}.$$

Moreover, the conductances $c(x_{\mathfrak{r}h+1h+1}, x_{\mathfrak{r}+1h+1h+1})$ are given by

$$c(x_{\mathfrak{r}h+1h+1}, x_{\mathfrak{r}+1h+1h+1}) = -c(x_{\mathfrak{r}+1h+1h+1}, x_{\mathfrak{r}+1h+2h+1}) \frac{u_{\mathfrak{r}01}(x_{\mathfrak{r}+1h+2h+1})}{u_{\mathfrak{r}01}(x_{\mathfrak{r}h+1h+1})}$$

$$- c(x_{\mathfrak{r}+1h+1h}, x_{\mathfrak{r}+1h+1h+1}) \frac{u_{\mathfrak{r}01}(x_{\mathfrak{r}+1h+1h})}{u_{\mathfrak{r}01}(x_{\mathfrak{r}h+1h+1})}.$$

Proof. We fix the indices $h \in \{0, \ldots, \ell_3 - 1\}$, $\mathfrak{r} = 1, \ldots, \ell_1$ and $\mathfrak{s} = 1, \ldots, \ell_2 - h - 1$. Then, using Proposition 4.2 (i),

$$0 = \mathcal{L}_q(u_{\mathfrak{rs}0})(x_{\mathfrak{r}+1\mathfrak{s}+h+1h+1}) = -c(x_{\mathfrak{rs}+h+1h+1}, x_{\mathfrak{r}+1\mathfrak{s}+h+1h+1})u_{\mathfrak{rs}0}(x_{\mathfrak{rs}+h+1h+1})$$

$$- c(x_{\mathfrak{r}+1\mathfrak{s}+h+1h+1}, x_{\mathfrak{r}+1\mathfrak{s}+h+2h+1})u_{\mathfrak{rs}0}(x_{\mathfrak{r}+1\mathfrak{s}+h+2h+1})$$

$$- c(x_{\mathfrak{r}+1\mathfrak{s}+h+1h}, x_{\mathfrak{r}+1\mathfrak{s}+h+1h+1})u_{\mathfrak{rs}0}(x_{\mathfrak{r}+1\mathfrak{s}+h+1h}).$$

The value $c(x_{\mathfrak{rs}+h+1h+1}, x_{\mathfrak{r}+1\mathfrak{s}+h+1h+1})$ is the unique unknown term of this equality and by Proposition 4.2(i) we know that $u_{\mathfrak{rs}0}(x_{\mathfrak{rs}+h+1h+1}) \neq 0$.

Moreover, we fix the indices $h \in \{0, \ldots, \ell_3 - 1\}$, $\mathfrak{r} = 1, \ldots, \ell_1$. Then, using Proposition 4.2 (ii),

$$0 = \mathcal{L}_q(u_{\mathfrak{r}01})(x_{\mathfrak{r}+1h+1h+1}) = -c(x_{\mathfrak{r}h+1h+1}, x_{\mathfrak{r}+1h+1h+1})u_{\mathfrak{r}01}(x_{\mathfrak{r}h+1h+1})$$

$$- c(x_{\mathfrak{r}+1h+1h+1}, x_{\mathfrak{r}+1h+2h+1})u_{\mathfrak{r}01}(x_{\mathfrak{r}+1h+2h+1})$$

$$- c(x_{\mathfrak{r}+1h+1h}, x_{\mathfrak{r}+1h+1h+1})u_{\mathfrak{r}01}(x_{\mathfrak{r}+1h+1h}).$$

The value $c(x_{\mathfrak{r}h+1h+1}, x_{\mathfrak{r}+1h+1h+1})$ is the unique unknown term of this equality and by Proposition 4.2(ii) we know that $u_{\mathfrak{r}01}(x_{\mathfrak{r}h+1h+1}) \neq 0$. □

When $h = 0$, Proposition 4.5 shows that $c(x_{\mathfrak{r}\mathfrak{s}+11}, x_{\mathfrak{r}+1\mathfrak{s}+11})$ is known for all $\mathfrak{r} = 1, \ldots, \ell_1$ and $\mathfrak{s} = 1, \ldots, \ell_2 - 1$ and the same happens with $c(x_{\mathfrak{r}11}, x_{\mathfrak{r}+111})$ for all $\mathfrak{r} = 1, \ldots, \ell_1$. See Figure 6 (A) and (B) in order to see all the information gathered at the end of this step.

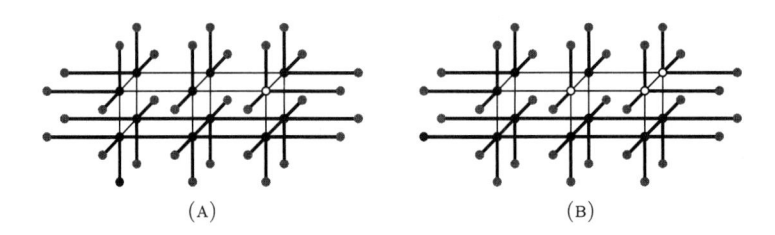

(A)   (B)

FIGURE 6. The bold red items are the ones known at the end of Step 5 for the case $\mathfrak{r} = \mathfrak{s} = 1$.

Step 6. Let us define the linear operator $\wp \colon \mathcal{C}(\bar{F}) \longrightarrow \mathcal{C}(F)$ given by the values

$$\wp(v)(x_{ijk}) = c(x_{ijk-1}, x_{ijk})v(x_{ijk-1}) + c(x_{i-1jk}, x_{ijk})v(x_{i-1jk}) + c(x_{ijk}, x_{i+1jk})v(x_{i+1jk})$$

$$+ c(x_{ij-1k}, x_{ijk})v(x_{ij-1k}) + c(x_{ijk}, x_{ij+1k})v(x_{ij+1k})$$

for all $v \in \mathcal{C}(\bar{F})$ and $x_{ijk} \in F$. This operator will be useful in this step and also in the following ones because of its relation with the Schrödinger operator for any $x_{ijk} \in F$

(4.1)
$$\mathcal{L}_q(v)(x_{ijk}) = \frac{v(x_{ijk})}{\omega(x_{ijk})}\wp(\omega)(x_{ijk}) - \wp(v)(x_{ijk})$$
$$- c(x_{ijk}, x_{ijk+1})v(x_{ijk+1}) + \frac{\omega(x_{ijk+1})}{\omega(x_{ijk})}c(x_{ijk}, x_{ijk+1})v(x_{ijk}).$$

In this step we give the conductances of all the edges joining the vertices from D_1 and D_2. Furthermore, we state a more general result.

PROPOSITION 4.6. *Let* $h \in \{0, \ldots, \ell_3 - 1\}$. *For every* $\mathfrak{r} = 1, \ldots, \ell_1$ *and* $\mathfrak{s} = 1, \ldots, \ell_2$, *let us suppose that we know the values of* $u_{\mathfrak{r}\mathfrak{s}0}$ *on* D_h *and* D_{h+1}. *Also, let us suppose that we know the conductances of all the edges joining vertices from* D_h *and* D_{h+1}, *and the ones of the edges with both ends in* D_{h+1}. *Then, for every* $\mathfrak{r} = 1, \ldots, \ell_1$ *and* $\mathfrak{s} = 1, \ldots, \ell_2 - h - 1$ *the conductances* $c(x_{\mathfrak{r}\mathfrak{s}+h+1h+1}, x_{\mathfrak{r}\mathfrak{s}+h+1h+2})$ *are given by*

$$c(x_{\mathfrak{r}\mathfrak{s}+h+1h+1}, x_{\mathfrak{r}\mathfrak{s}+h+1h+2}) = \frac{\wp(u_{\mathfrak{r}\mathfrak{s}0})(x_{\mathfrak{r}\mathfrak{s}+h+1h+1})\omega(x_{\mathfrak{r}\mathfrak{s}+h+1h+1})}{u_{\mathfrak{r}\mathfrak{s}0}(x_{\mathfrak{r}\mathfrak{s}+h+1h+1})\omega(x_{\mathfrak{r}\mathfrak{s}+h+1h+2})} - \frac{\wp(\omega)(x_{\mathfrak{r}\mathfrak{s}+h+1h+1})}{\omega(x_{\mathfrak{r}\mathfrak{s}+h+1h+2})}.$$

Proof. We fix the indices $h \in \{0, \ldots, \ell_3 - 1\}$, $\mathfrak{r} = 1, \ldots, \ell_1$ and $\mathfrak{s} = 1, \ldots, \ell_2 - 1$. Observe that $\wp(\omega)(x_{\mathfrak{r}\mathfrak{s}+h+1h+1})$ and $\wp(u_{\mathfrak{r}\mathfrak{s}0})(x_{\mathfrak{r}\mathfrak{s}+h+1h+1})$ are already known. Then, from Equation (4.1)

$$0 = \mathcal{L}_q(u_{\mathfrak{r}\mathfrak{s}0})(x_{\mathfrak{r}\mathfrak{s}+h+1h+1})$$

$$= u_{\mathfrak{r}\mathfrak{s}0}(x_{\mathfrak{r}\mathfrak{s}+h+1h+1})\wp(\omega)(x_{\mathfrak{r}\mathfrak{s}+h+1h+1}) - \omega(x_{\mathfrak{r}\mathfrak{s}+h+1h+1})\wp(u_{\mathfrak{r}\mathfrak{s}0})(x_{\mathfrak{r}\mathfrak{s}+h+1h+1})$$

$$+ \omega(x_{\mathfrak{r}\mathfrak{s}+h+1h+2})c(x_{\mathfrak{r}\mathfrak{s}+h+1h+1}, x_{\mathfrak{r}\mathfrak{s}+h+1h+2})u_{\mathfrak{r}\mathfrak{s}0}(x_{\mathfrak{r}\mathfrak{s}+h+1h+1})$$

and hence $c(x_{\mathfrak{r}\mathfrak{s}+h+1h+1}, x_{\mathfrak{r}\mathfrak{s}+h+1h+2})$ is the only unknown term of this equality. $\square$

In particular, when $h = 0$, Proposition 4.6 shows that $c(x_{\mathfrak{r}\mathfrak{s}+11}, x_{\mathfrak{r}\mathfrak{s}+12})$ is known for all $\mathfrak{r} = 1, \ldots, \ell_1$ and $\mathfrak{s} = 1, \ldots, \ell_2 - 1$. See Figure 7 in order to see all the information obtained at the end of this step.

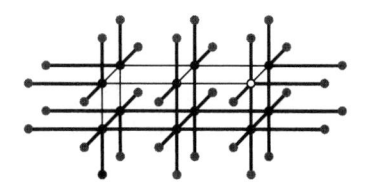

FIGURE 7. The bold red items are the ones known at the end of Step 6 for the case $\mathfrak{r} = \mathfrak{s} = 1$.

Step 7. In this step we are able to obtain the unknown values of $u_{\mathfrak{r}\mathfrak{s}0}$ and $u_{\mathfrak{r}01}$ on D_2 for all $\mathfrak{r} = 1, \ldots, \ell_1$ and $\mathfrak{s} = 1, \ldots, \ell_2$. In fact, let us state a more general result.

PROPOSITION 4.7. *Let $h \in \{0, \ldots, \ell_3 - 1\}$. For every $\mathfrak{r} = 1, \ldots, \ell_1$ and $\mathfrak{s} = 1, \ldots, \ell_2$, let us suppose that we know the values of $u_{\mathfrak{r}\mathfrak{s}0}$ and of $u_{\mathfrak{r}01}$ on D_h and D_{h+1}. Also, let us suppose that we know the conductances of all the edges joining vertices from D_h and D_{h+1}, from D_{h+1} and D_{h+2} and the ones of the edges with both ends in D_{h+1}. Then, for every $\mathfrak{r} = 1, \ldots, \ell_1$ and $\mathfrak{s} = 1, \ldots, \ell_2 - h - 2$ the values of $u_{\mathfrak{r}\mathfrak{s}0}$ on D_{h+2} are given by*

$$u_{\mathfrak{r}\mathfrak{s}0}(x_{ijh+2}) = -\frac{\wp(u_{\mathfrak{r}\mathfrak{s}0})(x_{ijh+1})}{c(x_{ijh+1}, x_{ijh+2})} + \frac{\wp(\omega)(x_{ijh+1})}{\omega(x_{ijh+1})c(x_{ijh+1}, x_{ijh+2})} u_{\mathfrak{r}\mathfrak{s}0}(x_{ijh+1})$$

$$+ \frac{\omega(x_{ijh+2})}{\omega(x_{ijh+1})} u_{\mathfrak{r}\mathfrak{s}0}(x_{ijh+1})$$

for all $i = 1, \ldots, \ell_1$ and $j = \mathfrak{s} + h + 2, \ldots, \ell_2$. Moreover,

$$u_{\mathfrak{r}01}(x_{ijh+2}) = -\frac{\wp(u_{\mathfrak{r}01})(x_{ijh+1})}{c(x_{ijh+1}, x_{ijh+2})} + \frac{\wp(\omega)(x_{ijh+1})}{\omega(x_{ijh+1})c(x_{ijh+1}, x_{ijh+2})} u_{\mathfrak{r}01}(x_{ijh+1})$$

$$+ \frac{\omega(x_{ijh+2})}{\omega(x_{ijh+1})} u_{\mathfrak{r}01}(x_{ijh+1})$$

for all $i = 1, \ldots, \ell_1$ and $j = h + 2, \ldots, \ell_2$.

Proof. Fixed three indices $h \in \{0, \ldots, \ell_3 - 1\}$, $\mathfrak{r} = 1, \ldots, \ell_1$ and $\mathfrak{s} = 1, \ldots, \ell_2 - h - 2$, let $i \in \{1, \ldots, \ell_1\}$ and $j \in \{\mathfrak{s} + h + 2, \ldots, \ell_2\}$. Observe that $\wp(\omega)(x_{ijh+1})$

and $\wp(u_{\mathfrak{rs}0})(x_{ijh+1})$ are already known. Then,

$$0 = \mathcal{L}_q(u_{\mathfrak{rs}0})(x_{ijh+1}) = \frac{u_{\mathfrak{rs}0}(x_{ijh+1})}{\omega(x_{ijh+1})}\wp(\omega)(x_{ijh+1}) - \wp(u_{\mathfrak{rs}0})(x_{ijh+1})$$
$$- c(x_{ijh+1}, x_{ijh+2})u_{\mathfrak{rs}0}(x_{ijh+2}) + \frac{\omega(x_{ijh+2})}{\omega(x_{ijh+1})}c(x_{ijh+1}, x_{ijh+2})u_{\mathfrak{rs}0}(x_{ijh+1})$$

and hence $u_{\mathfrak{rs}0}(x_{ijh+2})$ is the unique unknown term of this equality. The identity for $u_{\mathfrak{r}01}$ can be proved analogously. □

In particular, when $h = 0$, Proposition 4.7 shows that $u_{\mathfrak{rs}0}$ and $u_{\mathfrak{r}01}$ are known on D_2 for all $\mathfrak{r} = 1, \ldots, \ell_1$ and $\mathfrak{s} = 1, \ldots, \ell_2 - 2$. Figure 8 (A) and (B) shows the information obtained until this step.

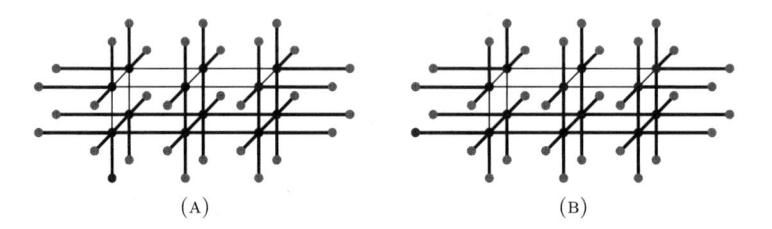

FIGURE 8. The bold red items are the ones known at the end of Step 7 for the case $\mathfrak{r} = \mathfrak{s} = 1$.

Step 8 and beyond. We keep repeating the same process to obtain more conductances, that is, we keep applying Proposition 4.4 from Step 4, then Proposition 4.5 from Step 5, Proposition 4.6 from Step 6 and then Proposition 4.7 from Step 7 for each $h = 1, \ldots, \ell_3 - 1$. We stop when we have obtained all the conductances between and all the vertices in $\bigcup_{k=1}^{\ell_3} D_k$, see Figure 9 (A), (B) and (C).

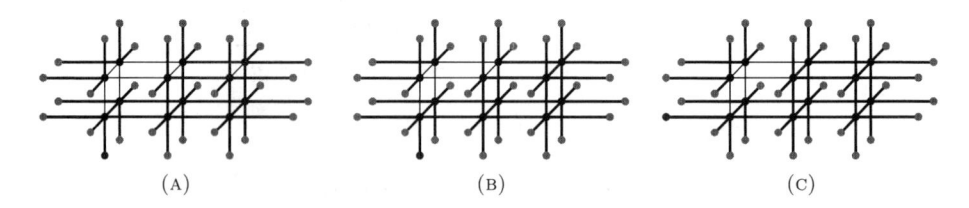

FIGURE 9. The bold red items are the ones known at the end of Step 7 for the case $\mathfrak{r} = \mathfrak{s} = 1$.

The final step left is to consider the pole of the overdetermined partial Poisson Kernel on D_{ℓ_3+1}; or equivalently to rotate the network 180^0. By proceeding analogously to the last steps, we obtain the lacking conductances of the 3–dimensional grid , see the rest of figures in Figure 10.

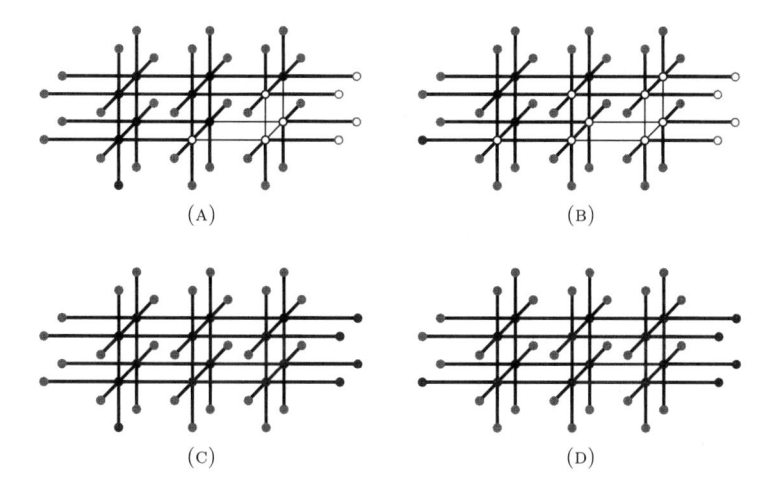

FIGURE 10. All the conductances have been recovered

References

[A1] C. Araúz, A. Carmona, and A. M. Encinas, *Dirichlet–to–Robin map on finite networks*, Appl. Anal. Discrete Math. **9** (2015), 85–102, DOI 10.2298/AADM150207004A. MR3362699.
[A2] C. Araúz, A. Carmona, and A. M. Encinas, *Overdetermined partial boundary value problems on finite networks*, J. Math. Anal. Appl. **423** (2015), no. 1, 191–207, DOI 10.1016/j.jmaa.2014.09.025. MR3273175
[A3] C. Araúz, A. Carmona, and A. M. Encinas, *Overdetermined partial resolvent kernels for networks*, J. Math. Anal. Appl. **435** (2016), no. 1, 96–111, DOI 10.1016/j.jmaa.2015.10.028. MR3423386.
[A4] C. Araúz, A. Carmona, and A. M. Encinas, *Recovering the conductances on grids*, Electron. Notes Discrete Math. **46** (2014), 11–18.
[C1] Edward B. Curtis and James A. Morrow, *The Dirichlet to Neumann map for a resistor network*, SIAM J. Appl. Math. **51** (1991), no. 4, 1011–1029, DOI 10.1137/0151051. MR1117430 (92g:31003)
[C2] E. B. Curtis, D. Ingerman, and J. A. Morrow, *Circular planar graphs and resistor networks*, Linear Algebra Appl. **283** (1998), no. 1-3, 115–150, DOI 10.1016/S0024-3795(98)10087-3. MR1657214 (99k:05096)
[C3] E. B. Curtis, and J. A. Morrow, *Inverse Problems for Electrical Networks*, Series on Applied Mathematics vol. 13. World Scientific 2000.
[D1] Yves Colin de Verdière, *Réseaux électriques planaires. I* (French), Comment. Math. Helv. **69** (1994), no. 3, 351–374, DOI 10.1007/BF01585564. MR1289333 (96k:05131)
[D2] Yves Colin de Verdière, Isidoro Gitler, and Dirk Vertigan, *Réseaux électriques planaires. II* (French), Comment. Math. Helv. **71** (1996), no. 1, 144–167, DOI 10.1007/BF02566413. MR1371682 (98a:05054)
[H] R.A. Horn, and F. Zhang, *Basic Properties of the Schur Complement, The Schur Complement and Its Applications*, Numerical Methods and Algorithms vol. 13. vol. 4: 17–46. Springer 2005.
[L] Thomas Lam and Pavlo Pylyavskyy, *Inverse problem in cylindrical electrical networks*, SIAM J. Appl. Math. **72** (2012), no. 3, 767–788, DOI 10.1137/110846476. MR2968749
[O] R. Oberlin, *Discrete Inverse Problems for Schrödinger and Resistor Networks*, (2000) unpublished, accessible at http://www.math.washington.edu/~morrow/papers/roberlin.pdf.
[S] John Sylvester and Gunther Uhlmann, *A global uniqueness theorem for an inverse boundary value problem*, Ann. of Math. (2) **125** (1987), no. 1, 153–169, DOI 10.2307/1971291. MR873380 (88b:35205)

[U] Gunther Uhlmann and Jenn-Nan Wang, *Reconstructing discontinuities using complex geometrical optics solutions*, SIAM J. Appl. Math. **68** (2008), no. 4, 1026–1044, DOI 10.1137/060676350. MR2390978 (2009d:35347)

Institut Cerdà, C Numància 185, 08034 Barcelona
E-mail address: crisaralom@gmail.com

Departament de Matemàtiques, Universitat Politècnica de Catalunya, 08034 Barcelona
E-mail address: angeles.carmona@upc.edu

Departament de Matemàtiques, Universitat Politècnica de Catalunya, 08034 Barcelona
E-mail address: andres.marcos.encinas@upc.edu

Departament de Matemàtiques, Universitat Politècnica de Catalunya, 08034 Barcelona
E-mail address: margarida.mitjana@upc.edu

The unified transform in two dimensions

Athanassios S. Fokas

ABSTRACT. There exists a particular class of nonlinear PDEs in two dimensions called *integrable*. The defining property of these equations is the fact that they can be written as the compatibility condition of two linear eigenvalue equations called a *Lax pair*. For integrable evolution PDEs the existence of a Lax pair gives rise to a novel method for solving the *initial-value* problem for these equations called the *inverse scattering transform*. In the late nineties the author introduced a new method for analyzing *boundary-value* problems. This method, in contrast to the inverse scattering transform which analyzes only *one* of the two equations defining the Lax pair, is based on the simultaneous spectral analysis of *both* equations defining the Lax pair. This implies that in contrast to the inverse scattering transform which still follows the philosophy of "separation of variables", albeit in a non-linear setting, the new method is based on the "synthesis of variables". In this sense, it is not surprising that the implementation of this method to *linear* PDEs has led to a new approach which differs drastically from the usual application of the well-known transforms. In particular, whereas the latter approach requires the given PDE, domain, and boundary conditions to be separable, and also it is usually not applicable to non-self-adjoint problems, the new method overcomes some of these difficulties. Furthermore, whereas the solution obtained from the usual transforms takes the form of either an integral or an infinite series neither of which are uniformly convergent on the boundary of the domain (for non-vanishing boundary conditions) and this renders such expressions unsuitable for numerical computations, the new method always yields representations which are uniformly convergent. This has led to new efficient numerical techniques. Here we review some of the most important implications of this new method, which is usually referred to as the "unified transform" or the "Fokas method".

1. Introduction

The application of appropriate transform pairs, such as the Fourier, the Laplace, the sine, the cosine and the Mellin transforms, provides the most well known method for constructing analytical solutions to a large class of physically significant boundary value problems. However, this approach which was inaugurated by Fourier in

---

2010 *Mathematics Subject Classification*. Primary 35A22.
A. S. Fokas acknowledges support from the EPSRC, UK.
This research has been co-financed by the European Union (European Social Fund - ESF) and Greek national funds through the Operational Program "Education and Lifelong Learning" of the National Strategic Reference Framework (NSRF) - Research Funding Program: Thales. Investing in knowledge society through the European Social Fund.

1807 has several limitations. In particular, it is based on "separation of variables" and thus requires the given PDE, domain and boundary conditions to be separable. Also it is usually *not* applicable to non-self-adjoint boundary value problems. Furthermore, it expresses the solution as either an integral or an infinite series, neither of which are uniformly convergent on the boundary of the domain (for nonvanishing boundary conditions). This very serious limitation, which has *not* been emphasized in the literature, renders such expressions unsuitable for numerical computations.

The next important step in the analysis of linear PDEs was taken by Green in 1828 who introduced the powerful approach of integral representations that can be obtained via Green's functions. However, these integral representations in general involve some unknown boundary values, thus they do not provide an effective analytic solution. For example, for the Laplace equation, the integral representations obtained by the above approach involve both the Dirichlet and the Neumann boundary values, and for a well posed problem only one of these boundary values is known. Thus, deriving an effective solution requires the determination of the (generalized) Dirichlet to Neumann map, *i.e.* the determination of the unknown boundary values in terms of the given boundary data. For certain simple domains Lord Kelvin introduced an ingenious technique, the method of images, for eliminating the unknown boundary values, and thus for obtaining an effective solution.

In summary, for linear PDEs there exist two important methods for obtaining analytic solutions: (i) The method of applying appropriate transforms, which was inaugurated by Fourier and which is based on separation of variables. This approach yields a representation of the solution which is formulated in the spectral (Fourier) plane. (ii) The method of Green and Kelvin which does *not* involve separation of variables and which is formulated in the physical plane.

It should be emphasized that although the problem of analyzing the Dirichlet to Neumann problem is usually mentioned in connection with elliptic PDEs, this problem also arises in evolution PDEs. For example, the Stokes equation on the half-line

(1.1) $$u_t + u_{xxx} + u_x = 0, \quad 0 < x < \infty, \quad 0 < t < T, \quad T > 0,$$

is well posed by prescribing $u(x,0)$ and $u(0,t)$. However, a standard integral representation in addition to $u(x,0)$ and $u(0,t)$, it also involves the unknown boundary values $u_x(0,t)$ and $u_{xx}(0,t)$. Thus, in this case the generalized Dirichlet to Neumann map consists of determining $u_x(0,t)$ and $u_{xx}(0,t)$ in terms of $u(0,t)$ and $u(x,0)$.

For PDEs involving only a second order spatial derivative, the choice of an appropriate transform is precisely dictated by the requirement of eliminating the single unknown boundary value.

1.1. Integrable Nonlinear PDEs. In the second half of the 20th century it was realized that certain *nonlinear evolution* PDEs, called *integrable*, can be formulated as the compatibility condition of two linear eigenvalue equations called a *Lax pair*, and that this formulation gives rise to a method for solving the initial-value problem for these equations, called the *inverse scattering transform* method. The author has emphasized that this method is based on a deeper form of separation of variables [1]. Indeed, the spectral analysis of one of the equations defining the Lax pair yields an appropriate *nonlinear Fourier transform pair*. In this sense, in spite of the fact that the inverse scattering transform is applicable to nonlinear PDEs, this method still follows the logic of separation of variables.

1.2. From Nonlinear to Linear PDEs. It was first realized by the late I. M. Gelfand and the author [2] that linear PDEs with constant coefficients also possess a Lax pair formulation. For example, the heat equation is the compatibility condition of the following pair of equations for the scalar function $\mu(x,t,k)$:

$$\mu_x - i\lambda\mu = u \tag{1.2}$$
$$\mu_t + \lambda^2\mu = u_x + i\lambda u, \quad \lambda \in \mathbb{C}. \tag{1.3}$$

Indeed, equations (1.2) and (1.3) can be rewritten in the form

$$\left(e^{-i\lambda x + \lambda^2 t}\mu\right)_x = e^{-i\lambda x + \lambda^2 t}u,$$
$$\left(e^{-i\lambda x + \lambda^2 t}\mu\right)_t = e^{-i\lambda x + \lambda^2 t}\left(u_x + i\lambda u\right).$$

Thus, equations (1.2) and (1.3) are compatible if and only if

$$\left(e^{-i\lambda x + \lambda^2 t}u\right)_t - \left(e^{-i\lambda x + \lambda^2 t}(u_x + i\lambda u)\right)_x = 0, \quad \lambda \in \mathbb{C}, \tag{1.4}$$

which is equivalent to the heat equation

$$u_t = u_{xx}. \tag{1.5}$$

Equations (1.2) and (1.3) should be contrasted with the following pair of ODEs obtained from the heat equation via separation of variables, $u = X(x,\lambda)T(t,\lambda)$:

$$\frac{d^2 X}{dx^2} + \lambda^2 X = 0, \tag{1.6}$$
$$\frac{dT}{dt} + \lambda^2 T = 0, \quad \lambda \in \mathbb{C}. \tag{1.7}$$

Let us consider the Dirichlet problem for the heat equation on the half-line $0 < x < \infty$. The spectral analysis of equation (1.6) gives rise to the sine transform,

$$\hat{f}(\lambda) = \int_0^\infty \sin(\lambda x) f(x) dx, \quad \lambda > 0; \quad f(x) = \frac{2}{\pi}\int_0^\infty \sin(\lambda x)\hat{f}(\lambda)d\lambda, x > 0. \tag{1.8}$$

Alteratively, the same transform can be obtained via the spectral analysis of equation (1.2). The latter analysis is simpler, since equation (1.2) is a first order ODE. Actually, the Lax pair formulation has an additional more important advantage: Equations (1.2) and (1.3) are valid for any solution u of the heat equation (in contrast to equations (1.6) and (1.7) which are valid only for a particular solution). Thus, the Lax pair provides the possibility of performing the *simultaneous spectral analysis* of equations (1.2) and (1.3). This gives rise to a novel integral representation which is based on the "synthesis" as opposed to "separation" of variables:

$$u(x,t) = \frac{1}{2\pi}\int_{-\infty}^\infty e^{i\lambda x - \lambda^2 t}\hat{u}_0(\lambda)d\lambda - \frac{1}{2\pi}\int_{\partial D^+} e^{i\lambda x - \lambda^2 t}[G_1(\lambda^2) + i\lambda G_0(\lambda^2)]d\lambda, \tag{1.9}$$

where $\hat{u}_0(\lambda)$, G_0, G_1 are defined by

(1.10) $\quad \hat{u}_0(\lambda) = \int_0^\infty e^{-i\lambda x} u(x,0) dx, \quad \text{Im}\lambda \leq 0,$

(1.11) $\quad G_0(\lambda) = \int_0^T e^{\lambda s} u(0,s) ds, \quad G_1(\lambda) = \int_0^T e^{\lambda s} u_x(0,s) ds, \quad \lambda \in \mathbb{C},$

and ∂D^+ is the union of the rays making an angle $\pi/4$ and $3\pi/4$ with the real axis of the complex λ-plane, see Fig. 1.

The new expressions for the solution of linear PDEs were first obtained by the above method [3, 4]. However, it was later realized that for linear PDEs it is possible to avoid the spectral analysis of the associated Lax pair and to obtain these expressions directly; for evolution PDEs this involves using the Fourier transform and appropriate contour deformation [5], and for elliptic PDEs this involves using Green's functions [6–8].

The latter approach is consistent with the following fact: The new method for linear PDEs provides the analogue of Green's approach but the relevant formulation takes place in the spectral instead of the physical plane [9]. Thus, the new method combines features of the two most well known methods for solving linear PDEs summarized earlier. Furthermore, it can be "nonlinearized" via the simultaneous spectral analysis of the associated Lax pair [5].

In what follows we will review the most important results obtained via the new method for both linear and integrable nonlinear PDEs.

2. Linear Evolution PDEs

In order to introduce the new method we consider the simplest possible initial-boundary value (IBV) problem for an evolution PDE, namely the heat equation on the half-line:

(2.1) $\quad u_t = u_{xx}, \quad 0 < x < \infty, \quad 0 < t < T, \quad T > 0,$

where

(2.2) $\quad u(x,0) = u_0(x), \quad 0 < x < \infty, \quad u(0,t) = g_0(t), \quad 0 < t < T.$

The functions $u_0(x)$ and $g_0(t)$ are given functions with appropriate smoothness and $u_0(x)$ decays as $x \to \infty$.

Employing the sine transform we find

(2.3) $\quad u(x,t) = \frac{2}{\pi} \int_0^\infty e^{-\lambda^2 t} \sin(\lambda x) \left[\int_0^\infty \sin(\lambda\xi) u_0(\xi) d\xi - \lambda \int_0^t e^{\lambda^2 s} g_0(s) ds \right] d\lambda.$

The above representation suffers from the generic disadvantage that is associated with *every* representation obtained via a classical transform, namely it is *not* uniformly convergent at the boundary. Indeed, if the right-hand-side of (2.3) converged uniformly at $x = 0$, then one could take the limit $x \to 0$ inside the integral, and then one would obtain $u(0,t) = 0$ instead of $u(0,t) = g_0(t)$.

It will be shown below that the new method yields

(2.4) $$u(x,t) = \frac{1}{2\pi} \int_{-\infty}^\infty e^{i\lambda x - \lambda^2 t} \hat{u}_0(\lambda) d\lambda - \frac{1}{2\pi} \int_{\partial D^+} e^{i\lambda x - \lambda^2 t} \left[\hat{u}_0(-\lambda) + 2i\lambda G_0(\lambda^2) \right] d\lambda,$$

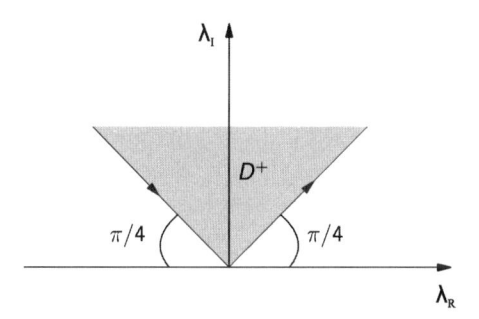

FIGURE 1. The domain D^+ and the contour ∂D^+ for the heat equation.

where the functions $\hat{u}_0(\lambda)$ and $G_0(\lambda)$ are defined by

$$(2.5) \quad \hat{u}_0(\lambda) = \int_0^\infty e^{-i\lambda x} u_0(x)dx, \quad \operatorname{Im}\lambda \leq 0, \quad G_0(\lambda) = \int_0^T e^{\lambda s} g_0(s)ds, \quad \lambda \in \mathbb{C},$$

and the contour ∂D^+ is the union of the rays making an angle $\pi/4$ and $3\pi/4$ with the real axis of the complex λ-plane, see Fig. 1.

It is straightforward to show [5] that the right-hand-side of (2.4) is indeed uniformly convergent at $x = 0$. Furthermore, the only (x,t) dependence of the right-hand-side of (2.4) is in the form $e^{i\lambda x - \lambda^2 t}$, thus it is immediately obvious that this representation satisfies the heat equation.

The experienced reader may worry about the dependence of $G_0(\lambda)$ on T, which contradicts causality (the solution of an evolution equation *cannot* depend on future data). However, using analyticity arguments, it can be shown [5] that $G_0(\lambda)$ can be replaced by the function $G_0(\lambda, t)$, where

$$(2.6) \quad G_0(\lambda, t) = \int_0^t e^{\lambda s} g_0(s)ds, \quad \lambda \in \mathbb{C}.$$

The new method involves the following three steps:
Step 1: *Rewrite the given PDE in a divergence form.*
For the heat equation we find equation (1.4). This equation can be derived as follows: Let $\tilde{u}$ be a solution of the formal adjoint of the heat equation

$$(2.7) \quad \tilde{u}_t + \tilde{u}_{xx} = 0.$$

Combining this equation with the heat equation (1.5) we find

$$(2.8) \quad (\tilde{u}u)_t - (\tilde{u}u_x - u\tilde{u}_x)_x = 0.$$

Choosing the particular solution

$$\tilde{u} = e^{-i\lambda x + \lambda^2 t},$$

equation (2.8) becomes (1.4).

If the PDE is valid in a given domain Ω then Green's theorem immediately implies the following *global relation* (GR):

$$(2.9) \quad \int_{\partial \Omega} e^{-i\lambda x + \lambda^2 t}[udx + (u_x + i\lambda u)dt] = 0.$$

For the heat equation on the half-line the GR becomes:

(2.10) $$e^{\lambda^2 T}\hat{u}(\lambda, T) = \hat{u}_0(\lambda) - i\lambda G_0(\lambda^2) - G_1(\lambda^2), \quad \text{Im}\lambda \leq 0,$$

where $\hat{u}_0(\lambda)$ and $G_0(\lambda)$ are defined in (2.5), the function $G_1(\lambda)$ is defined in (1.11) and $\hat{u}(\lambda, T)$ is defined by

(2.11) $$\hat{u}(\lambda, T) = \int_0^\infty e^{-i\lambda x} u(x, T) dx, \quad \text{Im}\lambda \leq 0.$$

Step 2: *Integral representation of the solution.*
The simplest way to obtain such a representation is to use the Fourier transform on the half-line and then to deform the relevant integral from the real line to the complex λ-plane [5]. For the heat equation this yields equation (1.9).

Step 3: *Elimination of the unknown boundary values.*
This can be achieved by using the GR and by employing all transformations in the complex λ-plane which leave the associated symbol, denoted by $\omega(\lambda)$, invariant. For the heat equation, $\omega(\lambda) = \lambda^2$, thus replacing in the GR λ with $-\lambda$, we find

(2.12) $$e^{\lambda^2 T}\hat{u}(-\lambda, T) = \hat{u}_0(-\lambda) + i\lambda G_0(\lambda^2) - G_1(\lambda^2), \quad \text{Im}\lambda \geq 0.$$

Solving this equation for G_1 and using the fact that $\hat{u}(-\lambda, T)$ does not contribute to the solution $u(x, t)$ [9], we find that (1.9) becomes (2.4).

REMARK 2.1. The limited applicability of the standard transforms becomes evident by considering the Stokes equation on the half-line

(2.13) $$u_t + u_{xxx} + u_x = 0, \quad 0 < x < \infty, \quad 0 < t < T, \quad T > 0,$$

with the initial and boundary conditions defined in (2.2).

It can be rigorously established [10, 11] that there does *not* exist an appropriate x-transform for this problem, *i.e.* there does *not* exist the analogue of the sine transform for a linear evolution PDE involving a third order derivative. One may attempt to solve the above IBV problem with the Laplace transform in t. But then one has to make the unnatural assumption of $T = \infty$, and furthermore one has to solve a cubic algebraic equation. Indeed, if $\tilde{u}(x, s)$ denotes the Laplace transform of $u(x, s)$, then $\tilde{u}$ satisfies

(2.14) $$\tilde{u}_{xxx}(x, s) + \tilde{u}_x(x, s) + s\tilde{u}(x, s) = u_0(x), \quad 0 < x < \infty.$$

Thus, in order to construct an appropriate Green's function for this ordinary differential equation, one seeks a solution of the homogeneous version of (2.14), which yields the *cubic* equation

$$\lambda(s)^3 + \lambda(s) + s = 0.$$

The unified method yields

(2.15) $$u(x, t) = \frac{1}{2\pi}\int_{-\infty}^\infty e^{i\lambda x - (i\lambda - i\lambda^3)t}\hat{u}_0(\lambda)d\lambda - \frac{1}{2\pi}\int_{\partial D^+} e^{i\lambda x - (i\lambda - i\lambda^3)t}\tilde{g}(\lambda)d\lambda,$$

where the function $\hat{u}_0(\lambda)$ is defined in the first of equations (2.5), $\tilde{g}(\lambda)$ is defined by

(2.16) $$\tilde{g}(\lambda) = \frac{1}{\nu_1 - \nu_2}[(\nu_1 - \lambda)\hat{u}_0(\nu_2) + (\lambda - \nu_2)\hat{u}_0(\nu_1)] + (3\lambda^2 - 1)G_0(\omega(\lambda)),$$

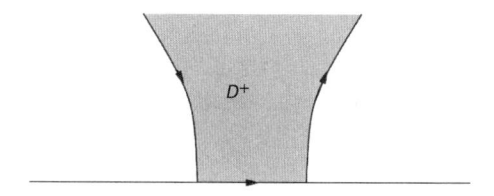

FIGURE 2. The domain D^+ for the Stokes equation.

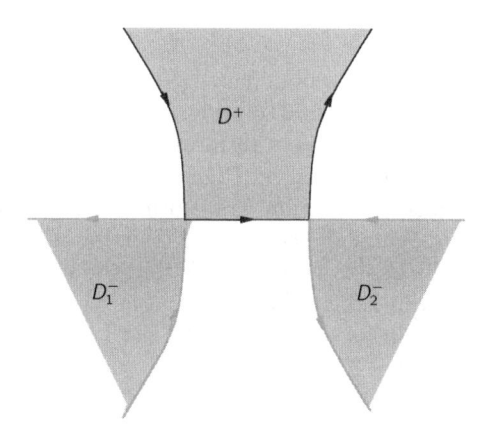

FIGURE 3. The analysis of the GR for the Stokes equation.

with $G_0(\lambda)$ defined in the second of equations (2.5) and $\omega(\lambda)$ given by $\omega(\lambda) = i\lambda - i\lambda^3$; the contour ∂D^+ is the boundary of the domain defined by

(2.17) $$D^+ = \{\operatorname{Re} \omega(\lambda) < 0\} \cap \mathbb{C}^+$$

and shown in Fig. 2. The complex numbers ν_1 and ν_2 are the two nontrivial transformations $\lambda \to \nu_1(\lambda)$, $\lambda \to \nu_2(\lambda)$ which leave $\omega(\lambda)$ invariant, i.e. they are the two nontrivial roots of the equation $\omega(\lambda) = \omega(\nu(\lambda))$:

(2.18) $$\nu_j^2 + \lambda\nu_j + \lambda^2 - 1 = 0, \quad j = 1, 2.$$

For the Stokes equation, the associated GR is

(2.19) $$G_2 + i\lambda G_1 = (\lambda^2 - 1)G_0 - \hat{u}_0(\lambda) - e^{(i\lambda - i\lambda^3)T}\hat{u}(\lambda, T), \quad \operatorname{Im}\lambda \leq 0.$$

Replacing in this equation λ by ν_1 and by ν_2 we find two equations, both of which are valid in D^+, see Fig. 3. Then, using the fact that the contribution of $\hat{u}(\nu_1, T)$ and $\hat{u}(\nu_2, T)$ to the solution vanishes [5], we find

(2.20)
$$G_1 \sim -i(\nu_1+\nu_2)G_0 + i\frac{\hat{u}_0(\nu_1) - \hat{u}_0(\nu_2)}{\nu_1 - \nu_2}, \quad G_2 \sim -(1+\nu_1\nu_2)G_0 + \frac{\nu_2\hat{u}_0(\nu_1) - \nu_1\hat{u}_0(\nu_2)}{\nu_1 - \nu_2}.$$

REMARK 2.2. It should be emphasized that *any* solution obtained via the standard transform approach suffers from lack of uniform converges at the boundary (for a non-homogeneous boundary condition). This fact, in addition to rendering such expressions unsuitable for numerical computations, also makes it difficult to

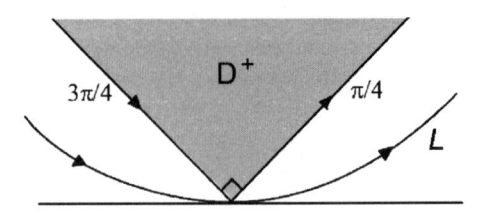

FIGURE 4. The contour L for the heat equation.

verify that the solution to an initial-boundary value problem (IBVP) obtained using an appropriate transform is indeed a solution. In this respect we note that the construction of the solution via any transform method *assumes* that a solution exists. Thus, unless one can appeal to PDE existence results, one *must* verify that the final formula obtained via any transform method *does* satisfy the PDE and the given initial and boundary conditions. For initial-value problems this is straightforward but for IBVPs, in order to verify that the given boundary conditions hold, one must overcome the difficulty of nonuniform convergence. It is interesting that this problem is *not* addressed in any of standard applied books on boundary value problems of linear PDEs.

Numerical implementation
For the case that $\hat{u}_0(\lambda)$ and $G_0(\lambda)$ can be computed explicitly, it is straightforward to compute numerically the solution. Consider, for example, the heat equation with the following initial and boundary conditions:

(2.21) $$u_0(x) = x\exp(-a^2 x), \quad g_0(t) = \sin bt, \quad a, b > 0.$$

Then, (2.4) becomes

(2.22) $$u(x,t) = \frac{1}{2\pi} \int_{\partial D^+} e^{i\lambda x - \lambda^2 t} \left[\frac{1}{(i\lambda + a)^2} - \frac{1}{(-i\lambda + a)^2} \right. \\ \left. - \lambda \left(\frac{e^{(\lambda + ib)t} - 1}{\lambda + ib} - \frac{e^{(\lambda - ib)t} - 1}{\lambda - ib} \right) \right] d\lambda.$$

On the contour ∂D^+, $e^{i\lambda x}$ decays exponentially for large λ, whereas $e^{-i\lambda^2 t}$ oscillates. However, deforming ∂D^+ to the contour L, see Fig. 4, we achieve decay as $\lambda \to \infty$ in both $e^{i\lambda x}$ and $e^{-i\lambda^2 t}$. Thus, the deformed integral can be computed numerically most efficiently [12].

Evolution PDEs on the interval
The heat and the Stokes equations with $\{0 < x < 1, 0 < x < T\}$ are analyzed in [13, 14]. For both these equations, the unified method yields $u(x,t)$ in terms of *integrals* in the complex λ-plane (as opposed to infinite series). It should be emphasized that it is *impossible* to express the solution of a typical IBV problem for the Stokes equation in terms of an infinite series. Therefore, the usual statement that a *finite* domain corresponds to a *discrete* spectrum is *not* valid in general (unless the associated problem is self-adjoint).

3. Linear elliptic PDEs in the interior of a polygon

For brevity of presentation we concentrate on the modified Helmholtz equation. Let

(3.1) $$u_{xx} + u_{yy} - k^2 u = 0, \quad (x,y) \in \Omega; \quad k > 0.$$

The global relation is given by

(3.2) $$\oint_{\partial \Omega} e^{\frac{ik}{2}\left[\frac{\bar{z}(t)}{\lambda} - \lambda z(t)\right]} \left[u^{\mathcal{N}} + \frac{ku}{2}\left(\lambda \frac{dz(t)}{dt} + \frac{1}{\lambda}\frac{d\bar{z}(t)}{dt}\right)\right] dt = 0, \quad k > 0, \quad \lambda \in \mathbb{C}\setminus\{0\},$$

where $u^{\mathcal{N}}$ denotes the normal derivative of the boundary and

$$z = x + iy, \quad \bar{z} = x - iy.$$

Let Ω denote the interior of the convex polygon formed by the complex numbers $\{z_1, z_2, \cdots, z_n\}$. Then, the global relation takes the form:

(3.3) $$\sum_{j=1}^{n} \hat{u}_j(\lambda) = 0, \quad \lambda \in \mathbb{C}\setminus\{0\},$$

where $\hat{u}_j(\lambda)$ is defined by

(3.4) $$\hat{u}_j(\lambda) = \int_{z_j}^{z_{j+1}} e^{-i\frac{k}{2}(\lambda z - \frac{\bar{z}}{\lambda})}\left[\left(u_z + i\frac{k}{2}\lambda u\right)dz - \left(u_{\bar{z}} + \frac{k}{2i\lambda}u\right)d\bar{z}\right], \quad j = 1, \cdots, n.$$

A second global relation is obtained from equation (3.3) by replacing λ with $1/\lambda$. If u is real, we can obtain the second global relation by taking the Schwarz conjugate of equation (3.3).

By employing either the classical Green's representation formula [6], or by performing the spectral analysis of the associated Lax pair [15], we find the following novel integral representation:

(3.5a) $$u(z, \bar{z}) = \frac{1}{4i\pi} \sum_{j=1}^{n} \int_{l_j} e^{i\frac{k}{2}(\lambda z - \frac{\bar{z}}{\lambda})} \hat{u}_j(\lambda) \frac{d\lambda}{\lambda}, \quad z \in \Omega,$$

here $\{l_j\}_1^n$ are the rays in the complex λ-plane oriented towards infinity and defined by

(3.5b) $$l_j = \{\lambda \in \mathbb{C} : \arg \lambda = -\arg(z_{j+1} - z_j), \quad j = 1, \cdots, n, \quad z_{n+1} = z_1\}.$$

For simple domains, it is possible, using the global relations and their invariant properties, to express all transforms in terms of the given boundary data, using only algebraic manipulations. This has led to the analytic solution of several BVPs for which the usual approaches apparently fail [16]–[31].

For more complicated domains, it is remarkable that the global relations suggest a novel numerical technique for the determination of the unknown boundary values, see for example [32]-[42]. This numerical technique can be viewed as the counterpart in the complex Fourier plane of the boundary integral method (which is formulated in the physical plane). Consider for example the Dirichlet problem. In this case, equation (3.3) and the equation obtained from (3.3) by replacing λ with $1/\lambda$ are two equations for the unknown Neumann boundary values. However, these equations are valid for *all* complex $\lambda \notin \{0\}$. Thus, by evaluating these equations at appropriate chosen λ's we can obtain numerically the unknown boundary values.

In order to illustrate the basic ideas, consider the simplest polygon, namely the square with corners at $(-1, 1)$, $(-1, -1)$, $(1, -1)$, $(1, 1)$. For the side $(-1, 1)$ we have $z = -1 + iy$, hence for $k = 2$ we find

$$(3.6) \qquad \hat{u}_1(\lambda) = e^{(i\lambda + \frac{1}{i\lambda})} \int_{+1}^{-1} e^{(\lambda + \frac{1}{\lambda})y} \left[u_x^{(1)} + \left(i\lambda + \frac{1}{i\lambda}\right) u^{(1)} \right] dy,$$

and similarly for the other sides.

Let $\hat{D}_j$ and $\hat{N}_j$ denote the parts of $\hat{u}_j$ corresponding to the Dirichlet and Neumann boundary values. Then,

$$(3.7) \qquad \hat{u}_1(\lambda) = -e^{(i\lambda + \frac{1}{i\lambda})} \hat{N}_1(\lambda) - \left(i\lambda + \frac{1}{i\lambda}\right) e^{(i\lambda + \frac{1}{i\lambda})} \hat{D}_1(\lambda),$$

and similarly for the other sides. Thus, for the Dirichlet problem, $\{\hat{D}_j(\lambda)\}_1^4$ are known and the global relation (3.3) together with its Schwartz conjugate yield two equations for the four unknown boundary values. One must now choose: (a) a suitable basis for expanding the unknown boundary values, and (b) an appropriate set of complex values at which to evaluate the global relations. Several such choices have already been used in the literature. It appears that the best choice is (a) to expand the unknown boundary values in terms of Legendre polynomials [32, 33], and (b) to choose collocation points on certain rays in the complex λ-plane [40–42]. In this connection we note the following analytical formula for the Fourier transform of the Legendre polynomials denoted by $P_m(x)$ [43]:

$$(3.8) \qquad \int_{-1}^{1} e^{-i\lambda x} P_m(x) dx = \sum_{n=0}^{m} \frac{(m+n)!}{2^n n! (m-n)!} \frac{(-1)^{m+n} e^{i\lambda} - e^{-i\lambda}}{(i\lambda)^{n+1}}, \quad n \in \mathbb{Z}.$$

It was shown in [42] that the following provides a convenient choice of collocation points:

$$(3.9) \qquad \lambda_r = -\frac{h_j R}{M} r, \quad 1 \leq j \leq n, \quad 1 \leq r \leq M, \quad R > 0, \quad M \subset \mathbb{Z}^+,$$

where h_j is defined by

$$(3.10) \qquad h_j = \frac{x_j^2 - x_j^1}{2} + i \frac{y_j^2 - y_j^1}{2}, \quad 1 \leq j \leq n,$$

and (x_j^1, y_j^1), (x_j^2, y_j^2) denote the coordinates of the endpoints of the side j.

The parameters M and R must be chosen in such a way that the linear system of $2nM$ equations has a low condition number. It was shown in [42] that the following choice of M and R achieves this goal:

$$(3.11) \qquad M \geq nN, \quad \frac{R}{M} \geq 2.$$

4. Integrable nonlinear evolution PDEs on the half-line

The unified transform yields novel integral representations formulated in the complex k-plane (the Fourier plane) [44, 45]. These representations are similar to the integral representations for the linearized versions of these nonlinear PDEs, but also contain the entries of a certain matrix-valued function, which is the solution of a matrix Riemann-Hilbert (RH) problem.

We consider the nonlinear Schrödinger equation (NLS)

$$(4.1) \qquad iq_t + q_{xx} - 2\nu |q|^2 q = 0,$$

where ν equals $+1$ or -1 identifies the defocusing and focusing cases respectively.

Define the functions $a(k)$ and $b(k)$ in terms of the initial datum $q_0(x) = q(x,0)$ via a system of linear Volterra integral equations:

$$\begin{pmatrix} b(k) \\ a(k) \end{pmatrix} = v(0,k), \quad \mathrm{Im}\, k \geq 0,$$

where the vector $v(x,k)$ is the unique solution of

$$\left(\frac{d}{dx} + 2ik\begin{pmatrix} 1 & 0 \\ 0 & 0 \end{pmatrix}\right)v(x,k) = \begin{pmatrix} 0 & q_0(x) \\ \nu\bar{q}_0(x) & 0 \end{pmatrix}v(x,k),$$

$$\lim_{x \to \infty} v(x,k) = \begin{pmatrix} 0 \\ 1 \end{pmatrix}, \quad 0 < x < \infty, \quad \mathrm{Im}\, k \geq 0.$$

Define the functions $A(k)$ and $B(k)$ via the following system of linear Volterra integral equations, which involves both boundary values $q(0,t)$, $q_x(0,t)$:

$$A(k) = e^{2ik^2 t}\overline{V_2(t,\bar{k})}, \quad B(k) = -e^{2ik^2 t}V_1(t,k),$$

where the vector $V(t,k)$ with components $V_1(t,k)$ and $V_2(t,k)$ is the following solution of

$$\left(\frac{d}{dt} + 2ik^2\begin{pmatrix} 1 & 0 \\ 0 & 0 \end{pmatrix}\right)V(t,k)$$

$$= \left[2k\begin{pmatrix} 0 & q(0,t) \\ \nu\bar{q}(0,t) & 0 \end{pmatrix} + i\nu\begin{pmatrix} -|q(0,t)|^2 & \nu q_x(0,t) \\ -\bar{q}_x(0,t) & |q(0,t)|^2 \end{pmatrix}\right]V(t,k),$$

$$V(0,k) = \begin{pmatrix} 0 \\ 1 \end{pmatrix}.$$

Then, the following results are valid:

THEOREM 4.1. *Define $M(x,t,k)$ in terms of $\{a(k), b(k), A(k), B(k)\}$ as the solution of the following RH problem:*

(i) M *is analytic in* $k \in \mathbb{C} \setminus L$ *where L denotes the union of the real and the imaginary axes.*
(ii) $M = \mathrm{diag}\,(1,1) + O(1/k)$ *as* $k \to \infty$.
(iii) $M^- = M^+ J$, $k \in L$, *where:*

$$J_1 = \begin{bmatrix} 1 & 0 \\ \Gamma(k)e^{2ikx+4ik^2 t} & 1 \end{bmatrix}, \quad \Gamma(k) := \frac{\lambda \overline{B(\bar{k})}}{d(k)A(k)}, \quad d(k) := a(k)\overline{A(\bar{k})} - \lambda b(k)\overline{B(\bar{k})},$$

$$J_2 = J_3 J_4^{-1} J_1, \quad J_3 = \begin{bmatrix} 1 & \overline{\Gamma(\bar{k})}e^{-2ikx-4ik^2 t} \\ 0 & 1 \end{bmatrix}$$

$$J_4 = \begin{bmatrix} 1 & -\gamma(k)e^{-2ikx-4ik^2 t} \\ \overline{\gamma(\bar{k})}e^{2ikx+4ik^2 t} & 1-|\gamma|^2 \end{bmatrix}, \quad \gamma(k) := \frac{b(k)}{a(\bar{k})}.$$

Then M exists and is unique.

THEOREM 4.2. *Define $q(x,t)$ in terms of $\{a(k), b(k), A(k), B(k)\}$ by*

$$q(x,t) = -\frac{1}{\pi}\left\{\int_{\partial I_3} \overline{\Gamma(\bar{k})}e^{-2ikx-4ik^2 t} M_{11}^+ \, dk \right.$$

$$\left. + \int_{-\infty}^{\infty} \gamma(k)e^{-2ikx-4ik^2 t} M_{11}^+ \, dk + \int_{0}^{\infty} |\gamma(k)|^2 M_{12}^+ \, dk\right\},$$

where ∂I_3 denotes the boundary of the third quadrant of the complex k-plane. Then $q(x,t)$ solves the defocusing NLS and also satisfies $q(x,0) = q_0(x)$, and attains at $x = 0$ the boundary values used to define $A(k)$ and $B(k)$.

The main advantage of the new method is the fact that the above RH problem involves jump matrices with *explicit* (x,t)-dependence, uniquely defined in terms of four scalar functions $\{a(k), b(k), A(k), B(k)\}$.

A major difficulty is that some of these boundary values are unknown. For example, for the Dirichlet problem of the NLS, the Neumann boundary value $q_x(0,t)$ is unknown. It turns out that the problem of characterizing the unknown boundary values can be addressed by utilizing the so-called *global relation*, which is a simple algebraic equation that couples the spectral functions.

Linearizable boundary conditions

For a particular class of boundary conditions called linearizable, by employing the global relation it is possible to solve the problem on the half-line as effectively as the analogous problem on the full line. Indeed, for linearizable boundary conditions, by utilizing the global relation, it is possible to determine the functions $A(k)$ and $B(k)$ *directly*, without the need of determining first the unknown boundary values.

For example, for the celebrated NLS any of the following three different boundary conditions are linearizable [46]:

(4.2a) $$q(0,t) = 0,$$

(4.2b) $$q_x(0,t) = 0,$$

(4.2c) $$q_x(0,t) - \alpha q(0,t) = 0, \quad \alpha \text{ real constant.}$$

The linerized version of NLS equation, namely the equation

(4.3) $$iu_t + u_{xx} = 0,$$

with either the boundary conditions (4.2a) or (4.2b), just like the heat equation, can be solved by either the sine or the cosine transform. It is well known that these transforms can be obtained from the Fourier transform on the line, by using either an odd or an even extension. Hence, an equivalent approach to solve these IBVPs is first to use an odd or an even extension to the full line and then to employ the Fourier transform. Thus, it is natural to expect that such extensions are also applicable for the NLS equation. This is indeed the case: The NLS (4.1) with either the boundary conditions (4.2a) or (4.2b) were solved in [47] using an odd or an even extension, whereas equation (4.1) with the boundary conditions (4.2c) was solved in [48]–[51] using a slightly more complicated extension.

On the other hand, as stated earlier, *linear* evolution PDEs formulated on the half-line involving a *third*-order spatial derivative, *cannot* be solved by an x-transform (or equivalently, by an extension to the full line). However, such problems *can* be solved by the unified transform. The situation is similar for integrable nonlinear evolution PDEs: for the KdV equation with dominant surface tension, namely for the equation

(4.4) $$q_t + q_x + 6qq_x - q_{xxx} = 0,$$

either of the following set of boundary conditions is linearizable [52]:

(4.5a) $$q(0,t) = q_{xx}(0,t) = 0,$$

and

(4.5b) $\quad q(0,t) = \alpha, \quad q_{xx}(0,t) = \alpha + 3\alpha^2, \quad \alpha$ real constant.

Equation (4.4) with either (4.5a) or (4.5b) *cannot* be solved by an extension to the full line, but can be solved by the unified transform.

In summary, for those linearizable IBVPs for which the associated linearized PDE can be solved by an x- transform (or equivalently, by an extension to the full line), there exist two alternative approaches: (a) extension to the full line; (b) the Fokas method. On the other hand, for those problems for which the associated linearized PDEs cannot be solved by an x-transform, there exists only the Fokas method. For further details see [53].

Non-Linearizable boundary conditions

For non-linearizable boundary conditions, the complete solution of a boundary value problem on the half-line requires the determination of the unknown boundary values, i.e. it requires the characterization of the Dirichlet to Neumann map. This problem was recently analyzed in [54,55] using two different formulations, both of which are based on the analysis of the global relation.

It must be emphasized that for non-linearizable boundary conditions which decay for large t, by utilizing the crucial feature of the new method that it yields RH problems with explicit (x,t)-dependence, it is possible to obtain useful asymptotic information about the solution *without* characterizing the spectral functions $\{A(k), B(k)\}$ in terms of the given initial and boundary conditions. This can be achieved by employing the Deift-Zhou method for the long-time asymptotics [56]–[59] and the Deift-Zhou-Venakides method for the zero-dispersion limit [60].

Asymptotically periodic boundary conditions

For the physically significant case of boundary conditions which are periodic in t, it is not possible to obtain the rigorous form of the long-time asymptotics of the solution, without first characterizing the Dirichlet to Neumann map, at least as $t \to \infty$. Pioneering results in this directions have been obtained in a series of paper by Boutet de Monvel and coauthors [61]–[64] for the particular case that the given Dirichlet datum consists of a *single* periodic exponential:

(4.6) $\quad q(0,t) = ae^{i\omega t}, \quad a,\omega \in \mathbb{R}, \quad t > 0.$

In particular, it was shown in [63] for the focusing (i.e. $\nu = -1$) NLS that if q satisfies (4.6), then the Neumann values of q has the asymptotics

(4.7) $\quad q_x(0,t) \sim ce^{i\omega t}, \quad t \to \infty, \quad c \in \mathbb{C},$

provided that the values of $\{a, \omega, c\}$ satisfy either

(4.8a) $\quad c = \pm a\sqrt{\omega - a^2} \quad \text{and} \quad \omega \geq a^2,$

or

(4.8b) $\quad c = ia\sqrt{|\omega| + 2a^2} \quad \text{and} \quad \omega \leq -6a^2.$

The Dirichlet datum (4.6) is complex-valued, thus it cannot be used for the KdV and modified KdV equations in real situations. Results valid for Dirichlet data more general than (4.6), including the case of the real-valued Dirichlet datum

(4.9) $\quad q(0,t) = a\sin t, \quad a \in \mathbb{R}, \quad t > 0,$

can be obtained using the perturbation expansion of [**55**]. In particular, it was shown in [**55**] for the NLS, and in [**65**] for the mKdV, that if $q(0,t)$ is given by the right-hand side of (4.9), then the function $q_x(0,t)$ for the NLS and the functions $\{q_x(0,t), q_{xx}(0,t)\}$ for the mKdV, respectively, can be computed *explicitly* at least up to and including terms of $O(a^3)$ and furthermore the above functions become *periodic* as $t \to \infty$. Unfortunately, the perturbative approach of [**55**] is quite cumbersome. Actually, it becomes practically impossible to go beyond terms of $O(a^3)$, thus this approach *cannot* be used to establish that $q(0,t)$ for the NLS and $\{q_x(0,t), q_{xx}(0,t)\}$ for the mKdV indeed approach periodic functions as $t \to \infty$. A new much simpler perturbative approach is presented in [**66**]. Using this new approach, it is possible to find explicit expressions for the unknown boundary values to any order. For example, for the nonlinear Schrödinger equation with the boundary condition (4.9), the unknown boundary value $q_x(0,t)$ is computed in [**66**] up to $O(a^8)$.

5. Rigorous Results for Evolution PDEs

The rigorous foundation of the new method for *linear forced* evolution PDEs in Sobolev spaces is presented in [**67**], [**68**]. These results actually lead to a new approach for proving well posedness for *nonlinear* IBVPs. The crucial ingredient of this approach is to use for the linear version of the given nonlinear PDE, the formulae obtained via the new method. The authors of [**69–71**] have been able to prove well posedness for IBVPs using ideas similar to those used in the treatment of initial-value problems. In particular, one first obtains a solution formula for the linear IBVP with forcing and then uses this formula to derive appropriate linear estimates. Subsequently, one replaces the forcing in the linear formula by the nonlinearity and uses the linear estimates together with a contraction mapping argument to deduce well-posedness of the nonlinear IBVP.

It is often the case, however, that even the derivation of the linear solution formula is somewhat technical and unintuitive, not to mention the derivation of the relevant linear estimates. The main advantage of the new method is that it yields explicit formulae for forced linear evolution equations with arbitrary number of derivatives. Thus, it is not surprising that these "naturally emerging" linear formulae can be used to establish local well-posedness of nonlinear evolution IBVPs through a contraction mapping approach.

So far the above approach has been implemented to the nonlinear Schrödinger and to the KdV equations. If it can also be implemented to other nonlinear PDEs, such as the equation

$$q_t + q_{xxx} + q^\alpha q_x = 0, \quad 1 \leq \alpha \leq 2,$$

then this will be a significant development in the general methodology of treating nonlinear evolution equations via PDE techniques.

6. Further Developments

Pedagogical aspects of the new method are discussed in [**72**] and the accompanying editorial [**73**].

Systems of evolution PDEs are discussed in [**74**] and [**75**].

Boundary value problems with variable coefficients are discussed in [**76**] and [**77**].

Moving boundary value problems for linear evolution PDEs are discussed in [**78**] and [**79**].

Linear evolution PDEs with time periodic boundary conditions are discussed in [**80**] and [**81**].

Anthony Ashton employing the new method has developed a remarkable formalism for the rigorous analysis of elliptic PDEs, see for example [**82, 83**].

For early results regarding the characterization of the Dirichlet to Neumann map for integrable nonlinear evolution PDEs see [**84**] and [**85**]. For more recent results see [**86**]–[**92**].

Integrable nonlinear elliptic PDEs are analyzed in [**93**] and [94].

Linear evolution PDEs with either non-separable or other complicated boundary conditions are analyzed in [**95**]–[**101**].

A new approach for analyzing problems in elasticity is introduced in [**102**].

The new method can be extended to three dimensions, see for example [**103**]–[**107**].

Reviews of the new method for linear and for integrable nonlinear PDEs are presented in [**108**] and [**109**], respectively.

References

[1] A. S. Fokas, *Lax pairs: a novel type of separability*, Inverse Problems **25** (2009), no. 12, 123007, 44, DOI 10.1088/0266-5611/25/12/123007. MR2565573 (2010k:37112)

[2] A. S. Fokas and I. M. Gel′fand, *Integrability of linear and nonlinear evolution equations and the associated nonlinear Fourier transforms*, Lett. Math. Phys. **32** (1994), no. 3, 189–210, DOI 10.1007/BF00750662. MR1299036 (96h:35206)

[3] A. S. Fokas, *A unified transform method for solving linear and certain nonlinear PDEs*, Proc. Roy. Soc. London Ser. A **453** (1997), no. 1962, 1411–1443, DOI 10.1098/rspa.1997.0077. MR1469927 (98e:35007)

[4] A. S. Fokas, *On the integrability of linear and nonlinear partial differential equations*, J. Math. Phys. **41** (2000), no. 6, 4188–4237, DOI 10.1063/1.533339. MR1768651 (2001k:37121)

[5] Athanassios S. Fokas, *A unified approach to boundary value problems*, CBMS-NSF Regional Conference Series in Applied Mathematics, vol. 78, Society for Industrial and Applied Mathematics (SIAM), Philadelphia, PA, 2008. MR2451953 (2010b:35038)

[6] A. S. Fokas and M. Zyskin, *The fundamental differential form and boundary-value problems*, Quart. J. Mech. Appl. Math. **55** (2002), no. 3, 457–479, DOI 10.1093/qjmam/55.3.457. MR1919978 (2003f:35267)

[7] E. A. Spence and A. S. Fokas, *A new transform method I: domain-dependent fundamental solutions and integral representations*, Proc. R. Soc. Lond. Ser. A Math. Phys. Eng. Sci. **466** (2010), no. 2120, 2259–2281, DOI 10.1098/rspa.2009.0512. MR2659494 (2011e:35004)

[8] E. A. Spence and A. S. Fokas, *A new transform method II: the global relation and boundary-value problems in polar coordinates*, Proc. R. Soc. Lond. Ser. A Math. Phys. Eng. Sci. **466** (2010), no. 2120, 2283–2307, DOI 10.1098/rspa.2009.0513. MR2659495 (2011f:35008)

[9] A. S. Fokas and E. A. Spence, *Synthesis, as opposed to separation, of variables*, SIAM Rev. **54** (2012), no. 2, 291–324, DOI 10.1137/100809647. MR2916309

[10] David A. Smith, *Well-posed two-point initial-boundary value problems with arbitrary boundary conditions*, Math. Proc. Cambridge Philos. Soc. **152** (2012), no. 3, 473–496, DOI 10.1017/S030500411100082X. MR2911141

[11] Vassilis G. Papanicolaou, *An example where separation of variables fails*, J. Math. Anal. Appl. **373** (2011), no. 2, 739–744, DOI 10.1016/j.jmaa.2010.07.035. MR2720717 (2012d:35040)

[12] N. Flyer and A. S. Fokas, *A hybrid analytical-numerical method for solving evolution partial differential equations. I. The half-line*, Proc. R. Soc. Lond. Ser. A Math. Phys. Eng. Sci. **464** (2008), no. 2095, 1823–1849, DOI 10.1098/rspa.2008.0041. MR2403130 (2009d:65135)

[13] A. S. Fokas and B. Pelloni, *A transform method for linear evolution PDEs on a finite interval*, IMA J. Appl. Math. **70** (2005), no. 4, 564–587, DOI 10.1093/imamat/hxh047. MR2156459 (2006e:35011)

[14] Beatrice Pelloni, *The spectral representation of two-point boundary-value problems for third-order linear evolution partial differential equations*, Proc. R. Soc. Lond. Ser. A Math. Phys. Eng. Sci. **461** (2005), no. 2061, 2965–2984, DOI 10.1098/rspa.2005.1474. MR2165521 (2006e:35043)

[15] A. S. Fokas, *Two-dimensional linear partial differential equations in a convex polygon*, R. Soc. Lond. Proc. Ser. A Math. Phys. Eng. Sci. **457** (2001), no. 2006, 371–393, DOI 10.1098/rspa.2000.0671. MR1848093 (2002j:35084)

[16] D. G. Crowdy and A. S. Fokas, *Explicit integral solutions for the plane elastostatic semi-strip*, Proc. R. Soc. Lond. Ser. A Math. Phys. Eng. Sci. **460** (2004), no. 2045, 1285–1309, DOI 10.1098/rspa.2003.1206. MR2066407 (2005c:74027)

[17] Daniel ben-Avraham and Athanassios S. Fokas, *The solution of the modified Helmholtz equation in a wedge and an application to diffusion-limited coalescence*, Phys. Lett. A **263** (1999), no. 4-6, 355–359, DOI 10.1016/S0375-9601(99)00698-2. MR1732097

[18] D. ben-Avraham and A. S. Fokas, *The modified Helmholtz equation in a triangular domain and an application to diffusion-limited coalescence*, Phys. Rev. E **64**, 016114, 2001.

[19] A. S. Fokas and A. A. Kapaev, *A Riemann-Hilbert approach to the Laplace equation*, J. Math. Anal. Appl. **251** (2000), no. 2, 770–804, DOI 10.1006/jmaa.2000.7052. MR1794770 (2001k:35241)

[20] G. Dassios and A. S. Fokas, *The basic elliptic equations in an equilateral triangle*, Proc. R. Soc. Lond. Ser. A Math. Phys. Eng. Sci. **461** (2005), no. 2061, 2721–2748, DOI 10.1098/rspa.2005.1466. MR2165508 (2006f:35048)

[21] Y. A. Antipov and A. S. Fokas, *The modified Helmholtz equation in a semi-strip*, Math. Proc. Cambridge Philos. Soc. **138** (2005), no. 2, 339–365, DOI 10.1017/S0305004104008205. MR2132175 (2005k:35058)

[22] A. S. Fokas and A. A. Kapaev, *On a transform method for the Laplace equation in a polygon*, IMA J. Appl. Math. **68** (2003), no. 4, 355–408, DOI 10.1093/imamat/68.4.355. MR1988152 (2004c:37176)

[23] G. Dassios and M. Doschoris, *On the global relation and the Dirichlet-to-Neumann correspondence*, Stud. Appl. Math. **126** (2011), no. 1, 75–102, DOI 10.1111/j.1467-9590.2010.00498.x. MR2724565 (2012c:35010)

[24] G. Dassios and M. Doschoris, *Axisymmetric Stokes' flow in a spherical shell revisited via the Fokas method. Part I: irrotational flow*, Math. Methods Appl. Sci. **34** (2011), no. 7, 850–868, DOI 10.1002/mma.1407. MR2815775 (2012d:76044)

[25] Darren G. Crowdy and Anthony M. J. Davis, *Stokes flow singularities in a two-dimensional channel: a novel transform approach with application to microswimming*, Proc. R. Soc. Lond. Ser. A Math. Phys. Eng. Sci. **469** (2013), no. 2157, 20130198, 14, DOI 10.1098/rspa.2013.0198. MR3078201

[26] A. S. Fokas and K. Kalimeris, *Eigenvalues for the Laplace operator in the interior of an equilateral triangle*, Comput. Methods Funct. Theory **14** (2014), no. 1, 1–33, DOI 10.1007/s40315-013-0038-7. MR3194311

[27] M. Dimakos and A. S. Fokas, *The Poisson and the Biharmonic Equations in the Interior of a Convex Polygon*, Stud. Appl. Math. **134** (2015), no. 4, 456–498.

[28] E. Spence, *Boundary value problems for linear elliptic PDEs*, PhD Thesis, Cambridge University, 2010.

[29] K. Kalimeris, *Initial and Boundary Value Problems in two and three dimensions*, PhD Thesis, Cambridge University, 2010.

[30] G. Baganis and M. Hadjinicolaou, *Analytic solution of an exterior Dirichlet problem in a non-convex domain*, IMA J. Appl. Math. **74** (2009), no. 5, 668–684, DOI 10.1093/imamat/hxp023. MR2549954 (2011a:35124)

[31] Min-Hai Huang and Yu-Qiu Zhao, *High-frequency asymptotics for the modified Helmholtz equation in a quarter-plane*, Appl. Anal. **90** (2011), no. 12, 1927–1938, DOI 10.1080/00036811.2010.534858. MR2847497 (2012k:35073)

[32] Bengt Fornberg and Natasha Flyer, *A numerical implementation of Fokas boundary integral approach: Laplace's equation on a polygonal domain*, Proc. R. Soc. Lond. Ser. A Math. Phys. Eng. Sci. **467** (2011), no. 2134, 2983–3003, DOI 10.1098/rspa.2011.0032. MR2835606

[33] Christopher-Ian Raphaël Davis and Bengt Fornberg, *A spectrally accurate numerical implementation of the Fokas transform method for Helmholtz-type PDEs*, Complex Var. Elliptic Equ. **59** (2014), no. 4, 564–577, DOI 10.1080/17476933.2013.766883. MR3177010

[34] C.-I. Davis, *Numerical tests of the Fokas method for Helmholtz-type partial differential equations: Dirichlet to Neumann Maps*, Masters Thesis, University of Colorado at Boulder, 2008.

[35] A. S. Fokas, N. Flyer, S. A. Smitheman, and E. A. Spence, *A semi-analytical numerical method for solving evolution and elliptic partial differential equations*, J. Comput. Appl. Math. **227** (2009), no. 1, 59–74, DOI 10.1016/j.cam.2008.07.036. MR2512760 (2010d:65276)

[36] S. R. Fulton, A. S. Fokas, and C. A. Xenophontos, *An analytical method for linear elliptic PDEs and its numerical implementation*, J. Comput. Appl. Math. **167** (2004), no. 2, 465–483, DOI 10.1016/j.cam.2003.10.012. MR2064703 (2005d:65216)

[37] Y. G. Saridakis, A. G. Sifalakis, and E. P. Papadopoulou, *Efficient numerical solution of the generalized Dirichlet-Neumann map for linear elliptic PDEs in regular polygon domains*, J. Comput. Appl. Math. **236** (2012), no. 9, 2515–2528, DOI 10.1016/j.cam.2011.12.011. MR2879717 (2012m:65077)

[38] A. G. Sifalakis, S. R. Fulton, E. P. Papadopoulou, and Y. G. Saridakis, *Direct and iterative solution of the generalized Dirichlet-Neumann map for elliptic PDEs on square domains*, J. Comput. Appl. Math. **227** (2009), no. 1, 171–184, DOI 10.1016/j.cam.2008.07.025. MR2512770 (2010d:35061)

[39] A. G. Sifalakis, E. P. Papadopoulou and Y. G. Saridakis, *Numerical study of iterative methods for the solution of the Dirichlet-Neumann map for linear elliptic PDEs on regular polygon domains*, Int. J. Appl. Math. Comput. Sci. **4**, pp. 173–178, 2007.

[40] A. G. Sifalakis, A. S. Fokas, S. R. Fulton, and Y. G. Saridakis, *The generalized Dirichlet-Neumann map for linear elliptic PDEs and its numerical implementation*, J. Comput. Appl. Math. **219** (2008), no. 1, 9–34, DOI 10.1016/j.cam.2007.07.012. MR2437692 (2009g:35042)

[41] S. A. Smitheman, E. A. Spence and A. S. Fokas, IMA J. Num. Anal., doi: 10.1093/imanum/drn079, 2009.

[42] P. Hashemzadeh, A. S. Fokas and S. A. Smitheman, *A Numerical Technique for Linear Elliptic PDEs in Polygonal Domains*, Proc. R. Soc. A **471** (2015), no. 2175.

[43] A. S. Fokas, A. Iserles, and S. A. Smitheman, *The Unified Method in Polygonal Domains via the Explicit Fourier Transform of Legendre Polynomials*, in A. S. Fokas and B. Pelloni (EDs), *Unified transform method for boundary value problems: applications and advances*, SIAM, 2014.

[44] A. S. Fokas, *Integrable nonlinear evolution equations on the half-line*, Comm. Math. Phys. **230** (2002), no. 1, 1–39, DOI 10.1007/s00220-002-0681-8. MR1930570 (2004d:37100)

[45] A. S. Fokas, A. R. Its, and L.-Y. Sung, *The nonlinear Schrödinger equation on the half-line*, Nonlinearity **18** (2005), no. 4, 1771–1822, DOI 10.1088/0951-7715/18/4/019. MR2150354 (2006c:37074)

[46] E. K. Sklyanin, *Boundary conditions for integrable equations* (Russian), Funktsional. Anal. i Prilozhen. **21** (1987), no. 2, 86–87. MR902305 (88g:35181)

[47] M. J. Ablowitz and H. Segur, *The inverse scattering transform: semi-infinite interval*, J. Math. Phys. **16**, pp. 1054–1056, 1975.

[48] R. F. Bikbaev and V. O. Tarasov, *Initial-boundary value problem for the nonlinear Schrödinger equation*, J. Phys. A **24** (1991), no. 11, 2507–2516. MR1117621 (92d:35267)

[49] A. S. Fokas, *An initial-boundary value problem for the nonlinear Schrödinger equation*, Phys. D **35** (1989), no. 1-2, 167–185, DOI 10.1016/0167-2789(89)90101-2. MR1004191 (90h:35206)

[50] I. T. Khabibullin, *Integrable initial-boundary value problems* (Russian, with English summary), Teoret. Mat. Fiz. **86** (1991), no. 1, 43–52, DOI 10.1007/BF01018494; English transl., Theoret. and Math. Phys. **86** (1991), no. 1, 28–36. MR1106826 (92a:35146)

[51] V. O. Tarasov, *The integrable initial-boundary value problem on a semiline: nonlinear Schrödinger and sine-Gordon equations*, Inverse Problems **7** (1991), no. 3, 435–449. MR1108285 (92f:35133)

[52] Burak Gürel, Metin Gürses, and Ismagil Habibullin, *Boundary value problems compatible with symmetries*, Phys. Lett. A **190** (1994), no. 3-4, 231–237, DOI 10.1016/0375-9601(94)90747-1. MR1285789 (95c:35215)

[53] G. Biondini, A. S. Fokas and D. Shepalsky, *Comparison of Two Approaches to the IBVP for the NLS Equation on the Half-Line with Robin Boundary Conditions*, in A. S. Fokas

and B. Pelloni (EDs), *Unified transform method for boundary value problems: applications and advances*, SIAM, 2014.

[54] A. S. Fokas and J. Lenells, *The unified method: I. Nonlinearizable problems on the half-line*, J. Phys. A **45** (2012), no. 19, 195201, 38, DOI 10.1088/1751-8113/45/19/195201. MR2924497

[55] J. Lenells and A. S. Fokas, *The unified method: II. NLS on the half-line with t-periodic boundary conditions*, J. Phys. A **45** (2012), no. 19, 195202, 36, DOI 10.1088/1751-8113/45/19/195202. MR2924498

[56] A. S. Fokas and A. R. Its, *An initial-boundary value problem for the sine-Gordon equation in laboratory coordinates* (English, with English and Russian summaries), Teoret. Mat. Fiz. **92** (1992), no. 3, 387–403, DOI 10.1007/BF01017074; English transl., Theoret. and Math. Phys. **92** (1992), no. 3, 964–978 (1993). MR1225785 (94f:35123)

[57] A. S. Fokas and A. R. Its, *Soliton generation for initial-boundary value problems*, Phys. Rev. Lett. **68** (1992), no. 21, 3117–3120, DOI 10.1103/PhysRevLett.68.3117. MR1163545 (93a:35130)

[58] A. S. Fokas and A. R. Its, *An initial-boundary value problem for the Korteweg-de Vries equation*, Math. Comput. Simulation **37** (1994), no. 4-5, 293–321, DOI 10.1016/0378-4754(94)00021-2. Solitons, nonlinear wave equations and computation (New Brunswick, NJ, 1992). MR1308105 (95m:35162)

[59] A. S. Fokas and A. R. Its, *The linearization of the initial-boundary value problem of the nonlinear Schrödinger equation*, SIAM J. Math. Anal. **27** (1996), no. 3, 738–764, DOI 10.1137/0527040. MR1382831 (97d:35204)

[60] A. S. Fokas and S. Kamvissis, *Zero-dispersion limit for integrable equations on the half-line with linearisable data*, Abstr. Appl. Anal. **5** (2004), 361–370, DOI 10.1155/S1085337504306093. MR2063331 (2005a:37121)

[61] Anne Boutet de Monvel, Alexander Its, and Vladimir Kotlyarov, *Long-time asymptotics for the focusing NLS equation with time-periodic boundary condition* (English, with English and French summaries), C. R. Math. Acad. Sci. Paris **345** (2007), no. 11, 615–620, DOI 10.1016/j.crma.2007.10.018. MR2371477 (2008m:35298)

[62] Anne Boutet de Monvel, Alexander Its, and Vladimir Kotlyarov, *Long-time asymptotics for the focusing NLS equation with time-periodic boundary condition on the half-line*, Comm. Math. Phys. **290** (2009), no. 2, 479–522, DOI 10.1007/s00220-009-0848-7. MR2525628 (2010i:37169)

[63] Anne Boutet de Monvel, Vladimir Kotlyarov, and Dmitry Shepelsky, *Decaying long-time asymptotics for the focusing NLS equation with periodic boundary condition*, Int. Math. Res. Not. IMRN **3** (2009), 547–577, DOI 10.1093/imrn/rnn139. MR2482124 (2010e:35254)

[64] Anne Boutet de Monvel, Vladimir P. Kotlyarov, Dmitry Shepelsky, and Chunxiong Zheng, *Initial boundary value problems for integrable systems: towards the long time asymptotics*, Nonlinearity **23** (2010), no. 10, 2483–2499, DOI 10.1088/0951-7715/23/10/007. MR2683777 (2011i:37094)

[65] Guenbo Hwang and A. S. Fokas, *The modified Korteweg-de Vries equation on the half-line with a sine-wave as Dirichlet datum*, J. Nonlinear Math. Phys. **20** (2013), no. 1, 135–157, DOI 10.1080/14029251.2013.792492. MR3202131

[66] A. S. Fokas and J. Lenells, *Perturbative and exact results on the Neumann value for the nonlinear Schrödinger equation on the half-line*, Journal of Physics: Conference Series **482**, 012015, 2014.

[67] A. S. Fokas, A. A. Himonas, and D. Mantzavinos, *The nonlinear Schrödinger equation on the half-line*, Trans. Amer. Math. Soc. (2015), DOI 10.1090/tran/6734.

[68] A. S. Fokas, A. A. Himonas, and D. Mantzavinos, *The Korteweg-de Vries equation on the half-line* (2015, submitted).

[69] Jerry L. Bona, S. M. Sun, and Bing-Yu Zhang, *A non-homogeneous boundary-value problem for the Korteweg-de Vries equation in a quarter plane*, Trans. Amer. Math. Soc. **354** (2002), no. 2, 427–490, DOI 10.1090/S0002-9947-01-02885-9. MR1862556 (2002h:35258)

[70] J. E. Colliander and C. E. Kenig, *The generalized Korteweg-de Vries equation on the half line*, Comm. Partial Differential Equations **27** (2002), no. 11-12, 2187–2266, DOI 10.1081/PDE-120016157. MR1944029 (2004m:35226)

[71] Justin Holmer, *The initial-boundary-value problem for the 1D nonlinear Schrödinger equation on the half-line*, Differential Integral Equations **18** (2005), no. 6, 647–668. MR2136703 (2006d:35258)

[72] Bernard Deconinck, Thomas Trogdon, and Vishal Vasan, *The method of Fokas for solving linear partial differential equations*, SIAM Rev. **56** (2014), no. 1, 159–186, DOI 10.1137/110821871. MR3246302

[73] L. F. Rossi, *Education (Editorial)*, SIAM Review **56**(1), pp. 157–158, 2014.

[74] P. A. Treharne and A. S. Fokas, *Boundary value problems for systems of linear evolution equations*, IMA J. Appl. Math. **69** (2004), no. 6, 539–555, DOI 10.1093/imamat/69.6.539. MR2101843 (2005h:35313)

[75] A. S. Fokas and B. Pelloni, *Boundary value problems for Boussinesq type systems*, Math. Phys. Anal. Geom. **8** (2005), no. 1, 59–96, DOI 10.1007/s11040-004-1650-6. MR2136653 (2006i:35312)

[76] A. S. Fokas, *Boundary-value problems for linear PDEs with variable coefficients*, Proc. R. Soc. Lond. Ser. A Math. Phys. Eng. Sci. **460** (2004), no. 2044, 1131–1151, DOI 10.1098/rspa.2003.1208. MR2133859 (2006a:35260)

[77] P. A. Treharne and A. S. Fokas, *Initial-boundary value problems for linear PDEs with variable coefficients*, Math. Proc. Cambridge Philos. Soc. **143** (2007), no. 1, 221–242, DOI 10.1017/S0305004107000084. MR2340985 (2009d:35274)

[78] A. S. Fokas and B. Pelloni, *Generalized Dirichlet-to-Neumann map in time-dependent domains*, Stud. Appl. Math. **129** (2012), no. 1, 51–90, DOI 10.1111/j.1467-9590.2011.00545.x. MR2946204

[79] A. S. Fokas and S. De Lillo, *The unified transform for linear, linearizable and integrable nonlinear partial differential equations*, Phys. Scr. **89**, 038004, 2013.

[80] Guillaume Michel Dujardin, *Asymptotics of linear initial boundary value problems with periodic boundary data on the half-line and finite intervals*, Proc. R. Soc. Lond. Ser. A Math. Phys. Eng. Sci. **465** (2009), no. 2111, 3341–3360, DOI 10.1098/rspa.2009.0194. With supplementary data available online. MR2545299 (2011d:35102)

[81] J. L. Bona and A. S. Fokas, *Initial-boundary-value problems for linear and integrable nonlinear dispersive partial differential equations*, Nonlinearity **21** (2008), no. 10, T195–T203, DOI 10.1088/0951-7715/21/10/T03. MR2439474 (2009j:35298)

[82] A. C. L. Ashton, *On the rigorous foundations of the Fokas method for linear elliptic partial differential equations*, Proc. R. Soc. Lond. Ser. A Math. Phys. Eng. Sci. **468** (2012), no. 2141, 1325–1331, DOI 10.1098/rspa.2011.0478. MR2910351

[83] A. C. L. Ashton, *The spectral Dirichlet-Neumann map for Laplace's equation in a convex polygon*, SIAM J. Math. Anal. **45** (2013), no. 6, 3575–3591, DOI 10.1137/13090523X. MR3134426

[84] A. S. Fokas, *The generalized Dirichlet-to-Neumann map for certain nonlinear evolution PDEs*, Comm. Pure Appl. Math. **58** (2005), no. 5, 639–670, DOI 10.1002/cpa.20076. MR2141894 (2006i:35311)

[85] A. Boutet de Monvel, A. S. Fokas, and D. Shepelsky, *Analysis of the global relation for the nonlinear Schrödinger equation on the half-line*, Lett. Math. Phys. **65** (2003), no. 3, 199–212, DOI 10.1023/B:MATH.0000010711.66380.77. MR2033706 (2005b:35254)

[86] Guenbo Hwang, *A perturbative approach for the asymptotic evaluation of the Neumann value corresponding to the Dirichlet datum of a single periodic exponential for the NLS*, J. Nonlinear Math. Phys. **21** (2014), no. 2, 225–247, DOI 10.1080/14029251.2014.905298. MR3206243

[87] K. Kalimeris, *Explicit soliton asymptotics for the nonlinear Schrödinger equation on the half-line*, J. Nonlinear Math. Phys. **17** (2010), no. 4, 445–452, DOI 10.1142/S1402925110000994. MR2771185 (2011m:35354)

[88] Jonatan Lenells, *The solution of the global relation for the derivative nonlinear Schrödinger equation on the half-line*, Phys. D **240** (2011), no. 6, 512–525, DOI 10.1016/j.physd.2010.11.004. MR2755178 (2011m:35358)

[89] A. S. Fokas and J. Lenells, *Explicit soliton asymptotics for the Korteweg-de Vries equation on the half-line*, Nonlinearity **23** (2010), no. 4, 937–976, DOI 10.1088/0951-7715/23/4/010. MR2608593 (2011b:35447)

[90] D. C. Antonopoulou and S. Kamvissis, *On the Dirichlet to Neumann problem for the 1D cubic NLS equation*, (preprint).

[91] J. Lenells and A. S. Fokas, *The nonlinear Schrodinger equation with t-periodic data: I. Exact results*, Proc. R. Soc. A **471** (2015), no. 2181.

[92] J. Lenells and A. S. Fokas, *The nonlinear Schrodinger equation with t-periodic data: II. Perturbative results*, Proc. R. Soc. A **471** (2015), no. 2181.

[93] B. Pelloni and D. A. Pinotsis, *The elliptic sine-Gordon equation in a half plane*, Nonlinearity **23** (2010), no. 1, 77–88, DOI 10.1088/0951-7715/23/1/004. MR2576374 (2011c:35195)

[94] A. S. Fokas, J. Lenells, and B. Pelloni, *Boundary value problems for the elliptic sine-Gordon equation in a semi-strip*, J. Nonlinear Sci. **23** (2013), no. 2, 241–282, DOI 10.1007/s00332-012-9150-5. MR3041625

[95] A. S. Fokas and D. T. Papageorgiou, *Absolute and convective instability for evolution PDEs on the half-line*, Stud. Appl. Math. **114** (2005), no. 1, 95–114, DOI 10.1111/j.0022-2526.2005.01541.x. MR2117328 (2006e:35334)

[96] Dionyssios Mantzavinos and Athanassios S. Fokas, *The unified method for the heat equation: I. Non-separable boundary conditions and non-local constraints in one dimension*, European J. Appl. Math. **24** (2013), no. 6, 857–886, DOI 10.1017/S0956792513000223. MR3181485

[97] B. Deconinck, B. Pelloni and N. E. Sheils, *Non-steady-state heat conduction in composite walls*, Proc. Roy. Soc. A **470**, 2165: 22pp., 2014.

[98] Natalie E. Sheils and Bernard Deconinck, *Heat conduction on the ring: interface problems with periodic boundary conditions*, Appl. Math. Lett. **37** (2014), 107–111, DOI 10.1016/j.aml.2014.06.006. MR3231736

[99] M. Asvestas, A. G. Sifalakis, E. P. Papadopoulou and Y. G. Saridakis, *Fokas method for a multi-domain linear reaction-diffusion equation with discontinuous diffusivity*, Journal of Physics: Conference Series **490**, 012143, doi:10.1088/1742-6596/490/1/012143, 2014.

[100] D. Mantzavinos, M. Papadomanolaki, Y. Saridakis and A. Sifalakis, *A novel transform approach for a brain tumor invasion model with heterogeneous diffusion in 1+1 dimensions*, Appl. Num. Math., doi: 10.1016/j.apnum.2014.09.006, 2014.

[101] Natalie E. Sheils and Bernard Deconinck, *Heat conduction on the ring: interface problems with periodic boundary conditions*, Appl. Math. Lett. **37** (2014), 107–111, DOI 10.1016/j.aml.2014.06.006. MR3231736

[102] A. Its, E. Its and J. Kaplunov, *Riemann-Hilbert Approach to the Elastodynamical Equation*, Letters in Mathematical Physics **96**, N 1–3, 2011.

[103] A. S. Fokas, *A new transform method for evolution partial differential equations*, IMA J. Appl. Math. **67** (2002), no. 6, 559–590, DOI 10.1093/imamat/67.6.559. MR1942267 (2003j:35005)

[104] David M. Ambrose and David P. Nicholls, *Fokas integral equations for three dimensional layered-media scattering*, J. Comput. Phys. **276** (2014), 1–25, DOI 10.1016/j.jcp.2014.07.018. MR3252567

[105] D. P. Nicholls, *A High-Order Perturbation of Surfaces (HOPS) Approach to Fokas Integral Equations: Three Dimensional Layered-Media Scattering*, Quarterly of Applied Mathematics, (to appear).

[106] M. J. Ablowitz, A. S. Fokas, and Z. H. Musslimani, *On a new non-local formulation of water waves*, J. Fluid Mech. **562** (2006), 313–343, DOI 10.1017/S0022112006001091. MR2263547 (2007k:76013)

[107] G. Dassios and A. S. Fokas, *Methods for solving elliptic PDEs in spherical coordinates*, SIAM J. Appl. Math. **68** (2008), no. 4, 1080–1096, DOI 10.1137/070679223. MR2390980 (2009h:35038)

[108] George Dassios, *What non-linear methods offered to linear problems? The Fokas transform method*, Internat. J. Non-Linear Mech. **42** (2007), no. 1, 146–156, DOI 10.1016/j.ijnonlinmec.2006.11.010. MR2313765 (2007m:35259)

[109] Beatrice Pelloni, *Advances in the study of boundary value problems for nonlinear integrable PDEs*, Nonlinearity **28** (2015), no. 2, R1–R38. MR3303170

DEPARTMENT OF APPLIED MATHEMATICS AND THEORETICAL PHYSICS, UNIVERSITY OF CAMBRIDGE, CAMBRIDGE, CB30WA, UNITED KINGDOM – AND – RESEARCH CENTER OF MATHEMATICS, ACADEMY OF ATHENS, SORANOU EFESSIOU 4, ATHENS 11527, GREECE

E-mail address: t.fokas@damtp.cam.ac.uk

Semi-homogeneous maps

Wen-Fong Ke, Hubert Kiechle, Günter Pilz, and Gerhard Wendt

ABSTRACT. Let V be a vector space over a field F. A functional $f : V \to F$ is called "homogeneous (of degree 1)" if $f(\lambda v) = \lambda f(v)$ holds for all $v \in V$ and all $\lambda \in F$, and "semi-homogeneous" if this equations only holds for all $\lambda \in \operatorname{Im}(f)$. If $F = \mathbb{R}$ and $\operatorname{Im}(f) = \mathbb{R}^+ = \{x \in \mathbb{R} \mid x \geq 0\}$, a semi-homogeneous map is usually called "positively homogeneous". Certainly the most famous result in this area is *Euler's Homogeneous Function Theorem* (Kudryavtsev (2002)): If $f : \mathbb{R}^n \to \mathbb{R}$ is continuously differentiable then it is semi-homogeneous of degree 1 if and only if $v \cdot \nabla f(v) = f(v)$ holds for all $v \in \mathbb{R}^n$. Observe that a semi-homogeneous functional on V is automatically linear if $\dim(V) = 1$ and "often" linear in the other cases (cf. Fuchs et al. (1991) and Maxson and Van der Merwe (2002)).

In this paper, we will completely characterize semi-homogeneous functionals. In fact, we will do this in a much more general context. That is, we will characterize semi-homogeneous maps from a set S to a certain monoid G acting on S. It will come as a pleasant surprise that this general view on semi-homogeneous maps gives information on two completely different topics, namely of algebraic structures called "planar nearrings" and of fixed point free monoid actions on sets. In a certain sense, semi-homogeneous maps, fixed point free monoid actions, and planar nearrings are basically the same.

1. Semi-homogeneous maps and monoid actions

DEFINITION 1.1. Consider a monoid G acting on a nonempty set S, where we assume that the unit element 1 fulfills $1 \cdot s = s$ for all $s \in S$. A map $f : S \to G$ is called *semi-homogeneous* if

$$f(g \cdot s) = gf(s) \quad \text{for all } g \in \operatorname{Im}(f) \text{ and } s \in S.$$

This can be rewritten into (and is equivalent to)

$$f(f(s) \cdot t) = f(s)f(t) \quad \text{for all } s, t \in S.$$

From this we can immediately derive that $\operatorname{Im}(f)$ is a sub-semigroup of G. In the case of a vector space V over a field F, $(F, \cdot)$ acts on the set V. Another example is the following.

2010 *Mathematics Subject Classification.* Primary 26E; Secondary 16Y30.
Key words and phrases. Semi-homogeneous map, G-act set, semigroup, planar nearring.
The first author was partially supported by the Ministry of Science and Technology, Taiwan, project no. MOST 103-2115-M-006-006-MY2.
The fourth author was supported by Austrian Science Fund FWF under Project no. 23689-N18.

EXAMPLE 1.2. For any nonempty set S, the monoid $G = (S^S, \circ)$ of all maps from S to itself acts naturally on $M := S$. Given an arbitrary map $f : S \to G$; $s \mapsto f_s$, we can turn S into a groupoid by defining $s * t = f_s(t)$ for $s, t \in S$. Furthermore, $*$ is associative if and only $f_{f_s(t)} = f_s \circ f_t$ for all $s, t \in S$, which is equivalent to saying that f is semi-homogeneous. Therefore, every semi-homogeneous map from a set S to S^S gives rise to a semigroup structure on S.

One can also go the other way:

EXAMPLE 1.3. Let $(S, *)$ be a monoid. Select some element $s_0 \in S$ and define a map $f : S \to S$ by $f(s) = s * s_0$. Then f is easily seen to be (semi-)homogeneous. Different choices of s_0 give, in general, different semi-homogeneous maps.

When G is a group, we can determine the semi-homogeneous maps in more detail.

LEMMA 1.4. Let the group G act on the set S, $f : S \to G$ a semi-homogeneous function, and set $H = \{h \in G \mid f(h \cdot s) = hf(s) \text{ for all } s \in S\}$. Then $\mathrm{Im}(f) = H$, and this is a subgroup of G which acts fixed point freely on S.

PROOF. First assume that $f : S \to G$ is a semi-homogeneous map with G being a group. Then H in the statement is clearly a semigroup. Moreover, take $h \in H$; we have $f(s) = f(hh^{-1} \cdot s) = hf(h^{-1} \cdot s)$, and so $f(h^{-1} \cdot s) = h^{-1}f(s)$. Since $s \in S$ is arbitrary, $h^{-1} \in H$. This shows that H is a subgroup of G. By definition, $\mathrm{Im}(f) \subseteq H$. On the other hand, let $h \in H$ and $s \in S$. Then

$$h = hf(s)^{-1}f(s) = f((hf(s)^{-1}) \cdot s) \in \mathrm{Im}(f).$$

Hence $H = \mathrm{Im}(f)$. Further, assume that $h \cdot s = s$ for some $s \in S$ and $h \in H$. Then $f(s) = f(h \cdot s) = hf(s)$. Thus $h = 1$. This means that if $h \neq 1$, then $h \cdot s \neq s$ for all $s \in S$, so H is a fixed point free subgroup of G. □

REMARK 1.5. If G is just a monoid, then H is a semigroup containing the inverses for all invertible elements from H. If $\mathrm{Im}(f)$ contains an invertible element, then the proof above shows $H = \mathrm{Im}(f)$. In this case, if H satisfies the cancellation law, then H acts fixed point freely on S.

A zero element 0 of a structure $(B, *)$ is one such that $0 * b = b * 0 = 0$ for all $b \in B$. For any subset A of B, we use the notation $A^* = A \setminus \{0\}$. A semigroup can contain at most one zero element, and such an element is also called *absorbing*.

Now, we assume that G is a group with zero, i.e., G is a monoid with zero such that G^* is a group. For example, any subgroup H of the multiplicative group of a field gives a group with zero $H_0 = H \cup \{0\}$. More generally, any group H can be simply made into a group with zero by attaching a new element 0 to H and extending the multiplication in the obvious way. We will denote this group with zero by H_0.

EXAMPLE 1.6. If the group H acts faithfully on an abelian group as a group of automorphisms, then H_0 can be viewed as a substructure of the endomorphism ring. In particular, this applies to any subgroup of a field acting on a vector space.

Let G be a group with zero acting on a set S. Put $Z = 0 \cdot S$ and $S^\times = S \setminus Z$. We collect some more or less trivial observations which can be proved directly.

LEMMA 1.7.   (1) G^* acts on S.

(2) Z consists of fixed points of G^*, i.e., $Z \subseteq \{s \in S \mid g \cdot s = s \text{ for all } g \in G^*\}$.
(3) G^* acts on $S^\times$.
(4) For every $s \in S^\times$ the orbit $G \cdot s$ has the form $G \cdot s = G^* \cdot s \cup \{0 \cdot s\}$.
(5) For every $z \in Z$ we have $G \cdot z = \{0 \cdot z\}$.

We say that G acts fixed point freely on S if the restricted action of G^* on $S^\times$ is fixed point free. To put it in another way,
$$g \cdot s = s \implies g = 1 \text{ or } s \in Z.$$
In this case S splits into the set of fixed points Z and the set $S^\times$.

On the other hand, if we have a group G acting on a set S, by attaching new elements from a nonempty set Z to S, we obtain an action of the group with zero G_0 on the set $S_0 = S \cup Z$. The action is given (besides the action of G on S) by choosing $0 \cdot s \in Z$ (arbitrarily!) for all $s \in S$, and set $g \cdot z = z$ for all $g \in G$, $z \in Z$.

2. Characterization of all semi-homogeneous maps on group-acts

We now show how to get all semi-homogeneous maps f from a set S to a group G (with or without zero) which acts on S. They all arise from maps defined on orbit representatives.

LEMMA 2.1. *Let H be a subgroup of G whose action on S is fixed point free, and R a complete set of representatives of the orbits of the group H inside S. Then for every $s \in S$ there exist unique $g_s \in H$ and $r_s \in R$ with $s = g_s \cdot r_s$.*

In view of Lemma 1.4 we can use this unique decomposition to describe semi-homogeneous maps. Indeed, with the notations from the previous Lemma, we immediately obtain,

LEMMA 2.2. *Every semi-homogeneous map $f : S \to G$, is uniquely determined by its restriction $\hat{f} = f|_R$, where R is a complete set of representative of the orbits of the subgroup $H = \text{Im}(f)$ inside S. Namely, for $s = g_s \cdot r_s \in S$, $f(s) = g_s \hat{f}(r_s)$.*

More precisely, we have

THEOREM 2.3. *Let G be a group acting on a set S. Take a fixed point free subgroup H of G. Let R be a set of representatives of the orbits of H inside S. Then every map $\hat{f} : R \to H$ can be uniquely extended to a semi-homogeneous map $f : S \to G$ via $f(s) = h_s \hat{f}(r_s)$ if $s = h_s \cdot r_s$ as in Lemma 2.1, and every semi-homogeneous map $f : S \to G$ arises in this way. We also have $H = \text{Im}(f)$.*

For groups with zero, the situation is only slightly more complicated.

THEOREM 2.4. *Let G be a group with zero acting on a set S and $Z = 0 \cdot S$. Then every semi-homogeneous map $f : S \to G$ can be constructed in one of the following ways:*

(1) *$\text{Im}(f) = \{0\}$ or $\text{Im}(f) = \{1\}$, i.e., f is constant onto $\{0\}$ or $\{1\}$.*
(2) *Take a fixed point free subgroup H of G^*. Let R be a complete set of representatives of the orbits of H inside $S^\times$. The map f is uniquely determined by its restriction $\hat{f}$ to R as $f(s) = h_s \hat{f}(r_s)$ if $s = h_s \cdot r_s$ and $f(z) = 0$ for all $z \in Z$.*

Furthermore, $H_0 = \text{Im}(f)$.

Notice that Theorem 2.4 is very similar to Theorem 2.3: Just take away zero and put $Z = \emptyset$ in Theorem 2.4 and we have Theorem 2.3. The proof of Theorem 2.4 will follow from the following key lemma.

LEMMA 2.5. *Let G be a group with zero acting on a set S, let H be a fixed point free subgroup of G^*, $Z = 0 \cdot S$, and let R be a complete set of representatives of the orbits of H inside $S^\times$. For an arbitrary non-zero map $\hat{f} : R \to H_0$ we get a well-defined map $f : S \to G$ by*

$$f(s) = \begin{cases} h_s \hat{f}(r_s) & \text{if } s = h_s \cdot r_s \in S^\times, \\ 0 & \text{if } s \in Z. \end{cases}$$

This map is semi-homogeneous and $H_0 = \mathrm{Im}(f)$.

PROOF. Since the representation $s = h_s \cdot r_s \in S^\times$ is unique, f is well-defined. The set $R \cup Z$ is a complete set of representatives of H inside S. For $s \in Z$ we have to have $f(s) = 0$. The resulting map $f : S \to G$ is easily seen to be semi-homogeneous. □

We are now ready for the

PROOF OF THEOREM 2.4. The constant maps in (1) of Theorem 2.4 are clearly semi-homogeneous. Therefore we restrict our considerations to non-constant maps.

Assume first that a non-constant, semi-homogeneous map $f : S \to G$ is given. If there is some $g \in \mathrm{Im}(f)^*$, $g \neq 1$, then for all $z \in Z$ we have $f(z) = f(g \cdot z) = gf(z)$. This yields $f(z) = 0$. Thus $0 \in \mathrm{Im}(f)$. If no such g exists, then $\mathrm{Im}(f) = \{0, 1\}$ as f is non-constant. From this one easily concludes $f(z) = 0$ for all $z \in Z$.

Take $s \in S$ with $f(s) = 0$. Then $0 = gf(s) = f(g \cdot s)$ for all $g \in \mathrm{Im}(f)$. So f is zero on the orbit $\mathrm{Im}(f) \cdot s$.

The set S splits into $Z' = f^{-1}(0)$ and $S' = S \setminus Z'$. Notice that $Z \subseteq Z'$. Moreover, $\mathrm{Im}(f)^*$ acts on S'. So $f|_{R'}$ is given by the restriction of f onto $R \cap S'$. As $f|_{Z'}$ is the zero map, we see that f is determined by its restriction to R.

The converse and the information $H_0 = \mathrm{Im}(f)$ is in Lemma 2.5. □

So, if $R = (r_i)_{i \in I}$, we can define a semi-homogeneous map $f : S \to G$ uniquely by its values $(g_i)_{i \in I}$, where $g_i = \hat{f}(r_i)$. In analogy to matrix representations of linear maps, we might call the family $(g_i)_{i \in I}$ the "vector representation" of f and $(r_i)_{i \in I}$ a "basis" for f.

Finally, we study the influence of a "change of the basis".

THEOREM 2.6. *Let G be a group (without or with zero) acting on a set S, and H a fixed point free subgroup of G. Let R and R' be sets of representatives of the orbits of H inside S (or H^* inside $S^\times$, respectively).*

If $f : S \to G$ is a semi-homogeneous map with $H = \mathrm{Im}(f)$, and $\hat{f} = f|_R$ and $\hat{f}' = f|_{R'}$, then

(1) *for every $x \in R'$ there exists a unique $h_x \in H$ such that $h_x \cdot x \in R$ represents the same orbit as x, i.e. $H \cdot x = H \cdot (h_x \cdot x)$, and*
(2) *for all $x \in R'$, $h_x \hat{f}'(x) = \hat{f}(h_x \cdot x)$.*

Conversely, assume the maps $\hat{f} = R \to G$ and $\hat{f}' = R' \to G$ are given, and construct the maps $f : S \to G$, and $f' : S \to G$ as in Lemma 2.5. If for all $x \in R'$ we have $h_x \hat{f}'(x) = \hat{f}(h_x \cdot x)$, then $f = f'$.

PROOF. (1) is clear; (2) is a direct consequence of the semi-homogeneous property.

For the converse, observe first that for $s \in S^\times$, $s = h_s \cdot r'_s$ with $h_s \in H$ and $r'_s \in R'$ gives the according decomposition with respect to R by $s = h_s h_{r_s}^{-1} \cdot (h_{r_s} \cdot r'_s)$ (see Lemma 2.1). Thus we find for all $s \in S^\times$

$$f'(s) = h_s h_{r_s}^{-1} h_{r_s} \cdot \hat{f}'(r_s) = h_s h_{r_s}^{-1} \cdot \hat{f}(h_{r_s} \cdot r'_s) = f(s).$$

The claim for $s \in Z$ is trivial. □

3. Left distributive systems

We pick up the idea mentioned in 1.2 to connect semi-homogeneous maps to semigroups. We have seen that every semi-homogeneous map from a set S to S^S gives rise to a semigroup structure on S, and that, conversely, semigroups give rise to semi-homogeneous maps. If the set S has already a group structure, we get a much richer structure.

A left distributive system is a triple $(L, +, *)$ where $(L, +)$ and $(L, *)$ are groupoids, and $a*(b+c) = a*b + a*c$ for all $a, b, c \in L$. When $(L, +)$ is a group, not necessarily abelian, and $(L, *)$ is a semigroup, the left distributive system $(L, +, *)$ is a (left) nearring. In L, we define the relation $=_m$ of being "equivalent multipliers" by $a =_m b \iff \forall x \in L : a*x = b*x$. If $(L, +, *)$ is a nearring, it is called planar when the equation $a*x = b*x + c$ has exactly one solution for all $a, b, c \in L$ with $a \neq_m b$. We note that this is not exactly the definition of planar nearrings used in nearring theory where planarity requires that $|L/=_m| \geq 3$.

EXAMPLE 3.1. Let $(G, +)$ be a groupoid. Then the set $(G^G, +, \circ)$ of all maps from G into itself, with pointwise addition and composition $\circ$ of maps, is a left distributive system if the composition $f_1 \circ f_2$ is defined by $x \mapsto x f_1 f_2$. Furthermore, if $(G, +)$ is a group, then $(G^G, +, \circ)$ forms a (left) nearring. If we use the composition $f_1 \circ f_2 : x \mapsto f_1(f_2(x))$ instead, we would get a right distributive system.

Let $(L, +, *)$ be a left distributive system, and denote by $\text{End}\, L$ the set of all endomorphisms of the groupoid $(L, +)$, which is a monoid acting on L. For every $v \in L$, let $\lambda_v : L \to L; w \mapsto v * w$ denote the left translation by $v \in L$. Clearly, $*$ is uniquely determined by the map $\lambda : L \to \text{End}\, L; v \mapsto \lambda_v$.

Conversely, start with a groupoid $(L, +)$, every map $\lambda : L \to \text{End}\, L; v \mapsto \lambda_v$ gives a left distributive system $(L, +, *)$ when we define $v * w = \lambda_v(w)$.

REMARK 3.2. If the maps $*$ and λ are viewed as subsets of $L \times L \times L$, then actually $* = \lambda$.

The following is immediate.

THEOREM 3.3. *Let the left distributive system $(L, +, *)$ be given by the map $\lambda : L \to \text{End}\, L$, and let $\text{Trans}(L)$ be the collection $\{\lambda_v \mid v \in L\}$ of all left translations. Then $*$ is associative if and only if $\lambda_{\lambda_v(w)} = \lambda_v \circ \lambda_w$ for all $v, w \in L$. In other words, if and only if λ is semi-homogeneous (where $\text{Trans}(L)$ acts naturally on L). In this case, $(L, +, *)$ is a nearring if and only if $(L, +)$ is a group.*

When we have a left distributive system $(L, +, *)$ given by the map $\lambda : L \to \text{End}\, L$, and an index set I, then L acts on L^I naturally as endomorphisms via the componentwise action $g \cdot w = (\lambda_g(w_i))_{i \in I}$ for $g \in L$ and $w = (w_i)_{i \in I} \in L^I$.

Thus, it is natural to say that a map $f : L^I \to L$ is semi-homogeneous when $f(f(w) \cdot w') = f(w) * f(w')$ for all $w, w' \in L^I$.

Now, let $(N, +)$ be a group and let the group Φ act on N as a group of automorphisms of N. Let $f : N \to \Phi_0$ be a semi-homogeneous map with $\text{Im}(f) = \Phi_0$ and define the associative and distributive multiplication via $a * b = f(a) \cdot b$ for all $a, b \in N$. The next result shows that $(N, +, *)$ is "almost" a planar nearring:

THEOREM 3.4. *For $(N, +, *)$ as above, we get*
  (1) $a =_m b \iff f(a) = f(b)$.
  (2) *If $a \neq_m b$, then the equation $a * x = b * x + c$ has at most one solution.*
  (3) *If for every $\phi \in \Phi \setminus \{1\}$ the map $-1 + \phi$ is surjective, then $(N, +, *)$ is a planar nearring.*

PROOF. We have shown above that $(N, +, *)$ is a left nearring. (1) is trivial by construction.

(2) If $f(a) = 0$ or $f(b) = 0$, then the statement is trivial. Thus we can assume $f(a) \neq 0$ and $f(b) \neq 0$. This yields

$$(3.1) \qquad f(a) \cdot x = f(b) \cdot x + c \iff -x + f(b)^{-1} f(a) \cdot x = f(b)^{-1} \cdot c.$$

If this has more than one solution, then there are distinct $x, y \in N$ such that with $\phi = f(b)^{-1} f(a) \neq 1$

$$-x + \phi \cdot x = -y + \phi \cdot y \iff \phi \cdot (x - y) = x - y.$$

As Φ acts fixed point freely, this implies $x = y$, a contradiction.

(3) The extra assumption shows that the right hand equation of (3.1) always has a solution. $\square$

If (3) also holds, we call (N, Φ) a "Ferrero pair", after Celestina and Giovanni Ferrero who found the connection between planar nearrings and these pairs of groups $(N, +)$ and $(\Phi, \circ)$ (see [8]). There are two interesting cases when Theorem 3.4(3) is automatically satisfied: If N is finite, and if N is a finite dimensional vector space over some field F and Φ acts F-linearly.

THEOREM 3.5. *Let (N, Φ) be a Ferrero pair and choose a subset R of a set of orbit representatives inside N^* under the action of Φ. Put $Z = N \setminus (\Phi \cdot R)$. For each $a \in N \setminus Z$ there exist unique $r_a \in R$ and $\lambda_a \in \Phi$ such that $\lambda_a(r_a) = a$. If $\lambda_a = 0$ for all $a \in Z$, then the map $\lambda : N \to \Phi_0; a \mapsto \lambda_a$ is semi-homogeneous.*

Conversely, for each semi-homogeneous map $\mu : N \to \text{Aut}(N)_0$ we get a pair (N, Φ) where Φ is a fixed point free automorphism group of $(N, +)$ and $\Phi_0 = \text{Im}(\mu)$.

PROOF. As in Lemma 2.5 we define λ on the set R by $\lambda_r = 1$. For $a \in N \setminus Z$ we find the decomposition according to Lemma 2.1 as $a = \lambda_a(r_a)$. Therefore $\lambda_a = \lambda_a \circ \lambda_{r_a}$; and $\lambda_z = 0$ for $z \in Z$.

The converse is clear from the fact that Φ is fixed point free (see Lemma 1.4). $\square$

Theorem 3.4(3) is not always true. We get many examples easily from any non-planar nearfield (cf. [5]).

EXAMPLE 3.6. Let N be a non-planar nearfield. By [5, I.(9.1)], there exists some $c \in N \setminus \{1\}$ such that the equation $1 + x = cx$ has no solution. If we choose any subgroup Φ of N^* that contains c, e.g. $\Phi = \langle c \rangle$, then Φ is a fixed point free automorphism group on $(N, +)$, but (N, Φ) is not a Ferrero pair. Indeed, the map $x \mapsto -x + cx$ will be injective, but not surjective. In particular, 1 is not in the image. Note that N^* is non-commutative, so $\langle c \rangle$ is a proper subgroup.

4. Semi-homogeneous maps on vector spaces

Now we get to the point of characterizing all semi-homogeneous functionals on a given vector space mentioned in the abstract. But this is an easy task with what we have developed above.

Let V be a vector space over the field F. We view F as a multiplicative group with 0 acting on the additive group $(V, +)$ via the scalar multiplication. In this case, we have $V^\times = V \setminus \{0\}$, and a semi-homogeneous map on V is a functional $f : V \to F$ such that $f(av) = af(v)$ for all $a \in \text{Im}(f)$ and $v \in V$. Note that the action of F on V is fixed point free; so every subgroup of F^* qualifies for $\text{Im}(f) \setminus \{0\}$.

Let H be a multiplicative subgroup of F^*. Once we fix a complete set of orbit representatives of H in $V^\times$, we can get all (non-constant) semi-homogeneous maps on V with image H_0 using Theorem 2.4.

EXAMPLE 4.1 (cf. [2]). Consider the complex number field $\mathbb{C}$ as a vector space over $\mathbb{R}$ and let the multiplicative group $\mathbb{R}^+$ of positive reals act on $\mathbb{C}$ by multiplication. The orbits are $\{0\}$ and all rays (half lines) starting from 0. The set $C = \{e^{i\theta} \mid \theta \in [0, 2\pi)\} \cup \{0\}$ (the unit circle together with 0) is a complete set of orbit representatives. Any nonzero map $\hat{f} : C \to \mathbb{R}_0^+$ gives a well-defined non-constant semi-homogeneous map f on V with $f(re^{i\theta}) = r\hat{f}(e^{i\theta})$ for all $r \in \mathbb{R}_0^+$, $\theta \in [0, 2\pi)$, and $\text{Im}(f) = \mathbb{R}_0^+$.

One can "dualize" this example:

EXAMPLE 4.2. Consider again the complex number field $\mathbb{C}$ as a vector space over $\mathbb{R}$ and let now the circle group C act on $\mathbb{C}$ by multiplication. Now the orbits are all circles around the origin, and any ray starting from 0 is a complete set of orbit representatives.

5. Canonical Orbit Representatives

As we have seen, the orbit representatives play crucial roles in the characterization of semi-homogeneous maps. In many important cases, there are "canonical" choices of orbit representatives. We discuss in the following some of such cases.

5.1. The Case F^k, F a Field. Let F be a field, S be a multiplicative subgroup of F, and let C be a complete set of orbit representatives of S in F^*. First we take all $(c, x_2, \ldots, x_k)$ ($c \in C$ and $x_j \in F$ for $j = 2, \ldots, k$), and put them into R_S. These must be in different orbits since $s(c, x_2, \ldots, x_k) = (d, y_2, \ldots, y_k)$, $s \in S$, forces $c = d$ and $s = 1$. Next, if some vector $v = (0, 0, \ldots, 0, x_t, \ldots, x_k)$ in F^k starts with some zero components and if $x_t \neq 0$, we find the orbit representative $c \in C$ of x_t and v is in the orbit of $(0, \ldots, 0, c, \frac{cx_{t+1}}{x_t}, \ldots, \frac{cx_k}{x_t})$. The zero vector is an orbit of its own.

So the collection R_S consists of all $(0, \ldots, 0, c, y_{t+1}, \ldots, y_k)$ with $1 \leq t \leq k$, $c \in C$, $y_{t+1}, \ldots, y_k \in F$, and is a complete set of orbit representatives of S in F^k, referred to as the canonical orbit representatives with respect to S.

Any other system R' of orbit representatives relates to R_S as follows.

- If $r = (y_1, \ldots, y_k) \in R'$ starts with a non-zero y_1, we look up the orbit representative $c \in C$ of y_1. Then r is in the orbit of $\frac{y_1}{c} r = (c, \frac{cy_2}{y_1}, \ldots, \frac{cy_k}{y_1})$. So each element of the orbit of $(x_1, x_2, \ldots, x_k) \in F^k$, $x_1 \neq 0$, can be identified by its "polar coordinates" $(c, \frac{cx_2}{x_1}, \ldots, \frac{cx_k}{x_1})$, where the first component plays the role of the "radius", while the others are the "spherical coordinates".
- If $r = (0, \ldots, 0, y_t, \ldots, y_k) \in R'$, proceed similarly to get the representative
$$(0, \ldots, 0, c, \frac{cy_{t+1}}{y_t}, \ldots, \frac{cy_k}{y_t}) \in R_S$$

5.2. Special Case 1: F a finite field. Let F be a finite field of order $q = p^m$ and S be a multiplicative subgroup of F^*. Let b be the number of cosets of S in F^*, each coset having $|S|$ elements. We then have $\ell = \frac{q^k - 1}{|S|}$ nonzero orbits in F^k. There are bq^{k-1} orbit representatives of the first type, and $b(q^{k-2} + q^{k-3} + \ldots + q + 1)$ representatives of the second type, which add up to $b \frac{q^k - 1}{q - 1} = \frac{q-1}{|S|} \frac{q^k - 1}{q - 1} = \frac{q^k - 1}{|S|}$. So we see again that we have all representatives of nonzero orbits.

How can one find "coset leaders" c for S? If we have a generating element ζ of F^* then the size of S determines a generator ζ^ℓ of S. Then $1, \zeta, \zeta^2, \ldots, \zeta^{\ell-1}$ are convenient coset leaders. Recall that finding a generator is not an easy task in general.

5.3. Special Case 2: $\mathbb{R}^2$ and $G = \{x \in \mathbb{R} \mid x > 0\}$ or $G = \{x \in \mathbb{R} \mid x \neq 0\}$. In these cases, we also have other canonical choices: the non-zero orbits are the rays (lines, respectively) in the plane excluding $(0, 0)$ (see Example 4.1). Canonical orbit representatives are, e.g., the points on the unit sphere S (the upper half of the unit sphere with $(-1, 0)$ excluded, respectively). From another orbit representative we easily get this canonical one by reducing it to length 1.

5.4. Special Case 3: $\mathbb{R}^2$ and G = unit circle (i.e., rotations). The non-zero orbits are the circles around the origin. Canonical orbit representatives C are the points on the positive x-axis (see Example 4.2). If d is in another set D of orbit representatives, its canonical representative is just the intersection of the circle around the origin which passes through d with the positive x-axis.

5.5. Special Case 4: $F = \mathbb{R}^2$ and S = the logarithmic spiral. For every $\alpha \in \mathbb{R}^+$, the set $\mathscr{S} = \{x e^{i\alpha \log(x)} \mid x \in \mathbb{R}^+\}$ is a subgroup of $(\mathbb{C}^*, \cdot)$. The orbit of some $a = |a| e^{i\varphi} \in \mathbb{C}$ is given by

$$\{x e^{i\alpha \log(x)} \cdot a \mid x \in \mathbb{R}^*\} = \{|a| x e^{i(\alpha \log(x) + \varphi)} \mid x \in \mathbb{R}^*\}$$
$$= \{y e^{i(\alpha \log(y) - \alpha \log(|a|) + \varphi)} \mid y \in \mathbb{R}^*\},$$

and is hence a rotation of $\mathscr{S}$.

Since every (rotated) spiral intersects the unit circle exactly once, the unit circle can serve as a "distinguished" system C of representatives. Any other system D of representatives relates to C as follows. Take any point with polar coordinates (r, φ) in the complex plane. There is a unique rotation of $\mathscr{S}$ which meets the circle around $(0, 0)$ and radius r. Let this point be (r, ψ). Then the rotation by $\psi - \varphi$ indicates the orbit (r, φ) belongs to.

5.6. Other cases over fields. Of course, there are many more multiplicative subgroups in $F^k \setminus \{0\}$, for instance $\mathbb{Z} \times \mathbb{Q}$ in $\mathbb{R}^2$ (in polar coordinates (r, φ)), but there is clearly no hope to find canonical orbit representatives in each of these cases.

6. Planar nearrings

We close with some remarks on planar nearrings. As we have seen, given a Ferrero pair (N, Φ), we can use the construction from Lemma 2.5 to obtain a distributive system, in fact a planar nearring.

Now, let α be a cardinal and let N be a planar nearring. Then N acts on N^α componentwise. Recall the map $\lambda : N \to \Phi_0; a \mapsto \lambda_a$ from Theorem 3.5.

PROPOSITION 6.1. *If $f : N^\alpha \to N$ is semi-homogeneous, then so is $\lambda \circ f : N^\alpha \to \Phi_0$.*

PROOF. Directly we have

$$\lambda \circ f(\lambda_{f(u)}(v)) = \lambda_{f(f(u)v)} = \lambda_{f(u)f(v)} = \lambda_{f(u)}\lambda_{f(v)}.$$

$\square$

PROPOSITION 6.2. *Let $\tilde{f} : N^\alpha \to \Phi_0$ be semi-homogeneous. There exists a semi-homogeneous map $f : N^\alpha \to N$ such that $\tilde{f} = \lambda \circ f$.*

PROOF. Let e be a left identity in N. Then $B_e = \{a \in N \setminus Z \mid ae = a\} \cup \{0\}$ is a group with zero inside $(N, \cdot)$ and $\tilde{\lambda} : B_e \to \Phi_0; a \mapsto \lambda_a$ is an isomorphism. Put $f = \tilde{\lambda}^{-1} \circ \tilde{f}$. All we need to show is that f is semi-homogeneous. Notice first that $\text{Im}(f) \subseteq B_e$ by definition. We compute for $u, v \in N^\alpha$

$$\begin{aligned} f(f(u)v) &= \tilde{\lambda}^{-1}\big(\tilde{f}(\lambda_{f(u)}(v))\big) = \tilde{\lambda}^{-1}\big(\lambda_{f(u)} \circ \tilde{f}(v)\big) \\ &= \tilde{\lambda}^{-1}\big(\lambda_{f(u)} \circ \lambda_{f(v)}\big) = \tilde{\lambda}^{-1}\big(\lambda_{f(u)f(v)}\big) \quad \text{... by associativity} \\ &= f(u)f(v). \end{aligned}$$

$\square$

References

[1] James R. Clay, *Nearrings*, Oxford Science Publications, The Clarendon Press, Oxford University Press, New York, 1992. Geneses and applications. MR1206901 (94b:16001)

[2] Wen-Fong Ke, Hubert Kiechle, Günter Pilz, and Gerhard Wendt, *Planar nearrings on the Euclidean plane*, J. Geom. **105** (2014), no. 3, 577–599, DOI 10.1007/s00022-014-0221-7. MR3267561

[3] L. D. Kudryavtsev. *Homogeneous Functions*. In: Michael Hazewinkel (Editor), Encyclopaedia of Mathematics. Springer-Verlag, Berlin, 2002.

[4] K. D. Magill Jr., *Topological nearrings whose additive groups are Euclidean*, Monatsh. Math. **119** (1995), no. 4, 281–301, DOI 10.1007/BF01293589. MR1328819 (96c:16058)

[5] Heinz Wähling, *Theorie der Fastkörper* (German), Thales Monographs, vol. 1, Thales-Verlag, Essen, 1987. MR956467 (90e:12024)

[6] P. Fuchs, C. J. Maxson, and G. Pilz, *On rings for which homogeneous maps are linear*, Proc. Amer. Math. Soc. **112** (1991), no. 1, 1–7, DOI 10.2307/2048473. MR1042265 (91h:16054)

[7] C. J. Maxson and A. B. Van der Merwe, *Forcing linearity numbers for modules over rings with nontrivial idempotents*, J. Algebra **256** (2002), no. 1, 66–84, DOI 10.1016/S0021-8693(02)00128-X. MR1936879 (2003i:16002)

[8] Giovanni Ferrero, *Due generalizzazioni del concetto di anello e loro equivalenza nell'ambito degli "stems" finiti* (Italian, with English summary), Riv. Mat. Univ. Parma (2) **7** (1966), 145–150. MR0228549 (37 #4129)

Department of Mathematics and Research Center for Theoretical Sciences, National Cheng Kung University, Tainan 701, Taiwan
E-mail address: wfke@mail.ncku.edu.tw

Fachbereich Mathematik, Universität Hamburg, Bundesstr. 55, 20146 Hamburg, Germany
E-mail address: hubert.kiechle@uni-hamburg.de

Department of Algebra, Johannes Kepler Universität Linz, Altenberger Strasse 69, 4040 Linz, Austria
E-mail address: guenter.pilz@jku.at

Department of Algebra, Johannes Kepler Universität Linz, Altenberger Strasse 69, 4040 Linz, Austria
E-mail address: gerhard.wendt@jku.at

Adaptive numerical solution of eigenvalue problems arising from finite element models. AMLS vs. AFEM

C. Conrads, V. Mehrmann, and A. Międlar

ABSTRACT. We discuss adaptive numerical methods for the solution of eigenvalue problems arising either from the finite element discretization of a partial differential equation (PDE) or from discrete finite element modeling. When a model is described by a partial differential equation, the adaptive finite element method starts from a coarse finite element mesh which, based on a posteriori error estimators, is adaptively refined to obtain eigenvalue/eigenfunction approximations of prescribed accuracy. This method is well established for classes of elliptic PDEs, but is still in its infancy for more complicated PDE models. For complex technical systems, the typical approach is to directly derive finite element models that are discrete in space and are combined with macroscopic models to describe certain phenomena like damping or friction. In this case one typically starts with a fine uniform mesh and computes eigenvalues and eigenfunctions using projection methods from numerical linear algebra that are often combined with the algebraic multilevel substructuring method to achieve an adequate performance. These methods work well in practice but their convergence and error analysis is rather difficult. We analyze the relationship between these two extreme approaches. Both approaches have their pros and cons which are discussed in detail. Our observations are demonstrated with several numerical examples.

1. Introduction

Eigenvalue problems associated with complex mathematical models described by partial differential equations (PDEs) or very large finite element models arise in a number of applications ranging from quantum mechanical models, design of periodic structures for wave-guides, structural mechanics, stability analysis in dynamical systems, as well as model reduction. A typical real world example that was recently considered in [35] is the analysis and treatment of disc brake squeal. This

---

2010 *Mathematics Subject Classification.* Primary 65F15, 65N25, 65N30.

The second author's research was supported in the framework of MATHEON project *D-OT3, Adaptive finite element methods for nonlinear parameter-dependent eigenvalue problems in photonic crystals* supported by Einstein Foundation Berlin.

The third author's research was supported in the framework of MATHEON project *D-OT3, Adaptive finite element methods for nonlinear parameter-dependent eigenvalue problems in photonic crystals* supported by Einstein Foundation Berlin, and within a DFG Research Fellowship under the DFG GEPRIS Project *Adaptive methods for nonlinear eigenvalue problems with parameters* and by the Chair of Numerical Algorithms and High-Performance Computing, Mathematics Institute of Computational Science and Engineering, École Polytechnique Fédérale de Lausanne.

©2016 American Mathematical Society

phenomenon arises from self-excited vibrations caused by a flutter-type instability originating from friction forces at the pad-rotor interface [2] of the brake. Since a full atomistic modeling via the Langevin equation [63] is computationally infeasible, a commonly used approach in practice is to employ macro-scale approximations via multibody dynamics and finite element modeling (FEM) with very fine uniform meshes, see, e.g., [46, 56]. Using macroscopic models of material damping and friction forces, one obtains as model equations for the finite element coefficients of the position variables $Q = \sum_{i=1}^{n} Q_i(t)\varphi_i(x)$, a dynamical system

$$(1) \qquad M\ddot{Q} + C\dot{Q} + KQ = f,$$

where $M, C, K \in \mathbb{R}^{n,n}$ are large scale mass, damping, and stiffness matrices, respectively, and f is an external force. For finite element (FE) models of rotating machinery, the matrices C and K are typically nonsymmetric to incorporate gyroscopic and circulatory forces, and they depend on various parameters that include model operating conditions, material conditions, as well as the rotational speed of the disc. For self-excited vibrations, one also includes the excitation force via a nonsymmetric term added to the stiffness matrix. Furthermore, the mass matrix is often singular, usually due to mass lumping or due to explicit algebraic equations which constrain the dynamics of the system.

System (1) is a (space) discrete FE model and typically, since the macroscopic approximations do not hold in the continuous limit for mesh size equal to zero, a corresponding continuous model in the form of a PDE is not available. If the system would be valid in the limit, then it would be a hyperbolic PDE

$$(2) \qquad \ddot{U} + c\dot{U} + k\,\Delta U = f,$$

in a domain $\Omega \in \mathbb{R}^d, d = 1, 2, \ldots$, with given parameter dependent functions c, k, together with boundary conditions. We discuss here Dirichlet conditions $U = 0$ on the boundary $\partial \Omega$; the case of general boundary conditions can be reduced to this case [11].

Choosing the ansatz $Q = \exp(\mu t)\mathbf{q}$ or $U = \exp(\mu t)u$, respectively, then yields eigenvalue problems. In the (space) discrete case this is the finite dimensional quadratic eigenvalue problem:

Determine $\mu \in \mathbb{C}$ *and a nonzero* $\mathbf{q} \in \mathbb{C}^n$ *such that*

$$(3) \qquad L(\mu)\mathbf{q} := (\mu^2 M + \mu C + K)\mathbf{q} = 0.$$

A finite dimensional eigenvector $\mathbf{q} = [q_i]_{i=1}^n \in \mathbb{C}^n$ associated with an eigenvalue μ is then the coordinate vector of coefficients for the eigenfunction $q = \sum_{i=1}^{n} q_i \psi^{(i)}(x)$ in the FE basis $\{\psi^{(i)}(x)\}_{i=1}^n$, which has been used to generate the FE model.

In the (space) continuous case one obtains an infinite dimensional quadratic eigenvalue problem.

Determine $\mu \in \mathbb{C}$ *and a nonzero function u in an appropriate function space V such that*

$$(4) \qquad \mathcal{L}(\mu)u := (\mu^2 + c\mu + k\Delta)u = 0 \text{ in } \Omega, \quad u = 0 \text{ on } \partial\Omega.$$

In order to solve the continuous problem (4), one discretizes the variational form of the problem using a Galerkin approach. If an appropriate n_h dimensional FE subspace $V_h \subset V$ spanned by a FE basis $\{\varphi_h^{(i)}(x)\}_{i=1}^{n_h}$ is chosen, then (4) takes the form of the finite dimensional quadratic eigenvalue problem

$$(5) \qquad \mathcal{L}_h(\mu_h)u_h := (\mu_h^2 M_h + \mu_h C_h + K_h)u_h = 0,$$

and the eigenfunction u_h is represented as $u_h = \sum_{i=1}^{n_h} u_{h,i} \varphi_h^{(i)}(x)$, with a coordinate vector $\mathbf{u}_h = [u_{h,i}]_{i=1}^{n_h} \in \mathbb{C}^n$.

The two discrete finite dimensional problems (3) and (5) are very similar in nature. If a PDE model is available, then using the same basis functions $\psi^{(i)} = \varphi_h^{(i)}$, $i = 1, 2, \ldots, n_h$, on the same uniform grid without exploiting any macroscopic model approximations, would result in strongly related or even the same matrix coefficients.

In the continuous case, for some PDE operators, a priori and a posteriori error estimates can be obtained that permit assessment of the quality of the solution and allow for adaptive grid refinement. In the discrete case, such estimates are typically not available, and error control has to be based on algebraic techniques and comparisons with experiments. However, in both the discrete and continuous case, once approximations $\widetilde{\mu}$ of an eigenvalue μ and $\widetilde{q}(x) = \sum_{i=1}^{n} \widetilde{q}_i \psi^{(i)}(x)$, respectively, $\widetilde{u}_h(x) = \sum_{i=1}^{n_h} \widetilde{u}_{h,i} \varphi_h^{(i)}(x)$ to the associated eigenfunctions have been determined, then eigenvalue residuals can be formed

$$(6) \qquad r := (\widetilde{\mu}^2 M + \widetilde{\mu} C + K)\widetilde{\mathbf{q}},$$

$$(7) \qquad r_h(x) := (\widetilde{\mu}_h^2 M_h + C_h \widetilde{\mu}_h + K_h)\widetilde{u}_h,$$

respectively. Using backward error analysis [53], it follows that if the stability constant of the associated eigenvalue/eigenfunction pair (μ_h, u_h), respectively the condition number of the eigenvalue/eigenvector pair $(\mu, \mathbf{q})$ are not too large, i.e., small perturbations in the model do not lead to large perturbations in computed eigenvalue/eigenfunction approximations, then these residuals can be used to estimate the associated errors. This analysis will be discussed in Section 4.

There are essentially two extreme approaches to compute a specific eigenvalue/eigenvector or eigenvalue/eigenfunction pair. In the continuous case, when a priori and a posteriori error estimates are available, then one can apply the adaptive finite element method (AFEM), and, at least in some special cases, prove its reliability and efficiency [16]. Starting from a sufficiently fine initial mesh, AFEM uses local error estimates to adaptively refine the mesh, so that the resulting error in the finite element approximation is within a given tolerance. However, the analytic background of AFEM requires the presence of an associated PDE model which, as discussed before, is not always available. Furthermore, the method is currently restricted to some very special problem classes of elliptic problems with real or purely imaginary eigenvalues. Numerical methods that use AFEM for some more complex problems were discussed in [15, 39, 60], but no detailed analysis is available for problems with complex eigenvalues or eigenvalues with Jordan blocks. For multiple real eigenvalues of symmetric problems, there has been analysis in [28, 29, 71] recently. In contrast to this, in (3) the only available data are matrices associated with the fine mesh. A very common method that is frequently used in structural engineering is the automated multilevel substructuring method (AMLS) [9, 43]. Starting from (3) or (5) associated with the fine mesh, AMLS uses an algebraic substructuring technique (component mode synthesis [20]), local computations of eigenvalues and eigenvectors for algebraically constructed substructures in (3), as well as projections, to obtain a small projected eigenvalue problem which then can be solved by standard, dense eigenvalue methods [7, 37]. This algebraic approach is very flexible and can in principle be extended to treat damping terms as well as complex and multiple eigenvalues. However, the theoretical analysis of the method is rather limited [25, 38, 70], and the results may not always be satisfactory, see Section 6.

In this work, we discuss a common basis for both the AFEM and the AMLS method. We compare these concepts and point out their advantages and disadvantages. For a direct analytical comparison, we restrict ourself to consider (3) and (5) with symmetric positive definite mass and stiffness matrices and discard damping or other parts of the model, see Section 2. In this situation, setting $\lambda := -\mu^2$, all the eigenvalues are real and there exist orthonormal (in an appropriate inner product) sets of associated eigenvectors/eigenfunctions. In Section 3 we introduce the automated multilevel substructuring method (AMLS) and discuss its properties and several implementation details. Section 4 is dedicated to the adaptive finite element method (AFEM) and a recently developed variant called AFEMLA. We compare both methods, provide some common ground, and discuss some of their advantages and disadvantages in Section 5. Section 6 illustrates our observations with several numerical examples.

2. A model problem for comparison

In order to compare and relate the AFEM and the AMLS method, we use a simple elliptic PDE eigenvalue problem.

Determine $\lambda \in \mathbb{R}$ and $u \in V := H_0^1(\Omega)$ (the space of functions that vanish on the boundary and have a first derivative that is Lebesgue integrable) such that

$$(8) \quad \begin{aligned} \mathcal{L}u &= \lambda u & \text{in } \Omega, \\ u &= 0 & \text{on } \partial\Omega, \end{aligned}$$

where $\Omega \in \mathbb{R}^d, d = 1, 2, \ldots$ is a bounded, polyhedral Lipschitz domain and $\partial\Omega$ is its boundary.

Here
$$\mathcal{L}u(x) = -\mathrm{div}(A(x)\nabla u(x)),$$
and A is a real symmetric positive definite matrix, so that $\mathcal{L}$ is self-adjoint and elliptic. Introducing the bilinear forms

$$a : V \times V \to \mathbb{R}, \qquad a(u,v) := \int_\Omega (\nabla u)^T A(x) \nabla v \, dx,$$

$$b : V \times V \to \mathbb{R}, \qquad b(u,v) := \int_\Omega uv \, dx,$$

then one has the variational form of problem (8):

Determine $\lambda \in \mathbb{R}$ and $u \in V$ such that

$$(9) \quad a(u,v) = \lambda b(u,v) \quad \text{for all} \ v \in V.$$

In order to find an approximation to the exact solution of the variational problem (9), we attempt to represent the solution by an element from a given finite dimensional subspace $V_h \subset V$. This is known as the *Galerkin* method (*Bubnov-Galerkin* or *Ritz-Galerkin* method in the self-adjoint case) [18]. For a given finite dimensional subspace $V_h \subseteq V$ the variational form of the eigenvalue problem (9) then is the *discretized eigenvalue problem*:

Determine $\lambda_h \in \mathbb{R}$ and $u_h \in V_h$ such that

$$(10) \quad a(u_h, v_h) = \lambda_h b(u_h, v_h) \quad \text{for all} \ v_h \in V_h.$$

Since A is symmetric positive definite, it follows that $a(\cdot,\cdot)$ defines an inner product on V and $b(\cdot,\cdot)$ is also an inner product on V.

There are many possible choices for the space V_h, see e.g., [11, 18]. For simplicity, we discuss only the 2D case and let $\mathcal{T}_h$ be a partition (triangulation) of the domain Ω into elements (triangles) T,

$$\bigcup_{T \in \mathcal{T}_h} = \overline{\Omega},$$

such that any two distinct elements in $\mathcal{T}_h$ share at most a common edge or a common vertex. For each element $T \in \mathcal{T}_h$, we denote the set of corresponding edges and vertices by $\mathcal{E}(T)$ and $\mathcal{N}(T)$, respectively, and $\mathcal{E}_h$ and $\mathcal{N}_h$ denotes all edges and vertices in $\mathcal{T}_h$. Likewise, we define h_T as the *diameter* (the length of the longest edge) of an element. For each edge E, we denote its length by h_E and the unit normal vector by $\vec{n}_E$. We set $h := \max_{T \in \mathcal{T}_h} h_T$. We say that the triangulation is *regular*, see [18], if there exists a positive constant ρ such that

$$\frac{h_T}{d_T} < \rho,$$

with d_T being the diameter of the largest ball that may be inscribed in element T, i.e., the minimal angle of all triangles in $\mathcal{T}_h$ is bounded away from zero.

Consider a regular triangulation $\mathcal{T}_h$ of Ω and the set of polynomials $\mathbb{P}_p$ of total degree $p \geq 1$ on $\mathcal{T}_h$, which vanish on the boundary of Ω, see, e.g., [11]. Then the Galerkin discretization of (10) with $V_h^p \subset V$, dim $V_h^p = n_h$, chosen as

$$V_h^p(\Omega) := \{v_h \in C^0(\overline{\Omega}) : v_h|_T \in \mathbb{P}_p \text{ for all } T \in \mathcal{T}_h \text{ and } v_h = 0 \text{ on } \partial\Omega\},$$

is called *finite element discretization*. The Finite Element Method (FEM) [18] is a Galerkin method where V_h is the subspace of piecewise polynomial functions, i.e., functions that are continuous in Ω and that are polynomial on each $T \in \mathcal{T}_h$. To simplify the presentation, here we only consider $\mathbb{P}_1$ finite elements, i.e., $p = 1$, and use $V_h := V_h^1$. The motivation to use piecewise polynomials is that in this case, the computational work in generating the system matrices is small and they are sparse, since the space V_h then has a canonical basis of functions with small support. The basis $\left\{\varphi_h^{(1)}, \ldots, \varphi_h^{(n_h)}\right\}$ is then a *Lagrange* or *nodal basis* [18] and an eigenfunction u_h is determined by its values at the n_h grid points of $\mathcal{T}_h$ and it can be written as

$$u_h = \sum_{i=1}^{n_h} u_{h,i} \varphi_h^{(i)}$$

and the discretized problem (10) reduces to a *generalized algebraic eigenvalue problem* of the form

(11) $$K_h \mathbf{u}_h = \lambda_h M_h \mathbf{u}_h,$$

where the matrices

$$K_h := [a(\varphi_h^{(j)}, \varphi_h^{(i)})]_{1 \leq i,j \leq n_h}, \quad M_h := [b(\varphi_h^{(j)}, \varphi_h^{(i)})]_{1 \leq i,j \leq n_h}$$

are called *stiffness* and *mass* matrix, respectively. The coordinate vector $\mathbf{u}_h$ associated with the eigenfunction u_h is defined as

$$\mathbf{u}_h := [u_{h,i}]_{1 \leq i \leq n_h}.$$

Since $a(\cdot, \cdot)$ and $b(\cdot, \cdot)$ are bounded, symmetric, and associated with inner products, the resulting matrices K_h, M_h are symmetric and positive definite. Thus, see, e. g., [57, §15.3], for the discrete eigenvalue problem (11) there is a full set of real M_h-orthogonal eigenvectors given by the columns of a matrix $\mathbf{U}_h \in \mathbb{R}^{n_h, n_h}$, and all eigenvalues $\lambda_h^{(i)}$, $i = 1, 2, \ldots, n_h$, are real and positive, i. e., we have

$$\mathbf{U}_h^T K_h \mathbf{U}_h = \operatorname{diag}(\lambda_h^{(1)}, \ldots, \lambda_h^{(n_h)}), \quad \mathbf{U}_h^T M_h \mathbf{U}_h = I_{n_h}.$$

Furthermore, it is well-known, e. g., [4, Equation (8.42), p. 699], [64, Equation (23), p. 223], or [19, 69], that for *conforming approximations*, i. e., if $V_h \subset V$, then the *Courant-Fischer min-max characterization* implies that the exact eigenvalues are approximated from above, that is,

$$\lambda^{(i)} \leq \lambda_h^{(i)}, \quad i = 1, 2, \ldots, n_h.$$

For the comparison of the pros and cons of AFEM and AMLS, we restrict ourselves to the computation of a few of the smallest eigenvalues and their corresponding eigenvectors/eigenfunctions.

3. Automated Multilevel Substructuring

In this section we briefly review the Automated Multilevel Substructuring method (AMLS) [9, 43], including some small improvements of its original formulation. For simplicity of presentation, we consider a problem of the form (11) which either arises from discrete FE modeling or from the FE discretization of a 2D elliptic PDE problem. The AMLS method needs as an input just the two matrices K_h and M_h and a user-supplied cutoff value $\lambda_c > 0$ and then works purely algebraically to determine all approximate eigenvalue/eigenvector pairs $(\lambda_h^{(i)}, \mathbf{u}_h^{(i)})$ associated with eigenvalues $0 < \lambda_h^{(i)} \leq \lambda_c$. The AMLS method can be viewed as an enhancement of the well-established *component mode synthesis method* of [20].

The first step in AMLS is to compute a nested dissection reordering [30], which is based on the computation of a set of vertex separators in the unweighted graph induced by the stiffness matrix, and apply it to *both* stiffness and mass matrix. Formally this reordering yields a permutation matrix $P \in \mathbb{R}^{n_h, n_h}$ such that $K := P^T K_h P$, $M := P^T K_h P$ are block-structured, as illustrated in Figure 1. In the FE setting, one obtains the same block structure in both matrices. Setting $w = P^T \mathbf{u}_h$, we then have the block-structured (sparse) generalized eigenvalue problem

(12) $$Kw = \lambda_h Mw.$$

Note that in practice one does not apply a permutation matrix, but just relabels the indices in the eigenvectors and undoes this relabeling when the eigenvectors $\mathbf{u}_h$ of the original problem are needed. The exact choice and number of partitions in the nested dissection reordering is irrelevant for the following description of the AMLS algorithm, but the structure is used heavily to improve the performance of the method when it is applied to a concrete eigenvalue problem. The resulting matrices are sparse block matrices with the diagonal blocks called *substructure blocks* and *coupling blocks*, respectively. Coupling blocks correspond to the vertex separators in the graph described by the stiffness matrix, whereas the substructure blocks correspond to substructures in the graph. Since the mass and stiffness matrices arise from the assembly of the inner products between the locally (via the mesh structure) defined basis functions, the resulting block structure partitioning can be

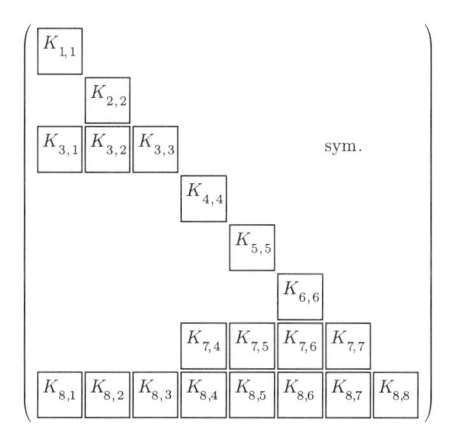

FIGURE 1. Symmetric stiffness matrix K with two levels of nested dissection reordering. The blocks $K_{1,1}, K_{2,2}, K_{4,4}, K_{5,5}$, and $K_{6,6}$ are substructure blocks, the blocks $K_{3,3}, K_{7,7}, K_{8,8}$ are coupling blocks. The first level of substructuring consists of the blocks $K_{1:3,1:3}$, $K_{4:7,4:7}$, and $K_{8,8}$.

viewed as a subdivision of the continuous domain [9]. However, since the method works in a completely algebraic fashion, it can also be applied if the matrices K, M are not associated with any finite element model if the graph partitioning software is applied to the combined graph of both matrices.

The second step in AMLS is to compute a block Cholesky decomposition [23, 33] $K = LDL^T$ of the reordered stiffness matrix, so that L is a block lower triangular and D is a block diagonal matrix. Since K is positive definite, this block Cholesky decomposition exists without employing any pivoting strategy.

REMARK 3.1. The second step of AMLS may be infeasible when K or some of the diagonal blocks are singular or close to being singular. Moreover, since this is a direct factorization there will be a non-negligible fill-in which may necessitate out-of-core algorithms in the implementation. It is possible to also include singular $K_{j,j}$ matrices by pivoting singular blocks of K to the bottom of the matrix. This will, however, partially destroy the sparsity and increase the size of the final block. For the comparison of the two methods, we restrict ourselves to the case of positive definite stiffness matrices.

Using the block Cholesky factors, another change of basis is performed. Setting $M_L := L^{-1}ML^{-T}$, $w_L := L^T w$ yields the transformed problem

(13) $$K_L w_L = \lambda_h M_L w_L,$$

with block diagonal matrix $K_L = \text{diag}(K_{1,1}, \ldots, K_{\ell,\ell})$ and blocks $K_{j,j}$, $j = 1, \ldots, \ell$, of sizes $n_1, \ldots, n_\ell$. Due to the structure of L, L^{-1}, it follows that $M_L = [M_{i,j}]$ retains the nested dissection reordering block structure if it is partitioned analogously. Furthermore, since the transformation is a congruence transformation, K_L and M_L are still symmetric positive definite due to Sylvester's law of inertia [33, Theorem 8.1.17].

In the third step of the AMLS algorithm one solves the local eigenvalue problems
$$K_{j,j}\widetilde{w}_j = \widetilde{\lambda}_j M_{j,j}\widetilde{w}_j, \quad j = 1, \ldots, \ell,$$
associated with the diagonal blocks of K_L, M_L, using any of the well-established methods for small dense eigenvalue problems [3, 10]. Let $\widetilde{\Lambda} = \mathrm{diag}(\widetilde{\Lambda}_1, \ldots, \widetilde{\Lambda}_\ell)$ be the block matrix with diagonal blocks containing all the computed eigenvalues $\widetilde{\lambda}_j^{(k)}$, $j = 1, \ldots, \ell$, $k = 1, \ldots, n_j$, of these subproblems, and let $\widetilde{W} = \mathrm{diag}(\widetilde{W}_1, \ldots, \widetilde{W}_\ell) \in \mathbb{R}^{n_h, n_h}$ be the corresponding block diagonal matrix of computed eigenvectors, i.e., $\widetilde{W}_j = [w_j^{(1)}, \ldots, w_j^{(n_j)}]$, $j = 1, \ldots, \ell$.

The fourth step in the AMLS method is to perform a *modal truncation* in the sub-blocks, i.e., for the given cutoff value λ_c, we select the subset of eigenvalues in the blocks $\widetilde{\Lambda}_j$ that fall below the cutoff λ_c, and set $\widehat{\Lambda}_j$, $j = 1, \ldots, \ell$, as the corresponding submatrices of $\widetilde{\Lambda}_j$. Moreover, we define $\widehat{\Lambda} = \mathrm{diag}(\widehat{\Lambda}_1, \ldots, \widehat{\Lambda}_\ell)$. Analogously, we define the block diagonal matrix $\widehat{W} = \mathrm{diag}(\widehat{W}_1, \ldots, \widehat{W}_\ell)$, where each block $\widehat{W}_j$ contains only those eigenvectors from $\widetilde{W}_j$ which correspond to the eigenvalues in $\widehat{\Lambda}_j$.

REMARK 3.2. It can be observed that surprisingly often the smallest values of $\widetilde{\lambda}_j^{(k)}$ are reasonable approximations of the smallest exact eigenvalues λ_h of (12). On the one hand this is motivated by the fact that the eigenvalue of the submatrices interlace those of the full problem by the Cauchy interlacing theorem, [33], even though by this argument the distances may be rather large, except if the diagonal blocks have eigenvalues that lie in the same range for many of the blocks. On the other hand this is motivated by the fact that the eigenvalues of the diagonal blocks are obtained by a perturbation (leaving out the off-diagonal blocks). For problems with submatrices that are associated with very different geometries or for nonsymmetric problem these motivating arguments may not justified.

The fifth step in the AMLS method is to project the matrix pencil (K_L, M_L) onto the space spanned by the columns of $\widehat{W}$. Let $\widehat{W}^\dagger$ be the Moore-Penrose pseudo-inverse of $\widehat{W}$ [33, p. 290], then we consider the projected eigenvalue problem

(14) $$(\widehat{W}^T K_L \widehat{W})\widehat{W}^\dagger w_L = \lambda_S (\widehat{W}^T M_L \widehat{W})\widehat{W}^\dagger w_L,$$

and we set $K_S := \widehat{W}^T K_L \widehat{W}$, $M_S := \widehat{W}^T M_L \widehat{W}$, $w_S := \widehat{W}^\dagger w_L$. Then the projected eigenvalue problem (14) has the form $K_S w_S = \lambda_S M_S w_S$. Note that the Moore-Penrose pseudo-inverse $\widehat{W}^\dagger$ does not have to be computed explicitly if we want to determine the eigenvectors of (13), since $w_L = \widehat{W} w_S$. Note further, that we can exchange the fourth and the fifth step in the AMLS method. Let Z be a matrix containing a subset of the columns of the identity matrix, then we can choose it such that
$$\widehat{W} = \widetilde{W} Z \text{ and } \widehat{\Lambda} = Z^T \widetilde{\Lambda} Z.$$
Accordingly, $K_S = Z^T \widetilde{W}^T K_L \widetilde{W}_d Z$ and we reversed the order of steps four and five. Moreover, notice that $K_S = \widehat{\Lambda}$.

The sixth step of the AMLS algorithm is to solve the projected eigenvalue problem (14), again with appropriate solvers from [3, 10] and to use the obtained m smallest eigenvalues $\widetilde{\lambda}_S^{(i)}, i = 1, \ldots, m$, as approximations to the smallest m eigenvalues $\lambda_h^{(i)}, i = 1, \ldots, m$, of the problem (12). The corresponding eigenvector

approximations $\widetilde{w}^{(i)}, i = 1, \ldots, m$, of the exact eigenvectors $w^{(i)}$ of (12) are then obtained from
$$\widetilde{w}^{(i)} = L^{-T}\widehat{W}\widetilde{w}_S^{(i)}, \quad i = 1, \ldots, m.$$
If the FE matrices are of a recursive multilevel structure, then the described approach can be carried out in a multilevel fashion, i.e., if the discussed diagonal blocks K_{jj}, M_{jj} again have the structure of a FE matrix, which is typically the case, then we can apply the same idea recursively for several levels.

The AMLS method contains several tuning parameters that are typically chosen based on heuristics, e.g., in the choice of the cutoff criterion in step four, quite often, a tolerance θ is added, see [43], to select all eigenvalues $\widetilde{\lambda}_j^{(k)}$ below $\theta\lambda_c$. The default value $\theta = 8.4^2$ is experimentally determined and we can confirm it as good choice based on our experiments. One can justify this heuristic choice using the eigenvalue error bounds of [50]. Let us, for simplicity, consider a single level of substructuring and partition the transformed matrices K_S, M_S conformably as

$$K_S = \begin{bmatrix} K_S^{11} & \\ & K_S^{22} \end{bmatrix}, \quad M_S = \begin{bmatrix} M_S^{11} & M_S^{12} \\ M_S^{12T} & M_S^{22} \end{bmatrix},$$

where M_S^{11} and M_S^{22} are identity matrices. Then K_S^{11}, M_S^{11} contain the substructure blocks, and K_S^{22}, M_S^{22} the coupling blocks. Theorem 2.5 in [50, p. 651] bounds the maximum difference between an exact eigenvalue $\lambda_h^{(i)}$ and an eigenvalue approximation $\widetilde{\lambda}_j^{(k)}$ that was computed by ignoring the off-diagonal blocks. If we assume that the modal truncation $\widetilde{\lambda}_j^{(k)} \leq \theta\lambda_c$ found all eigenvalue approximations corresponding to exact eigenvalues $\lambda_h^{(i)} \leq \lambda_c$, then we are implicitly assuming that $\left\|M_S^{12}\right\|_2 \leq \frac{\theta-1}{\theta}$ which for $\theta = 8.4^2$ yields $\left\|M_S^{12}\right\|_2 \leq 0.9858$. Since M_S contains identity matrices on its block diagonal, there is only one level of substructuring, and M_S is symmetric positive definite, $\left\|M_S^{12}\right\|_2 < 1$ holds and hence the choice of θ fits well.

It should be noted further, that making an efficient use of the extracted substructures in the eigenvalue problem (12) is an important part of the practical implementation of an AMLS method, since this will significantly reduce the fill-in in the off-diagonal blocks of the block Cholesky factorization.

4. The Adaptive Finite Element Method (AFEM)

The standard finite element method proceeds from the selection of a mesh and basis to the computation of a solution. However, it is well-known that the overall accuracy of the numerical approximation is determined by several factors: the regularity of the solution (smoothness of the eigenfunctions), the approximation properties of the finite element spaces, i.e., the search and test space, the accuracy of the eigenvalue solver and its influence on the total error. The most efficient approximations of smooth functions can be obtained using large higher-order finite elements (p-FEM), where the local singularities, arising e.g. from re-entrant corners, interior or boundary layers, can be captured by small low-order elements (h-FEM) [17]. Unfortunately, in real-world applications, these phenomena are typically not known a priori. Therefore, constructing an optimal finite dimensional space to improve the accuracy of the solution requires refining the mesh and (or) basis and performing the computations again. A more efficient procedure tries to decrease the mesh size (h-adaptivity) or (and) to increase the polynomial degree of the basis (p-refinement) automatically such that the accurate approximation can be obtained at a lower computational cost, retaining the overall efficiency. This adaptation is based on the local contributions of global *a posteriori error estimates*, the so-called *refinement indicators*, extracted from the numerical approximation. This algorithmic idea is called *Adaptive Finite Element Method (AFEM)* and can be described via the following loop

$$\text{SOLVE} \longrightarrow \text{ESTIMATE} \longrightarrow \text{MARK} \longrightarrow \text{REFINE}$$

The number of manuscripts addressing adaptive finite element methods is constantly growing, and its importance cannot be underestimated. On the other hand most of the publications do not deal with PDE eigenvalue problems but rather treat PDE boundary value problems. In the following sections we present a small fraction of material presented in [1, 4–6, 11, 13, 18, 24, 26, 27, 31, 32, 34, 36, 41, 45, 47, 48, 54, 55, 58, 59, 61, 62, 64, 67, 68].

The AFEM formulation that we will employ is based on the ansatz for the standard finite element method in Section 2. Since the AFEM will involve several levels of discretization with different mesh sizes h, we address this issue by slight modification of the notation introduced in Section 2. The label ℓ associated with the triangulation $\mathcal{T}_\ell$ indicates the refinement level of the mesh in the refinement hierarchy obtained by the adaptive FEM. We will assume that $\mathcal{T}_\ell \subset \mathcal{T}_{\ell+1}$, i.e., no coarsening is performed, and denote by N_ℓ and n_ℓ the maximal refinement level and the number of degrees of freedom associated with $\mathcal{T}_\ell$, respectively. We will denote the finite dimensional space over the partition $\mathcal{T}_\ell$ as V_ℓ and the associated Galerkin approximation as (λ_ℓ, u_ℓ). All other quantities are defined analogously as in Section 2 by the index h with ℓ.

The application of the adaptive FEM to the variationally stated eigenvalue problem (9) yields the following scheme: first the eigenvalue problem is solved on some initial mesh $\mathcal{T}_0$ to provide a finite element approximation (λ_ℓ, u_ℓ) of the continuous eigenpair (λ, u). Afterwards, the total error in the computed solution is estimated by some *error estimator* η_ℓ.

If the estimate for the global error is sufficiently small, then the adaptive algorithm terminates and returns (λ_ℓ, u_ℓ) as a final approximation, otherwise, the local contributions of the error are estimated on each element. A local *error indicator* (*refinement indicator*) for an element $T \in \mathcal{T}_\ell$ is usually denoted by η_T and related to a global error estimator η_ℓ through

$$\eta_\ell = \Big(\sum_{T \in \mathcal{T}_\ell} \eta_T^2 \Big)^{1/2}.$$

Based on these estimators, the elements for refinement are selected and form the set $\mathcal{M}_\ell \subset \mathcal{T}_\ell$ of *marked elements*.

The process of selecting the elements of $\mathcal{M}_\ell$ is called the *marking strategy*, and typical heuristic choices based on numerical experiments are discussed in [12, 13, 21]. It should be noted that marking an element actually means marking all its edges. The refinement of the finite element space can be performed using various techniques like moving grid points (*r-refinement*), subdividing elements of a fixed grid (*h-refinement*), applying locally higher-order basis functions (*p-refinement*) or any combinations of those [17]. For the sake of exposition, we discuss only the *h-refinement* of the elements, namely the longest-edge bisection [54, §4], and we do not discuss coarsening.

As we mentioned before, applying these refinement procedures may lead to nonconforming meshes with the so-called *hanging nodes*. Therefore, a *closure algorithm* [14] is applied to overcome this drawback and get a regular triangulation. For more details about adaptive refinement strategies see, e.g., [1, 17, 66].

In order to prove convergence of the AFEM one has to assume that the mesh refinement is done in such a way that a *saturation property* holds, as it has been proved for the Laplace eigenvalue problem in [16].

THEOREM 4.1 (Saturation property [16, Theorem 4.2]). *Let h_ℓ be the maximum mesh size on the ℓ-th mesh and let $\|\cdot\|_A := \sqrt{a(\cdot,\cdot)}$ denote the energy norm. Consider the adaptive FEM with sufficiently small maximal initial mesh size h_0 applied to the model problem (11). Then, there exists $0 \leq \rho < 1$ such that for all $\ell = 0, 1, \ldots, n_\ell - 1$ the following inequalities hold:*

$$\|u - u_{\ell+1}\|_A^2 \leq \rho \|u - u_\ell\|_A^2 + \lambda_{\ell+1}^3 h_\ell^4,$$

$$|\lambda - \lambda_{\ell+1}| \leq \rho |\lambda - \lambda_\ell| + \lambda_{\ell+1}^3 h_\ell^4,$$

where (λ, u) is an exact eigenpair of $\mathcal{L}$, λ_ℓ is an approximation of the eigenvalue λ on $\mathcal{T}_\ell$ and u_ℓ is the corresponding approximate eigenfunction.

With the saturation property given, a convergence proof for a specific AFEM in the computation of the smallest eigenvalues of (11) have been given in [16].

REMARK 4.2. Unfortunately, even the most accurate global error estimator itself does not guarantee the accuracy and efficiency of an adaptive algorithm. This can only be guaranteed if the desired eigenfunction has an approximate sparse representation in the used FE space, i.e., if they can be represented well by a small number of ansatz functions. This may, in particular, not be the case if several eigenfunctions are sought, which have singularities or oscillations in different regions of the spacial domain. Often the argument used in practice is that *nothing is better than a uniform mesh and brute force linear algebra* when studying eigenvalue problems. It should also be noted that almost all error estimators use inequalities that only hold up to an unknown constant and thus may lead to strong over- or underestimates of the true error. Finally, if the problem is sensitive to small perturbations (as it may be for non-self-adjoint non-normal problems), then even a very good and efficient error estimator may lead to large errors in eigenvalues and eigenfunctions. To include sensitivity of eigenvalues and eigenfunctions into the AFEM is currently a *very important open problem*.

The 'SOLVE step' in every step of the standard AFEM approach for eigenvalue problems (at least for reasonably fine levels) uses an iterative algebraic eigensolver, like an Arnoldi method [49]. Based on the computed eigenvalues/eigenvectors, then the a posteriori error estimates are determined and used to refine the grid. This approach, however, does not consider any influence of the errors in the algebraic eigensolver on the algorithm and most convergence or optimality results require that the eigenvalues and eigenvectors are computed exactly. Furthermore, since one may have to solve many algebraic eigenvalue problems related to finer and finer grids and information from the previous steps of the adaptive procedure, like previously well approximated eigenvalues, is not used on the next level, computational costs for the algebraic eigenvalue problem often dominate the total computational cost.

A more efficient approach is an AFEM variant called AFEMLA [14, 16, 51], which incorporates the information obtained during the iterative solving of algebraic eigenvalue problems into the error estimation and the refinement process. Since the accuracy of the computed eigenvalue approximation cannot be better than the quality of the discretization, there is no need to solve the intermediate algebraic eigenvalue problems (that are used to compute the error estimates) up to very high precision if the discretization scheme guarantees only small precision. And even in the final step it is enough to solve the problem within an accuracy that fits to the discretized system. Also nested iterations, i.e., using actual eigenvector approximation as a starting vector for the eigenvalue computation on the refined grid, reduce the total cost significantly. AFEMLA therefore follows exactly *the idea of adaptive methods to achieve a desired accuracy with the minimal computational effort.*

An additional advantage of the AFEMLA algorithm over the standard AFEM is that it can be applied even without any knowledge of the underlying PDE problem and even if one is not able to construct an appropriate a posteriori error estimator. AFEMLA allows to construct an adaptive algorithm nevertheless, since it can be based on the algebraic residual, provided the problem is such that the residual information is sufficient to characterize the error. We are not addressing this problem here, however, also in this case, performing the subspace adaptation requires information about underlying meshes and matrices obtained at different discretization levels [51–53].

Let us now consider two consecutive partitions $\mathcal{T}_\ell \subset \mathcal{T}_{\ell+1}$ and associated finite element spaces $V_\ell \subset V_{\ell+1}$ with a finite element basis $\{\varphi_\ell^{(1)}, \ldots, \varphi_\ell^{(n_\ell)}\}$ for V_ℓ and $\{\varphi_{\ell+1}^{(1)}, \ldots, \varphi_{\ell+1}^{(n_{\ell+1})}\}$ for $V_{\ell+1}$. Let us assume for simplicity, that the mesh $\mathcal{T}_{\ell+1}$ is obtained by a uniform refinement of $\mathcal{T}_\ell$.

With the Galerkin discretization followed by applying the Arnoldi process to the generalized eigenvalue problem

$$(15) \qquad K_\ell \mathbf{u}_\ell = \lambda_\ell M_\ell \mathbf{u}_\ell$$

we get approximations $\widetilde{\lambda}_\ell$ of the exact eigenvalues λ_ℓ on V_ℓ. With the approximation $\widetilde{\mathbf{u}}_\ell$ to the corresponding eigenvector $\mathbf{u}_\ell$, it follows that the corresponding approximate eigenfunction is given by

$$\widetilde{u}_\ell = \sum_{i=1}^{n_\ell} \widetilde{u}_{\ell,i} \varphi_\ell^{(i)},$$

where $\widetilde{u}_{\ell,i}$ are the coefficients of the eigenvector $\widetilde{\mathbf{u}}_\ell$, i.e., $\widetilde{\mathbf{u}}_\ell := [\widetilde{u}_{\ell,i}]_{i=1,\ldots,n_\ell}$. We can compute the residual for this approximation and use this information for adaptation.

From a geometric point of view, it is our goal to enrich the space V_ℓ corresponding to the coarse mesh $\mathcal{T}_\ell$ by some further functions. Since V_ℓ is a subspace of $V_{\ell+1}$ corresponding to the mesh $\mathcal{T}_{\ell+1}$, every function from V_ℓ can be expressed as a linear combination of functions from $V_{\ell+1}$. Thus,

$$\widetilde{u}_\ell = \sum_{i=1}^{n_\ell} \widetilde{u}_{\ell,i} \varphi_\ell^{(i)} = \sum_{i=1}^{n_{\ell+1}} \widehat{u}_{\ell+1,i} \varphi_{\ell+1}^{(i)},$$

with an appropriate coefficient vector $\widehat{\mathbf{u}}_{\ell+1} = [\widehat{u}_{\ell+1,i}]_{i=1,\ldots,n_{\ell+1}}$. The relationship between coefficient vectors $\widehat{\mathbf{u}}_{\ell+1}$ and $\widetilde{\mathbf{u}}_\ell$ can be described by multiplication with an easily constructed prolongation matrix P_ℓ. Therefore, the corresponding prolongated coordinate vector in the fine space $V_{\ell+1}$ associated with the computed

eigenvector $\widetilde{\mathbf{u}}_\ell$ is given as
$$\widehat{\mathbf{u}}_{\ell+1} = P_\ell \widetilde{\mathbf{u}}_\ell$$
Let us denote by $(\widehat{\lambda}_{\ell+1}, \widehat{\mathbf{u}}_{\ell+1})$ an approximate eigenpair obtained from the prolongation of the eigenvector $\widetilde{\mathbf{u}}_\ell$ on the finite space $V_{\ell+1}$, where $\widehat{\lambda}_{\ell+1}$ is a generalized Rayleigh quotient corresponding to $\widehat{\mathbf{u}}_{\ell+1}$.

REMARK 4.3. If the algebraic eigenvalue problem could be solved exactly, then $\widetilde{\lambda}_\ell$ and $\widehat{\lambda}_{\ell+1}$ would be equal. But, since eigenvalues usually cannot be computed exactly and since we work in finite precision arithmetic, roundoff errors, although not discussed here, have to be taken into account as well as an early termination of the iteration and therefore it is important to distinguish these values.

Based on $(\widehat{\lambda}_{\ell+1}, \widehat{\mathbf{u}}_{\ell+1})$ we can compute the corresponding algebraic residual associated with the Galerkin discretization of the original problem on the fine mesh $\mathcal{T}_{\ell+1}$, i.e.,

(16) $$\widehat{\mathbf{r}}_{\ell+1} = K_{\ell+1}\widehat{\mathbf{u}}_{\ell+1} - \widehat{\lambda}_{\ell+1} M_{\ell+1}\widehat{\mathbf{u}}_{\ell+1}.$$

This gives us a natural way of estimating the error in the computed eigenfunction using the coarse grid solution combined with the fine grid information, namely we can prolongate the already computed approximation $\widetilde{\mathbf{u}}_\ell$ from V_ℓ to $V_{\ell+1}$. Then every entry in the residual vector $\widehat{\mathbf{r}}_{\ell+1}$ in (16) corresponds to the appropriate basis function from the fine space. Furthermore, we know that if the i-th entry in the vector $\widehat{\mathbf{r}}_{\ell+1}$ is large, then the i-th basis function has a large influence on the solution, namely its support should be further investigated [42]. All these basis functions with large entries in the vector $\widehat{\mathbf{r}}_{\ell+1}$ together with all basis functions from the coarse space V_ℓ form a basis for the new refined space. The decision on whether an entry in the residual vector is small or large may again be based on different criteria, see [21]. When we have identified the basis functions that should be added to enrich our trial space, we start the marking procedure. Since every FEM basis function is associated with a specific node in the mesh, enriching the space by new basis function means marking the edge corresponding to its node. In order to avoid hanging nodes or irregular triangulations, we again mark some additional edges using a closure algorithm, i.e., if edge is marked and is not a reference edge (the longest edge) of the element, then we add the reference edge to the set of marked edges. After that we can perform the actual refinement to obtain a new mesh which will be an initial mesh for the next loop of our adaptive algorithm.

For more details on the AFEMLA algorithm and, in particular, for the error estimates involving the algebraic error for elliptic self-adjoint eigenvalue problems, we refer to [51, 53].

5. Comparison of AMLS and AFEM

In this section, we compare the discussed AMLS and AFEM methods. For this we consider the case, where the mass and stiffness matrices K_ℓ, M_ℓ resulting from the finest AFEM mesh $\mathcal{T}_{N_\ell}$ are given as input to the AMLS method. Let $\varphi_\ell^{(1)}, \varphi_\ell^{(2)}, \ldots, \varphi_\ell^{(n_\ell)}$, $\ell \in \{0, 1, \ldots, N_\ell\}$, denote the ansatz functions after the ℓ-th refinement. Moreover, since we do not perform a coarsening, we let

$$\varphi_\ell^{(i)} = \varphi_{\ell'}^{(i)}, \quad i = 1, 2, \ldots, n_\ell, \ \ell' > \ell.$$

We know that all eigenfunction approximations $\widetilde{u}_\ell$ computed by the AFEM will be linear combinations of the ansatz functions $\varphi_\ell^{(i)}, i = 1, 2, \ldots, n_\ell$, such that

$$\widetilde{u}_\ell = \sum_{i=1}^{n_\ell} \widetilde{u}_{\ell,i} \varphi_\ell^{(i)}, \quad \widetilde{u}_{\ell,i} \in \mathbb{R}.$$

Since $\mathcal{T}_\ell \subset \mathcal{T}_{\ell+1}$, we may as well write $\widetilde{u}_\ell, \ell = 0, 1, \ldots, N_{\ell-1}$, as

$$\widetilde{u}_\ell = \sum_{i=1}^{n_\ell} \widetilde{u}_{\ell,i} \varphi_{\ell'}^{(i)}, \ell' > \ell.$$

Since we have assumed linear ansatz functions, there will be one degree of freedom in the algebraic problem for every basis function $\varphi_\ell^{(i)}$ and hence $K_\ell, M_\ell \in \mathbb{R}^{n_\ell, n_\ell}$. For simplicity, let us order the degrees of freedom such that the i-th algebraic variable belongs to the ansatz function $\varphi_\ell^{(i)}$. Because there is no coarsening, we can compute the matrices associated to the coarser grids by either removing (in the case of a hierarchical FE method) the last rows and columns of matrices K_ℓ and M_ℓ or (in a more general FE method) by appropriate linear combinations.

If we apply the AMLS method to compute a set of eigenpairs of the matrix pencil (K_{N_ℓ}, M_{N_ℓ}), then as described before, this means the following steps:

(1) the computation of a nested dissection reordering,
(2) a congruence transformation using the block LDL^T decomposition,
(3) the computation of the eigenpairs of the diagonal pencils $(K_{j,j}, M_{j,j})$,
(4) modal truncation.

The first two steps are changes of basis in the algebraic problem, while the last two steps are the selection of a suitable linear combination of degrees of freedom such that the desired eigenspace is spanned. Since every degree of freedom corresponds to the coefficient of one basis function in the continuous problem, we can express aforementioned operations also in the continuous setting. Thus, the AFEM repeatedly refines the mesh only to assure that the algebraic eigensolver can discard certain linear combinations of ansatz functions.

To analyze the space of functions that is removed during the modal truncation and to study the relationship of AMLS and AFEM, we first have to keep in mind that the AMLS method does not have a direct access to the operator $\mathcal{L}$ that was discretized and sometimes, as we discussed in the introduction, there even is no such operator $\mathcal{L}$. Nevertheless, we know that the ansatz functions $\varphi_\ell^{(i)}$ exist and if we were to increase the number of degrees of freedom to infinity, the mass and stiffness matrices would be the exact representation of an (unknown) differential operator $\mathcal{L}$ associated with the limiting PDE, and for this operator AMLS is a method that computes approximate eigenvalues and eigenfunctions on a given mesh.

Within the AMLS method we repeatedly change bases in the coordinate vector spaces. If we assume that the 0-th level ansatz functions $\psi_0^{(i)} := \psi^{(i)}, i = 1, 2, \ldots, n$ are either arising from some FE or AFEM discretization with $\mathbb{P}_1$ ansatz functions $\psi^{(1)}, \psi^{(2)}, \ldots, \psi^{(n)}$ of the continuous problem associated with a PDE, then the algebraic changes of basis can be expressed via these ansatz functions. In the AFEM case we could start with some refinement level ℓ and set

$$\psi^{(i)} = \varphi_\ell^{(i)}, \quad i = 1, 2, \ldots, n_\ell.$$

But any other way of constructing the initial basis would be appropriate as well and then lead to an i-th basis function $\psi_k^{(i)}$ in the k-th AMLS step. To see what is happening, assume for simplicity a single level of substructuring with one coupling block of size n_2 and one substructure block of size n_1, such that $n = n_1 + n_2$. Let the corresponding index sets be denoted by $\mathcal{I}_1 := \{1, 2, \ldots, n_1\}$ and $\mathcal{I}_2 := \{n_1 + 1, \ldots, n\}$, respectively. Note that the AMLS method retains the nested dissection reordering computed in the first step. Thus, there is always one substructure block and one coupling block on the diagonal of the transformed matrices.

In the first step of the AMLS method a nested dissection reordering π is computed, such that
$$\psi_1^{(i)} = \psi_0^{(\pi(i))} = \psi^{(\pi(i))}.$$
Obviously, reordering the degrees of freedom in the discrete problem corresponds to a renumbering of the ansatz functions.

If in the second step of the AMLS method we denote the entries of the block matrix L^{-1} as $(L^{-1})_{ij}$, then
$$\psi_2^{(i)} = \sum_{j=1}^{n} (L^{-1})_{ij} \psi_1^{(j)}, i = 1, 2, \ldots, n.$$

Due to the block structure, $\psi_2^{(i)} = \psi_1^{(i)}, i \in \mathcal{I}_1$, for all ansatz functions of the substructure blocks. For the basis functions corresponding to the coupling block, we have
$$\psi_2^{(i)} = \psi_1^{(i)} + \sum_{j \in \mathcal{I}_1} (L^{-1})_{ij} \psi_1^{(j)}, \quad i \in \mathcal{I}_2,$$
i.e., the basis functions $\psi_2^{(i)}$ corresponding to the coupling block after the LDL^T decomposition are linear combinations of the basis functions $\psi_1^{(i)}, i \in \mathcal{I}_1$, corresponding to the substructure blocks *and* the basis functions $\psi_1^{(i)}, i \in \mathcal{I}_2$ corresponding to the coupling block.

If one uses an eigenvalue projection method, then the computation of the eigenvectors of the blocks in the third AMLS step corresponds to a change of basis with the new basis being the set of block diagonal eigenvectors contained in the matrix $\widetilde{W}$. Since $\widetilde{W}$ is block diagonal and partitioned conformably, we are computing linear combinations of the $\psi_2^{(i)}$ within their respective blocks:
$$\psi_3^{(i)} \in \mathrm{span}\{\psi_2^{(j)} \mid j \in \mathcal{I}_k, k = 1, 2\}, \quad i = 1, 2, \ldots, n.$$
The modal truncation within the AMLS method corresponds to the selection of transformed functions $\psi_3^{(i)}$:
$$\psi_4^{(i)} = \begin{cases} \psi_3^{(i)} & \lambda_i \leq \theta \lambda_c, \\ 0 & \text{otherwise}. \end{cases}$$
Thus whatever the initial set of ansatz function was, we have the following observations:

- Except for the LDL^T decomposition, the basis functions corresponding to one block do not appear in the linear combinations of the basis functions corresponding to the other blocks;
- the basis functions $\psi_\ell^{(i)}, \ell \geq 2$, corresponding to the coupling blocks depend on all basis functions;

- the basis functions $\psi_\ell^{(i)}$ corresponding to the substructure blocks depend only on the basis functions corresponding to the substructure blocks.

In any case, the AMLS method is computing linear combinations of the initial ansatz functions that approximate the desired eigenfunctions, by projecting into subspaces which are constructed using linear combinations spanned by global functions, which are still represented in the original basis. Thus all original fine basis functions participate in the representation and the final eigenfunctions are constructed using global functions in algebraically constructed substructures.

In contrast to this, the AFEM is selecting (by a projection with unit vectors), except for the very last step when the mesh refinement is terminated, local basis functions and a globalization is only done in the final eigenvalue/eigenvector computation.

From this point of view, one can say that AMLS starts with a fine mesh and uses globalization in substructures to reduce the system size, while AFEM starts with a similar fine mesh but uses selected local basis functions all the way to the final step.

5.1. Analysis of AMLS and AFEM. Suppose that we compute and assemble mass and stiffness matrices using an FE software. Depending on the software, this is done with or without explicitly using an available PDE. Nevertheless, the analytical theory of the finite element method holds as long as the FE software uses conforming finite elements. Therefore, we can apply the standard convergence results, e.g. [64, Theorem 3.7], and for the case of an infinite number of degrees of freedom the solution u_h of the discretized problem will be identical to the solution u of the continuous problem. Moreover, the theorem of Necas [54, Theorem 2] guarantees that u_h will be the unique solution to some (unknown to us) variationally stated problem. Therefore, we can always apply the finite element convergence theory even if we do not know the underlying PDE.

But these statements are made from a purely mathematical point of view. If we were able to discretize problems with an arbitrary fine step size, we might as well replace all practical models in engineering, physics, and chemistry with a model based on the interaction of elementary particles and thereby avoid the modeling error inherent to macroscopic models.

In practice, there are some other aspects of the AFEM and the AMLS method that are more relevant. At first, an AFEM implementation requires knowledge of the differential operator $\mathcal{L}$. It needs to be able to construct a partition of the domain Ω, discretize a given PDE, solve the algebraic eigenvalue problems, estimate the errors, and refine the mesh. All of these components must be integrated and cannot work independently of each other, but then if the corresponding theory is available, the convergence of the method can be proved. Furthermore, AFEM works with local basis functions until the very last step and thus the matrices stay sparse. However, many levels of refinement may be needed to achieve a desired level of accuracy. This is particularly difficult if several eigenfunctions have to be computed.

On the other hand, the AMLS method requires only the knowledge of the matrix pencil (K, M) and the cutoff parameter λ_c. This considerably simplifies the implementation, since only knowledge of standard black box numerical linear algebra tools is needed. Moreover, the AMLS method does not require the underlying PDE model, which means that the AMLS method is more suitable for an arbitrary generalized eigenvalue problems as long as the corresponding matrices have the required properties, i.e., symmetry and positive (semi-)definiteness. If this is not the case, then the method can still be applied but becomes partially heuristic.

Both methods are fully automatic. The AFEM does not require any user intervention to decide about a good mesh structure or the polynomial degree of the ansatz functions in different subdomains, instead it will infer this information from the problem when needed. The AMLS method requires a matrix pencil as input and determines all necessary information from the input as well. The fully automatic approach, though it may not always be fast, makes eigenvalue computations also accessible to non-experienced users.

It remains *an open question* how to optimally combine the two methods to obtain the best of both worlds. Both methods are in their current form not suited for more complex eigenvalue problems like the second order problems with damping discussed in the introduction. In AFEM no theory on error estimates is available for complex multiple eigenvalues or Jordan blocks, so even if one would apply the method it would be strongly heuristic. The only practical extension with some kind of error control is to use a homotopy approach as introduced in [15], but even there only very limited partial results are available. AMLS could in principle be applied but with the cutoff value being replaced by some other selection criterion. But there is no theory that this will work in the case of complex eigenvalues or Jordan blocks. See also the experiments in Section 6.

6. Numerical Experiments

Since all existing AMLS implementations are closed source, in the remainder of the paper, we consider our own MATLAB implementation of the method. In the current version we can treat symmetric generalized eigenvalue problems of the form (12) with positive definite stiffness matrices and positive (semi-)definite mass matrices, thus we can deal with problems having eigenvalues at or close to infinity. Moreover, we complement the AMLS method with a subspace iteration method (SIM) [8, 72] to turn the subspace generated by the AMLS method into a good approximation of the invariant subspace [72]. Our implementation does not require any additional assumptions on the matrices beyond symmetry and positive (semi-)definiteness, i.e., they do not need to be necessarily finite element matrices. The matrix reorderings are computed with METIS [44] which computes the vertex separators such that there are exactly two substructures on every level of substructuring.

In this section we will illustrate our AMLS implementation with several numerical examples and present some interesting observations about the method. From now on, we will refer to the original AMLS method described by [43] as the *vanilla AMLS method* (vAMLS) and our implementation as the subspace improved AMLS (sAMLS). We also investigate possible modifications and improvements which allow to turn AMLS into a stand-alone eigenvalue solver.

6.1. A Quadratic Eigenvalue Problem.
Consider a quadratic eigenvalue problem (QEP)

$$(\mu_h^2 M_h + \mu_h C_h + K_h)\mathbf{v}_h = 0. \tag{17}$$

M_h and K_h are real symmetric positive definite mass and stiffness matrices, C_h is a real symmetric positive semidefinite structural damping matrix. For a theoretical analysis of quadratic eigenvalue problems, see [65]. Many practitioners stress the ability of the AMLS method to generate from the undamped problem (17), i.e., $C_h = 0$, a subspace which is close to an invariant subspace associated with certain eigenvalues of the fully damped problem. As much as this statement is true for the case of proportional damping (Rayleigh damping), there is no mathematical reason why it should be true in the general case.

For this example and this example only, we will denote eigenvalue/eigenvector pairs of the generalized eigenvalue problem (GEP) (11) with $(\lambda_h, \mathbf{u}_h)$ and the eigenvalue/eigenvector pairs of (17) with $(\mu_h, \mathbf{v}_h)$.

In the case of proportional damping, it is assumed that there exist real parameters $\alpha, \beta > 0$ such that
$$C_h = \alpha K_h + \beta M_h.$$
Then the eigenvectors $\mathbf{u}_h$ of (11) are also eigenvectors of (17) and only the eigenvalues change, i.e., there exist eigenvalues $\mu_h \in \mathbb{C}$ such that
$$(\mu_h^2 M_h + \mu_h C_h + K_h)\mathbf{u}_h = 0.$$
Therefore, a good approximation to any invariant subspace of the GEP will also be a good approximation to an invariant subspace of the QEP. With general (real symmetric) damping matrices, this is in general not true and we will demonstrate this point using a simplified version of a real-life problem.

To assess the quality of an invariant subspace $\mathcal{S}$, we compute eigenvalue/eigenvector pair approximations within $\mathcal{S}$ and the error of each pair with respect to the whole space $\mathbb{R}^{n_h}$. Then we can evaluate the quality of the space by applying statistical measures to the set of errors, e.g., we can calculate minimum, maximum, or arithmetic mean, etc.

Let
$$P(t) := t^2 M_h + t C_h + K_h.$$
The relative backward normwise error for a computed pair $(\mu_h, \mathbf{v}_h)$ can be calculated, see e.g. [65, §4.2.1], via
$$\eta_P(\mu_h, \mathbf{v}_h) = \frac{\|P(\mu_h)\mathbf{v}_h\|_2}{\|\mathbf{v}_h\|_2 \left[|\mu_h|^2 \|M_h\|_2 + |\mu_h| \|C_h\|_2 + \|K_h\|_2\right]}.$$
It should be noted that an exact eigenvalue/eigenvector pair of the GEP (11) solves the QEP $(\mu_h^2 M_h + K_h)\mathbf{u}_h = 0$ if $\mu_h = \pm i\sqrt{\lambda_h}$, where i is the imaginary unit and we will use this fact to evaluate eigenvalue/eigenvector pairs of (11). Since we deal with relative errors, a small backward error has the same magnitude as the machine epsilon ε. The forward error can be computed by multiplying the backward error with the condition number and for an eigenvalue of (17), the condition number can be estimated via
$$\kappa_P(\mu_h, \mathbf{v}_h) = \frac{|\mu_h|^2 \|M_h\|_2 + |\mu_h| \|C_h\|_2 + \|K_h\|_2}{|\mu_h| \left|\mathbf{v}_h^T(2\mu_h M_h + C_h)\mathbf{v}_h\right|}.$$
If we consider an exact pair $(\lambda_h, \mathbf{u}_h)$ of (11) as an approximate eigenvalue/eigenvector pair for the QEP (17) (*with* damping) by setting
$$\widetilde{\mu}_h := i\sqrt{\lambda_h}, \widetilde{\mathbf{v}}_h := \mathbf{u}_h,$$
then, assuming that we can compute the required matrix 2-norms, we obtain
$$\eta_P(\widetilde{\mu}_h, \widetilde{\mathbf{v}}_h) = \frac{\|\widetilde{\mu}_h C_h \widetilde{\mathbf{v}}_h\|_2}{\|\widetilde{\mathbf{v}}_h\|_2 \left[|\widetilde{\mu}_h^2| \|M_h\|_2 + |\widetilde{\mu}_h| \|C_h\|_2 + \|K_h\|_2\right]}.$$

	$\|\cdot\|_2$	$\kappa_2(\cdot)$
K_h	$5.9803 \cdot 10^7$	$2.8114 \cdot 10^6$
M_h	$7.0909 \cdot 10^{-5}$	714.20
C_h	6.8600	∞

TABLE 1. Spectral norm $\|\cdot\|_2$ and condition number $\kappa_2(X) := \|X\|_2 \|X^{-1}\|_2$ for a coarse mesh brake model, with $n_h = 4669$.

This illustrates that we can cause a large backward error if $\|C_h \widetilde{\mathbf{v}}_h\|_2$ and $\|C_h\|_2$ are sufficiently large compared to $\|M_h\|$ and $\|K_h\|$, which confirms our claim that eigenspaces of (11) and (17) may be completely different.

Among other examples we consider as set of test matrices the Harwell-Boeing BCS structural engineering matrices which are a collection of real-world generalized eigenvalue problems [22]. For these problems, the stiffness matrix often has a norm that is several magnitudes larger than the norm of the mass matrix. If the damping matrix is small in norm compared to the stiffness matrix, then it does not have much influence on the backward error. To illustrate this, consider the coarse mesh FE model of an industrial disk brake with an external load from a brake pad [35]. Here, the matrices were obtained with the FEM software INTES PERMAS [40] and their properties are listed in Table 1. The damping matrix has only 342 columns and rows with nonzero entries and according to the MATLAB rank function, the damping matrix has numerical rank 50. Note this is not a realistic model, the fine mesh models in [35] have matrices of size 850 000, but here we use the small model to illustrate the properties.

In the following we show how well different invariant subspaces calculated based on (11) approximate invariant subspaces of (17) using the brake model matrices. Specifically, we will compare

- the subspace $\mathcal{S}_e$ spanned by the eigenvectors corresponding to the $k = 554$ smallest eigenvalues of (11),
- the subspace $\mathcal{S}_s$ calculated by the AMLS implementation (sAMLS, dimension 352),
- and the subspace $\mathcal{S}_v$ computed by vanilla AMLS (vAMLS, dimension 554).

The eigenvectors corresponding to the smallest eigenvalues of (11) were computed using the MATLAB function eigs. As a starting vector, opts.v0 in eigs, we used in all cases the vector of all ones in order to have reproducible results. In both AMLS variants, we used four levels of substructuring, i.e., there are 16 substructure blocks, and the cutoff $\lambda_c = 10^{10}$. sAMLS was run with the default settings whereas for vAMLS we set $\theta = 2.1^2$ (this value is used during modal truncation). Note that the default value for θ is 8.4^2 for both AMLS variants. The motivation in changing θ for vAMLS, was to reduce the excessive dimension of the vAMLS subspace when run with the default settings (3419). All computations were performed in double precision $\varepsilon = 2.2 \cdot 10^{-16}$. For every subspace, we will compare minimum, maximum, arithmetic mean, and median of the errors of all calculated eigenvalue/eigenvector pairs within a given subspace.

REMARK 6.1. The size of the subspace returned by vanilla AMLS is determined during the modal truncation step. Here, every eigenvalue $\widetilde{\lambda}_j, j = 1, \ldots, \ell$, of every

Space	Minimum	Maximum	Mean	Median
$\mathcal{S}_e$	$1.3 \cdot 10^{-17}$	$6.4 \cdot 10^{-16}$	$1.4 \cdot 10^{-16}$	$9.8 \cdot 10^{-17}$
$\mathcal{S}_s$	$1.5 \cdot 10^{-15}$	$3.1 \cdot 10^{-9}$	$1.2 \cdot 10^{-10}$	$3.3 \cdot 10^{-11}$
$\mathcal{S}_v$	$6.5 \cdot 10^{-7}$	$3.2 \cdot 10^{-2}$	$1.5 \cdot 10^{-3}$	$4.8 \cdot 10^{-4}$

TABLE 2. Relative normwise backward error with respect to the GEP (11) for eigenvalue/eigenvector pairs computed on different subspaces.

Space	Minimum	Maximum	Mean	Median
$\mathcal{S}_e$	$3.2 \cdot 10^{-14}$	$1.2 \cdot 10^{-10}$	$4.8 \cdot 10^{-13}$	$1.3 \cdot 10^{-13}$
$\mathcal{S}_s$	$8.1 \cdot 10^{-11}$	$1.1 \cdot 10^{-6}$	$1.6 \cdot 10^{-7}$	$8.3 \cdot 10^{-8}$
$\mathcal{S}_v$	$8.4 \cdot 10^{-2}$	$1.4 \cdot 10^{0}$	$5.2 \cdot 10^{-1}$	$4.5 \cdot 10^{-1}$

TABLE 3. Relative normwise forward error with respect to the GEP (11) for eigenvalue/eigenvector pairs computed on different subspaces.

diagonal block is compared individually to the cutoff λ_c. Clearly, the worst possible case is that all comparisons have the same result because then either *all* or *no* vectors are selected. As an example, consider a matrix pencil, for which for all j_1, j_2, k_1, k_2,

$$\widetilde{\lambda}_{j_1}^{(k_1)} = \widetilde{\lambda}_{j_2}^{(k_2)},$$

i.e., all diagonal blocks have only one eigenvalue (with corresponding algebraic multiplicity) and for all diagonal blocks it is the same value.

The coarse mesh disk brake problem is close to such a worst-case matrix pencil since 3116 out of $n_h = 4669$ possible distinct eigenvalues of the diagonal blocks are in the interval $[10^{11}, 10^{12}]$. Note that the largest generalized eigenvalue of (11) in this case is $\lambda_{h,\max} \approx 2 \cdot 10^{13}$, and the smallest eigenvalue is $\lambda_{h,\min} \approx 2 \cdot 10^6$.

In Table 2, we display the backward errors of the different spaces with respect to the GEP (11), i.e., we measure how well each subspace approximates an invariant subspace of (11). As expected, $\mathcal{S}_e$ is a very good approximation to an invariant subspace, the maximal relative backward error is comparable to the machine epsilon ε. The vAMLS subspace is a noticeably worse approximation to an invariant subspace than $\mathcal{S}_e$ but the subspace iterations of sAMLS make a real difference here, and decrease the mean and median backward error compared to $\mathcal{S}_v$ significantly. For completeness, we also list the relative forward errors for (11) in Table 3.

In Table 4, we display the relative normwise backward errors with respect to (17). The presence of the damping matrix leads to large changes and increases the backward error by several orders of magnitude for $\mathcal{S}_e$ and $\mathcal{S}_s$. For the minimum and maximum backward error, $\mathcal{S}_e$ contains considerably better eigenpair approximations but on average (mean, median), there is a hardly any difference between $\mathcal{S}_e$ and $\mathcal{S}_s$ with respect to our quantitative measures. Interestingly, the damping has hardly an impact on the vAMLS space; consequently, the average backward

Space	Minimum	Maximum	Mean	Median
$\mathcal{S}_e$	$3.5 \cdot 10^{-15}$	$5.8 \cdot 10^{-3}$	$2.6 \cdot 10^{-4}$	$8.5 \cdot 10^{-5}$
$\mathcal{S}_s$	$9.2 \cdot 10^{-13}$	$7.7 \cdot 10^{-3}$	$2.7 \cdot 10^{-4}$	$7.0 \cdot 10^{-5}$
$\mathcal{S}_v$	$7.1 \cdot 10^{-7}$	$3.1 \cdot 10^{-2}$	$1.6 \cdot 10^{-3}$	$5.8 \cdot 10^{-4}$

TABLE 4. Relative normwise backward error with respect to the QEP (17) for eigenpairs computed on different subspaces.

Space	Minimum	Maximum	Mean	Median
$\mathcal{S}_e$	$4.2 \cdot 10^{-10}$	$4.4 \cdot 10^{0}$	$2.0 \cdot 10^{-1}$	$1.2 \cdot 10^{-1}$
$\mathcal{S}_s$	$1.1 \cdot 10^{-7}$	$4.7 \cdot 10^{0}$	$3.5 \cdot 10^{-1}$	$2.1 \cdot 10^{-1}$
$\mathcal{S}_v$	$7.7 \cdot 10^{-2}$	$6.9 \cdot 10^{0}$	$5.9 \cdot 10^{-1}$	$4.8 \cdot 10^{-1}$

TABLE 5. Relative normwise forward error with respect to the QEP (17) for eigenpairs computed on different subspaces.

error difference to the other spaces shrinks from several magnitudes to a factor less than 10. Note that the backward errors of $\mathcal{S}_v$ are comparable for GEP and QEP.

Since we have used the invariant subspaces of (11) to find approximate invariant subspaces for (17), apparently a small backward error is not as important as other factors. To see whether we are least close to an exact solution, we also analyzed the forward error for (17), the results are shown in Table 5.

We see that on average, the spaces are hardly distinguishable using our measures. Especially surprising is the fact that the backward and the forward errors of $\mathcal{S}_v$ have the same order of magnitude with respect to (11) and (17).

From a mathematical point of view, there is no reason why an invariant subspace for equation (11) should resemble an invariant subspace for (17) and in large scale problems it has been observed that indeed this may not be the case [35]. Nevertheless, the projection of quadratic eigenvalue problems via AMLS subspaces is a standard procedure in engineering practice. Surprisingly, in many cases, the quantitative differences between the three spaces vanish when computing errors with respect to (17). In practice, AMLS seems to be able to handle larger problems than a standard Arnoldi-eigensolver and in conjunction with our examples, one can see why AMLS subspaces are so popular for dimensional reduction in mechanical engineering applications.

6.2. Isospectral Domains and AMLS. In practical engineering problems, the most efficient algorithms strongly benefit from exploiting symmetries which allow to reduce significantly the overall computational complexity. Since the AMLS method is a commonly used approach in structural mechanics, we were interested how it handles present symmetries. Ideally, one would expect the AMLS method to show the same sequence of operations in every step for every isospectral domain because it would simplify software development and mathematical analysis. Unfortunately this is not the case even for very simple problems. To illustrate this, we consider as a model problem the simple Laplace eigenvalue problem (8) defined over the rectangular domain $\Omega = (0, \alpha) \times (0, \beta)$. For this choice the exact eigenvalues

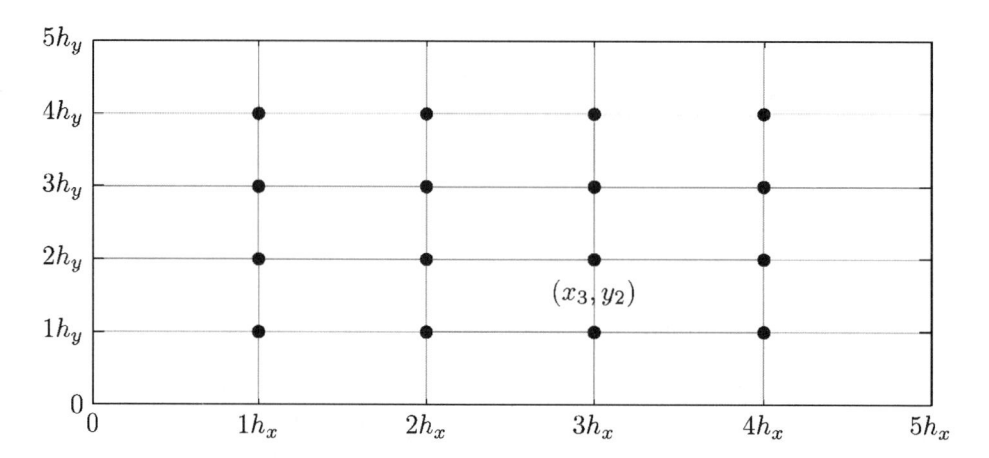

FIGURE 2. A grid over the rectangle $(0,2) \times (0,1)$ with $n = 4$ points on each axis. The inner nodes marked with black dots (•) are the degrees of freedom of the discrete problem.

and eigenvectors are known [19, Chapter VI, §4.1] and explicitly given as

$$(\lambda_{i,j}, u_{i,j}) = \left(\left(\frac{i^2}{\alpha^2} + \frac{j^2}{\beta^2}\right)\pi^2, \frac{2}{\sqrt{\alpha\beta}} \sin \frac{i\pi x}{\alpha} \sin \frac{j\pi y}{\beta}\right), \quad i,j = 1,2,\ldots.$$

Obviously, the problems with domains $(0,\alpha) \times (0,\beta)$ and $(0,\beta) \times (0,\alpha)$ possess the same set of eigenvalues. We discretized the problem with axis parallel quadrilateral elements and piecewise linear ansatz functions over n inner nodes on each axis with the boundary points given by (x,y) with $x \in \{0,\alpha\}$ or $y \in \{0,\beta\}$ and step sizes $h_x := \alpha(n+1)^{-1}$ and $h_y := \beta(n+1)^{-1}$ in x- and y-direction, respectively. For this discretization the eigenvectors will be pointwise exact in the grid nodes. In Figure 2 we show the domain $(0,\alpha) \times (0,\beta)$ with $\alpha = 2, \beta = 1$ and the associated mesh.

For the following experiments we chose $n_h = 32$, $\lambda_c = 100$, $\theta = 8.4^2$, $\alpha = 1, \beta = 32$, and $\alpha = 32, \beta = 1$, respectively. The analyzed problem has $n_c = 91$ eigenvalues $\lambda_h \leq \lambda_c$. From the numerical analysis point of view, this discretization corresponds to the problem

$$-\frac{\beta}{\alpha}\frac{\partial^2 u}{\partial x^2} - \frac{\alpha}{\beta}\frac{\partial^2 u}{\partial y^2} = \lambda u$$

on the unit square. Nevertheless, the two domains are isospectral and the resulting matrix pencils have the same multi-set of eigenvalues. Ideally, one would expect the AMLS method to behave identically for the two problems.

Let us investigate the properties of the subspace generated by the vAMLS method. At first, we notice that the matrix substructures obtained with METIS [44] for both domains are the same. Since, in both cases, the sparsity pattern of the stiffness matrix is the same and the METIS graph partitioner works with the *unweighted* induced graph, the fact of different directional scaling is not taken into account. Figure 3 shows the cosines of the angles between the subspace generated by the vAMLS method and the set of exact eigenvectors. The eigenvectors are sorted according to their corresponding eigenvalues in ascending order. Ideally, the vAMLS method computes the subspace spanned by the first $n_c = 91$ eigenvectors, i.e., for $i = 1, 2, \ldots, n_c$, the cosine of the angles is one and zero otherwise. As we can see in Figure 3, the vAMLS method does not compute the desired eigenspace in the second case and more important, it clearly does not recognize the isospectral property of the two domains.

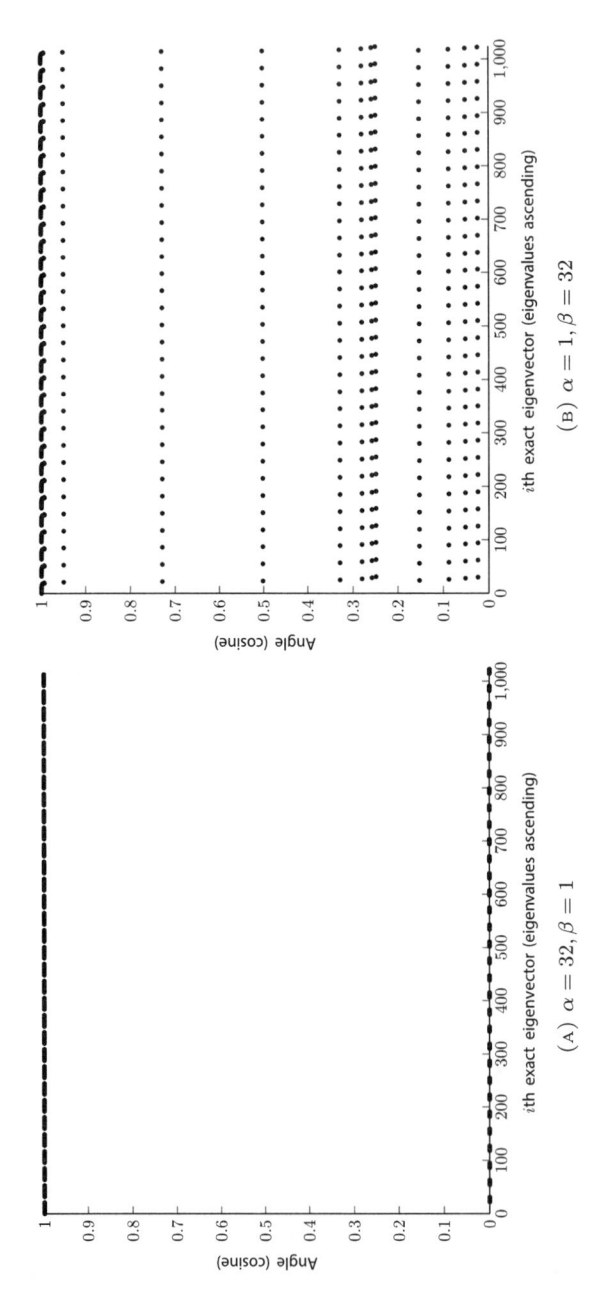

FIGURE 3. Subspace generated by vanilla AMLS with $n_h = 32$, $\lambda_c = 100$, and one level of substructuring with two substructure blocks.

7. Conclusion

We have discussed two of the most widely used methods for the solution of PDE eigenvalue problems as well as their advantages and disadvantages. The adaptive finite element method AFEM and its variant AFEMLA are provably efficient for standard elliptic problems. The automated multilevel substructuring method (AMLS) is widely used in practice for problems that require only approximate solutions. Since it does not compute a space with prescribed properties, its success is partially heuristic. In principle, both methods work with approximations of fine grid solutions. The difference is that in AFEM the final approximation is an approximation on a coarse mesh obtained by leaving out nodes in the fine mesh together with the associated basis functions, while in AMLS the reduction is achieved by choosing a basis of local eigenfunctions (which are represented by the fine mesh basis functions). Both methods work well on elliptic problems, but no complete analysis is available for damped problems. AMLS is fast, compared to the standard eigensolvers, applicable to very large scale problems, and often gives surprisingly good results even for damped problems. *However, what is needed is a major research effort to study AFEM or AMLS methods for practical PDE eigenvalue problems, with or without the given PDE model.*

References

[1] M. Ainsworth and J. T. Oden, *A Posteriori Error Estimation in Finite Element Analysis*, Pure and Applied Mathematics (New York), Wiley-Interscience [John Wiley & Sons], New York, 2000. MR1885308 (2003b:65001)

[2] A. Akay. Acoustics of friction. *J. of the Acoustical Society of America*, 111(4):1525–1548, 2002.

[3] E. Anderson, Z. Bai, C. Bischof, L. S. Blackford, J. Demmel, J. Dongarra, J. Du Croz, A. Greenbaum, S. Hammarling, A. McKenney, and D. Sorensen, editors. *LAPACK Users' Guide*. Software, Environments and Tools. Society for Industrial and Applied Mathematics, Philadelphia, PA, USA, 1999.

[4] I. Babuška and J. Osborn, *Eigenvalue problems*, Handbook of Numerical Analysis, Vol. II, Handb. Numer. Anal., II, North-Holland, Amsterdam, 1991, pp. 641–787. MR1115240

[5] I. Babuška and T. Strouboulis, *The Finite Element Method and Its Reliability*, Numerical Mathematics and Scientific Computation, The Clarendon Press, Oxford University Press, New York, 2001. MR1857191 (2002k:65001)

[6] I. Babuška, J. R. Whiteman, and T. Strouboulis, *Finite Elements: An Introduction to the Method and Error Estimation*, Oxford University Press, Oxford, 2011. MR2857237 (2012g:65001)

[7] *Templates for the Solution of Algebraic Eigenvalue Problems*, Software, Environments, and Tools, vol. 11, Society for Industrial and Applied Mathematics (SIAM), Philadelphia, PA, 2000. A practical guide; Edited by Zhaojun Bai, James Demmel, Jack Dongarra, Axel Ruhe and Henk van der Vorst. MR1792141 (2001k:65003)

[8] K.-J. Bathe. *Finite Element Procedures*. Prentice-Hall, Upper Saddle River, NJ, USA, 1996.

[9] J. K. Bennighof and R. B. Lehoucq, *An automated multilevel substructuring method for eigenspace computation in linear elastodynamics*, SIAM J. Sci. Comput. **25** (2004), no. 6, 2084–2106 (electronic), DOI 10.1137/S1064827502400650. MR2086832 (2005c:74030)

[10] L. S. Blackford, J. Choi, A. Cleary, E. D'Azevedo, J. Demmel, I. Dhillon, J. Dongarra, S. Hammarling, G. Henry, A. Petitet, K. Stanley, D. Walker, and R. C. Whaley, editors. *ScaLAPACK Users' Guide*. Software, Environments and Tools. Society for Industrial and Applied Mathematics, Philadelphia, PA, USA, 1997.

[11] D. Braess, *Finite Elements: Theory, Fast solvers, and Applications in Elasticity Theory*, 3rd ed., Cambridge University Press, Cambridge, 2007. Translated from the German by Larry L. Schumaker. MR2322235 (2008b:65142)

[12] S. C. Brenner and C. Carstensen. Finite element methods. In E. Stein, R. de Borst, and T. J. R. Huges, editors, *Encyclopedia of Computational Mechanics*, volume 1, pages 73–114. Wiley-Interscience, New York, NY, USA, 2004.

[13] S. C. Brenner and L. R. Scott, *The Mathematical Theory of Finite Element Methods*, 3rd ed., Texts in Applied Mathematics, vol. 15, Springer, New York, 2008. MR2373954 (2008m:65001)

[14] C. Carstensen and J. Gedicke, *An adaptive finite element eigenvalue solver of asymptotic quasi-optimal computational complexity*, SIAM J. Numer. Anal. **50** (2012), no. 3, 1029–1057, DOI 10.1137/090769430. MR2970733

[15] C. Carstensen, J. Gedicke, V. Mehrmann, and A. Międlar, *An adaptive homotopy approach for non-selfadjoint eigenvalue problems*, Numer. Math. **119** (2011), no. 3, 557–583, DOI 10.1007/s00211-011-0388-x. MR2845628 (2012j:65378)

[16] C. Carstensen, J. Gedicke, V. Mehrmann, and A. Międlar, *An adaptive finite element method with asymptotic saturation for eigenvalue problems*, Numer. Math. **128** (2014), no. 4, 615–634, DOI 10.1007/s00211-014-0624-2. MR3276869

[17] Z. Chen, *Finite Element Methods and Their Applications*, Scientific Computation, Springer-Verlag, Berlin, 2005. MR2158541 (2006d:65001)

[18] P. G. Ciarlet, *The Finite Element Method for Elliptic Problems*, Classics in Applied Mathematics, vol. 40, Society for Industrial and Applied Mathematics (SIAM), Philadelphia, PA, 2002. Reprint of the 1978 original [North-Holland, Amsterdam; MR0520174 (58 #25001)]. MR1930132

[19] R. Courant and D. Hilbert, *Methods of Mathematical Physics. Vol. I*, Interscience Publishers, Inc., New York, N.Y., 1953. MR0065391 (16,426a)

[20] R. R. Craig, Jr. and M. C. C. Bampton. Coupling of substructures for dynamic analyses. *AIAA Journal*, 6(7):1313–1319, 1968.

[21] W. Dörfler, *A convergent adaptive algorithm for Poisson's equation*, SIAM J. Numer. Anal. **33** (1996), no. 3, 1106–1124, DOI 10.1137/0733054. MR1393904 (97e:65139)

[22] I. Duff, R. Grimes, and J. Lewis. Sparse matrix test problems. *ACM Trans. Math. Software*, 15(1):1–14, 1989.

[23] I. S. Duff, A. M. Erisman, and J. K. Reid, *Direct Methods for Sparse Matrices*, 2nd ed., Monographs on Numerical Analysis, The Clarendon Press, Oxford University Press, New York, 1989. Oxford Science Publications. MR1029273 (90m:65061)

[24] H. C. Elman, D. J. Silvester, and A. J. Wathen, *Finite Elements and Fast Iterative Solvers: With Applications in Incompressible Fluid Dynamics*, 2nd ed., Numerical Mathematics and Scientific Computation, Oxford University Press, Oxford, 2014. MR3235759

[25] K. Elssel and H. Voss, *An a priori bound for automated multilevel substructuring*, SIAM J. Matrix Anal. Appl. **28** (2006), no. 2, 386–397 (electronic), DOI 10.1137/040616097. MR2255335 (2007h:65030)

[26] A. Ern and J.-L. Guermond, *Theory and Practice of Finite Elements*, Applied Mathematical Sciences, vol. 159, Springer-Verlag, New York, 2004. MR2050138 (2005d:65002)

[27] L. C. Evans, *Partial Differential Equations*, 2nd ed., Graduate Studies in Mathematics, vol. 19, American Mathematical Society, Providence, RI, 2010. MR2597943 (2011c:35002)

[28] D. Gallistl, *Adaptive nonconforming finite element approximation of eigenvalue clusters*, Comput. Methods Appl. Math. **14** (2014), no. 4, 509–535, DOI 10.1515/cmam-2014-0020. MR3259027

[29] D. Gallistl, An optimal adaptive FEM for eigenvalue clusters. *Numer. Math.*, 2015.

[30] A. George, *Nested dissection of a regular finite element mesh*, SIAM J. Numer. Anal. **10** (1973), 345–363. Collection of articles dedicated to the memory of George E. Forsythe. MR0388756 (52 #9590)

[31] M. S. Gockenbach, *Understanding and Implementing the Finite Element Method*, Society for Industrial and Applied Mathematics (SIAM), Philadelphia, PA, 2006. MR2256926 (2007e:65002)

[32] M. S. Gockenbach, *Partial Differential Equations: Analytical and Numerical Methods*, 2nd ed., Society for Industrial and Applied Mathematics (SIAM), Philadelphia, PA, 2011. MR2743564 (2011j:35001)

[33] G. H. Golub and C. F. Van Loan. *Matrix Computations*. Johns Hopkins University Press, Baltimore, MD, USA, 4 edition, 2012.

[34] C. Grossmann and H.-G. Roos, *Numerical Treatment of Partial Differential Equations*, Universitext, Springer, Berlin, 2007. Translated and revised from the 3rd (2005) German edition by Martin Stynes. MR2362757 (2009b:65004)

[35] N. Gräbner, S. Quraishi, C. Schröder, V. Mehrmann, and U. von Wagner. New numerical methods for the complex eigenvalue analysis of disk brake squeal. In *EuroBrake 2014 Conference Proceedings*. FISITA, London, UK, 2014. EB2014-SA-007.

[36] W. Hackbusch, *Elliptic Differential Equations: Theory and Numerical Treatment*, Springer Series in Computational Mathematics, vol. 18, Springer-Verlag, Berlin, 1992. Translated from the author's revision of the 1986 German original by Regine Fadiman and Patrick D. F. Ion. MR1197118 (94b:35001)

[37] S. Hammarling, C. J. Munro, and F. Tisseur, *An algorithm for the complete solution of quadratic eigenvalue problems*, ACM Trans. Math. Software **39** (2013), no. 3, 18:1–18:19, DOI 10.1145/2450153.2450156. MR3094974

[38] U. L. Hetmaniuk and R. B. Lehoucq, *Multilevel methods for eigenspace computations in structural dynamics*, Domain Decomposition Methods in Science and Engineering XVI, Lect. Notes Comput. Sci. Eng., vol. 55, Springer, Berlin, 2007, pp. 103–113, DOI 10.1007/978-3-540-34469-8_9. MR2334095 (2008e:65122)

[39] V. Heuveline and R. Rannacher, *A posteriori error control for finite approximations of elliptic eigenvalue problems*, Adv. Comput. Math. **15** (2001), no. 1-4, 107–138 (2002), DOI 10.1023/A:1014291224961. MR1887731 (2002m:65120)

[40] Ingenieurgesellschaft für technische Software mbH. PERMAS product description version 15, 2014.

[41] C. Johnson, *Numerical Solution of Partial Differential Equations by the Finite Element Method*, Dover Publications, Inc., Mineola, NY, 2009. Reprint of the 1987 edition. MR2553737 (2010h:65002)

[42] C. Kamm. A posteriori error estimation in numerical methods for solving self-adjoint eigenvalue problems. Master's thesis, Technische Universität Berlin, Berlin, Germany, 2007.

[43] M. F. Kaplan. *Implementation of Automated Multilevel Substructuring for Frequency Response Analysis of Structures*. PhD thesis, University of Texas at Austin, Austin, TX, USA, 2001.

[44] G. Karypis and V. Kumar, *A fast and high quality multilevel scheme for partitioning irregular graphs*, SIAM J. Sci. Comput. **20** (1998), no. 1, 359–392 (electronic), DOI 10.1137/S1064827595287997. MR1639073 (99f:68158)

[45] T. Kato, *Perturbation Theory for Linear Operators*, Classics in Mathematics, Springer-Verlag, Berlin, 1995. Reprint of the 1980 edition. MR1335452 (96a:47025)

[46] N. M. Kinkaid, O. M. O'Reilly, and P. Papadopoulos. Automotive disc brake squeal. *J. Sound & Vibration*, 267(1):105–166, 2003.

[47] M. G. Larson and F. Bengzon, *The Finite Element Method: Theory, Implementation, and Applications*, Texts in Computational Science and Engineering, vol. 10, Springer, Heidelberg, 2013. MR3015004

[48] S. Larsson and V. Thomée, *Partial Differential Equations with Numerical Methods*, Texts in Applied Mathematics, vol. 45, Springer-Verlag, Berlin, 2003. MR1995838 (2004f:65001)

[49] R. B. Lehoucq, D. C. Sorensen, and C. Yang, *ARPACK Users' Guide: Solution of Large-Scale Eigenvalue Problems with Implicitly Restarted Arnoldi Methods*, Software, Environments, and Tools, vol. 6, Society for Industrial and Applied Mathematics (SIAM), Philadelphia, PA, 1998. MR1621681

[50] R.-C. Li, Y. Nakatsukasa, N. Truhar, and S. Xu, *Perturbation of partitioned Hermitian definite generalized eigenvalue problems*, SIAM J. Matrix Anal. Appl. **32** (2011), no. 2, 642–663, DOI 10.1137/100808356. MR2817508

[51] V. Mehrmann and A. Miȩdlar, *Adaptive computation of smallest eigenvalues of self-adjoint elliptic partial differential equations*, Numer. Linear Algebra Appl. **18** (2011), no. 3, 387–409, DOI 10.1002/nla.733. MR2760060 (2012e:65255)

[52] A. Miȩdlar. Functional perturbation results and the balanced AFEM algorithm for self-adjoint PDE eigenvalue problems. Preprint 817, DFG Research Center Matheon, Berlin, Germany, 2011.

[53] A. Miȩdlar. *Inexact Adaptive Finite Element Methods for Elliptic PDE Eigenvalue Problems*. PhD thesis, Technische Universität Berlin, Berlin, Germany, 2011.

[54] R. H. Nochetto, K. G. Siebert, and A. Veeser, *Theory of adaptive finite element methods: an introduction*, Multiscale, Nonlinear and Adaptive Approximation, Springer, Berlin, 2009, pp. 409–542, DOI 10.1007/978-3-642-03413-8_12. MR2648380 (2011k:65164)

[55] R. H. Nochetto and A. Veeser, *Primer of adaptive finite element methods*, Multiscale and Adaptivity: Modeling, Numerics and Applications, Lecture Notes in Math., vol. 2040, Springer, Heidelberg, 2012, pp. 125–225, DOI 10.1007/978-3-642-24079-9. MR3076038

[56] H. Ouyang, W. Nack, Y. Yuan, and F. Chen. Numerical analysis of automotive disc brake squeal: a review. *International J. of Vehicle Noise & Vibration*, 1(3/4):207–231, 2005.

[57] B. N. Parlett, *The Symmetric Eigenvalue Problem*, Classics in Applied Mathematics, vol. 20, Society for Industrial and Applied Mathematics (SIAM), Philadelphia, PA, 1998. Corrected reprint of the 1980 original. MR1490034 (99c:65072)

[58] A. Quarteroni, *Numerical Models for Differential Problems*, MS&A. Modeling, Simulation and Applications, vol. 2, Springer-Verlag Italia, Milan, 2009. Translated from the 4th (2008) Italian edition by Silvia Quarteroni. MR2522375 (2010h:65004)

[59] A. Quarteroni and A. Valli, *Numerical Approximation of Partial Differential Equations*, Springer Series in Computational Mathematics, vol. 23, Springer-Verlag, Berlin, 1994. MR1299729 (95i:65005)

[60] R. Rannacher, A. Westenberger, and W. Wollner, *Adaptive finite element solution of eigenvalue problems: balancing of discretization and iteration error*, J. Numer. Math. **18** (2010), no. 4, 303–327, DOI 10.1515/JNUM.2010.015. MR2747810 (2011k:65165)

[61] P.-A. Raviart and J.-M. Thomas, *Introduction à l'Analyse Numérique des Équations aux Dérivées Partielles* (French), Collection Mathématiques Appliquées pour la Maîtrise. [Collection of Applied Mathematics for the Master's Degree], Masson, Paris, 1983. MR773854 (87a:65001a)

[62] S. Repin, *A Posteriori Estimates for Partial Differential Equations*, Radon Series on Computational and Applied Mathematics, vol. 4, Walter de Gruyter GmbH & Co. KG, Berlin, 2008. MR2458008 (2010b:35004)

[63] J. P. Sethna, *Statistical Mechanics*, Oxford Master Series in Physics, vol. 14, Oxford University Press, Oxford, 2006. Entropy, order parameters, and complexity; Oxford Master Series in Statistical Computational, and Theoretical Physics. MR2292303 (2008b:82001)

[64] G. Strang and G. J. Fix, *An Analysis of the Finite Element Method*, Prentice-Hall, Inc., Englewood Cliffs, N. J., 1973. Prentice-Hall Series in Automatic Computation. MR0443377 (56 #1747)

[65] L. Taslaman. *Algorithms and Theory for Polynomial Eigenvalue Problems*. PhD thesis, The University of Manchester, Manchester, UK, 2014.

[66] R. Verfürth. *A Review of A Posteriori Error Estimation and Adaptive Mesh-Refinement Techniques*. Wiley-Teubner, Stuttgart, Germany, 1996.

[67] R. Verfürth, *A Posteriori Error Estimation Techniques for Finite Element Methods*, Numerical Mathematics and Scientific Computation, Oxford University Press, Oxford, 2013. MR3059294

[68] P. Šolín, *Partial Differential Equations and the Finite Element Method*, Pure and Applied Mathematics (New York), Wiley-Interscience [John Wiley & Sons], Hoboken, NJ, 2006. MR2180081 (2006f:35004)

[69] H. F. Weinberger, *Variational Methods for Eigenvalue Approximation*, Society for Industrial and Applied Mathematics, Philadelphia, Pa., 1974. Based on a series of lectures presented at the NSF-CBMS Regional Conference on Approximation of Eigenvalues of Differential Operators, Vanderbilt University, Nashville, Tenn., June 26–30, 1972; Conference Board of the Mathematical Sciences Regional Conference Series in Applied Mathematics, No. 15. MR0400004 (53 #3842)

[70] C. Yang, W. Gao, Z. Bai, X. S. Li, L.-Q. Lee, P. Husbands, and E. Ng, *An algebraic substructuring method for large-scale eigenvalue calculation*, SIAM J. Sci. Comput. **27** (2005), no. 3, 873–892 (electronic), DOI 10.1137/040613767. MR2199912 (2006j:65114)

[71] Y. Yang, L. Sun, H. Bi, and H. Li, *A note on the residual type a posteriori error estimates for finite element eigenpairs of nonsymmetric elliptic eigenvalue problems*, Appl. Numer. Math. **82** (2014), 51–67, DOI 10.1016/j.apnum.2014.02.015. MR3212379

[72] J. Yin, H. Voss, and P. Chen. Improving eigenpairs of automated multilevel substructuring with subspace iterations. *Comput. & Structures*, 119:115–124, 2013.

Technische Universität Berlin, Institut für Mathematik, MA 4-5, Strasse des 17. Juni 136, 10623 Berlin, Germany
E-mail address: conrads@math.tu-berlin.de

Technische Universität Berlin, Institut für Mathematik, MA 4-5, Strasse des 17. Juni 136, 10623 Berlin, Germany
E-mail address: mehrmann@math.tu-berlin.de

University of Minnesota, Department of Computer Science & Engineering, 200 Union Street, S. E., RM4-192 Keller Hall, Minneapolis, MN 55455
E-mail address: amiedlar@umn.edu

Interpreting the von Neumann entropy of graph Laplacians, and coentropic graphs

Niel de Beaudrap, Vittorio Giovannetti, Simone Severini, and Richard Wilson

ABSTRACT. For any graph, we define a rank-1 operator on a bipartite tensor product space, with components associated to the set of vertices and edges respectively. We show that the partial traces of the operator are the Laplacian and the edge-Laplacian. This provides an interpretation of the von Neumann entropy of the (normalized) Laplacian as the amount of quantum entanglement between two systems corresponding to vertices and edges. In this framework, cospectral graphs correspond exactly to local unitarily equivalent pure states. Finally, we introduce the notion of coentropic graphs, that is, graphs with equal von Neumann entropy. The smallest coentropic (but not cospectral) graphs that we are able to construct have 8 vertices. The number of equivalence classes of coentropic graphs with n vertices and m edges is a lower bound to the number of (pure) bipartite entanglement classes with subsystems of corresponding dimension.

Our references on algebraic graph theory and quantum information are [2] and [6], respectively. Let $G = (V, E)$ be a simple graph with n vertices and m edges. We consider a representation of graphs using certain vectors in a bipartite tensor product space. Specifically, we define two configuration (Hilbert) spaces: $\mathcal{H}_V \cong \mathbb{C}^V$ with orthonormal basis $\mathbf{a}_v$ running over $v \in V$; $\mathcal{H}_E \cong \mathbb{C}^E$ with orthonormal basis $\mathbf{b}_e$ running over $e \in E$. We assume a commutative formal product from pairs of vertices to unordered pairs, $uv = \{u, v\}$. We also assume without loss of generality that there are total orders defined on V and E (which we denote by $\leq$ in both cases).

The graph representation which we consider is a vector in $\mathcal{H}_V \otimes \mathcal{H}_E$ related to graph Laplacians. The *Laplacian* of the graph G is an operator

(1) $$L(G) = D(G) - A(G)$$

acting on $\mathcal{H}_V$, where $D(G) = \mathrm{Diag}(\deg(v_1), \deg(v_2), \ldots, \deg(v_n))$ is the *degree matrix* of G and $A(G)$ is the adjacency matrix. There is an equivalent definition of a Laplacian, in terms of incidence matrices. An *orientation* of G is a collection $\mathcal{F} = \{f_e\}_{e \in E}$ of bijections $f_e : e \longrightarrow \{1, -1\}$ such that that for each $uv \in E$, we have $f_{uv}(u) = -f_{uv}(v) \in \{1, -1\}$. For an orientation $\mathcal{F}$ of G, the *incidence matrix* of G is the $n \times m$ matrix $M_{\mathcal{F}}$ — with rows indexed by vertices and columns by edges — such that

(2) $$(M_\mathcal{F})_{v,e} = \begin{cases} f_e(v), & \text{if } v \in e; \\ 0, & \text{otherwise.} \end{cases}$$

Independently of the chosen orientation $\mathcal{F}$, we then have

(3) $$L(G) = M_\mathcal{F} M_\mathcal{F}^\dagger.$$

This shows that $L(G)$ is positive semidefinite. We may equivalently formulate the incidence matrix as a sum of outer products,

(4) $$M_\mathcal{F} = \sum_{e \in E} \sum_{v \in e} f_e(v) \mathbf{a}_v \mathbf{b}_e^\dagger = \sum_{uv \in E} f_{uv}(u)(\mathbf{a}_u - \mathbf{a}_v)\mathbf{b}_{uv}^\dagger;$$

the middle expression is just another presentation of the definition in Eq. (2); the right-hand expression follows from $f_{uv}(u) = -f_{uv}(v)$.

Even though the Laplacian itself is the same for all orientations of G, the formulation of the Laplacian in Eq. (3) is orientation-dependent, essentially because we are not considering the graph G but rather a digraph D such that $A(G) = A(D) + A(D^T)$. Considering incidence matrices as a property of *digraphs* motivates the following definition. The *incidence matrix* of a directed graph G is the $n \times m$ matrix $\bar{M}$ — with rows indexed by vertices and columns by arcs — such that

$$\bar{M}_{v,\alpha} = \begin{cases} +1, & \text{if } v \text{ is a source of } \alpha; \\ -1, & \text{if } v \text{ is a sink of } \alpha; \\ 0, & \text{otherwise.} \end{cases}$$

The matrix $M_\mathcal{F}$ is then the resulting *directed* incidence matrix $\bar{M}$ for the directed graph in which we replace uv with the arc $u \to v$ if $f_{uv}(u) = +1$, and with the arc $v \to u$ if $f_{uv}(u) = -1$ (*i.e.*, taking $\mathcal{F}$ literally as a specification of how to uniquely orient the edges of G). Having a definition of incidence matrices on digraphs, we can describe the Laplacian of G in terms of the incidence matrix of G, interpreted as a symmetric digraph containing both the arc $u \to v$ and the arc $v \to u$ for each edge $uv \in E(G)$. We replace each $\mathbf{b}_{uv}$ with two vectors $\mathbf{d}_{u,v}$ and $\mathbf{d}_{v,u}$ corresponding to the arcs $u \to v$ and $v \to u$. For instance, we may do this by redefining $\mathcal{H}_E = \text{span}\{\mathbf{d}_{u,v} : uv \in E(G)\} \subseteq \mathcal{H}_V \otimes \mathcal{H}_V$, letting $\mathbf{d}_{u,v} = \mathbf{a}_u \otimes \mathbf{a}_v$. We thereby obtain

(5a) $$\bar{M} := \sum_{u \in V} \sum_{uv \in E} \mathbf{a}_u (\mathbf{d}_{u,v} - \mathbf{d}_{v,u})^\dagger = \sum_{u \in V} \sum_{uv \in E} (\mathbf{a}_u - \mathbf{a}_v) \mathbf{d}_{v,u}^\dagger$$

(5b) $$= \sum_{uv \in E} (\mathbf{a}_u - \mathbf{a}_v)(\mathbf{d}_{u,v} - \mathbf{d}_{v,u})^\dagger,$$

restricting in this case to graphs G containing no isolated vertices. We may then easily show that

$$L(G) = \tfrac{1}{2} \bar{M} \bar{M}^\dagger.$$

(The factor of $1/2$ may be seen to arise from doubling the edge-space by introduction of arc vectors rather than edge vectors.) We may thereby describe the Laplacian using incidence matrices, but without reference to any particular orientation $\mathcal{F}$ of the edges.

Implicitly, the latter formulation of the Laplacian also describes a way in which it may be formed as the partial trace of a rank-1 operator on $\mathcal{H}_V \otimes \mathcal{H}_E$ which is determined by the graph G. The rank-1 operator we may represent as an outer product $\psi_G \psi_G^\dagger$, where $\psi_G \in \mathcal{H}_V \otimes \mathcal{H}_E$ is a vectorization of the incidence matrix $\bar{M}$. Consider the vector

$$\psi_G = \tfrac{1}{\sqrt{2}} \sum_{uv \in E} (\mathbf{a}_u - \mathbf{a}_v) \otimes (\mathbf{d}_{u,v} - \mathbf{d}_{v,u}) \in \mathcal{H}_V \otimes \mathcal{H}_E; \tag{6}$$

With the use of the partial trace operation, and letting $\mathbf{X}_{E(G)}$ be the characteristic function of $E(G)$, we can describe a precise relationship between ψ_G and the Laplacian:

$$\begin{aligned}
\operatorname{tr}_E \left(\psi_G \psi_G^\dagger \right) &= \sum_{u,v \in V} \left(\mathbf{1}_V \otimes \mathbf{d}_{u,v}^\dagger \right) \psi_G \psi_G^\dagger \left(\mathbf{1}_V \otimes \mathbf{d}_{u,v} \right) \\
&= \tfrac{1}{2} \sum_{u,v \in V} \mathbf{X}_{E(G)}(uv) \, (\mathbf{a}_u - \mathbf{a}_v)(\mathbf{a}_u - \mathbf{a}_v)^\dagger \\
&= \tfrac{1}{2} \sum_{u,v \in V} \mathbf{X}_{E(G)}(uv) \, (\mathbf{a}_u - \mathbf{a}_v) \, \mathbf{d}_{u,v}^\dagger \mathbf{d}_{u,v} \, (\mathbf{a}_u - \mathbf{a}_v)^\dagger ;
\end{aligned} \tag{7}$$

Introducing separate terms involving $\mathbf{d}_{u,v}^\dagger \mathbf{d}_{u,v}$ and $\mathbf{d}_{v,u}^\dagger \mathbf{d}_{v,u}$ for each undirected edge $uv \in E(G)$, we may then obtain

$$\begin{aligned}
\operatorname{tr}_E \left(\psi_G \psi_G^\dagger \right) &= \tfrac{1}{2} \sum_{uv \in E(G)} \left[(\mathbf{a}_u - \mathbf{a}_v) \mathbf{d}_{u,v}^\dagger \mathbf{d}_{u,v} (\mathbf{a}_u - \mathbf{a}_v)^\dagger + (\mathbf{a}_v - \mathbf{a}_u) \mathbf{d}_{v,u}^\dagger \mathbf{d}_{v,u} (\mathbf{a}_v - \mathbf{a}_u)^\dagger \right] \\
&= \tfrac{1}{2} \sum_{st,uv \in E(G)} \left[(\mathbf{a}_s - \mathbf{a}_t) \otimes (\mathbf{d}_{s,t} - \mathbf{d}_{t,s})^\dagger \right] \left[(\mathbf{d}_{u,v} - \mathbf{d}_{v,u}) \otimes (\mathbf{a}_u - \mathbf{a}_v)^\dagger \right] \\
&= \tfrac{1}{2} \bar{M} \bar{M}^\dagger = L(G).
\end{aligned} \tag{8}$$

Note that $\psi_G \in \mathcal{H}_V \otimes \mathcal{H}_E$, as it has been defined above, may not be a unit vector; its normalization has been chosen specifically so that $\|\psi_G\|_2 = \sqrt{\operatorname{tr}(L(G))} = \sqrt{2|E|}$, which will differ from 1. If we wish, we may renormalize it, and retain the relation $L(G) = \operatorname{tr}_E(\psi_G \psi_G^\dagger)$ by dividing the Laplacian through by $2|E|$ to obtain an operator with unit trace.

Having obtained $L(G)$ as a partial trace of a rank-1 operator, we may ask the following: what is the result of taking the other partial trace? This is (a normalized version of) what is known in the literature as the *edge Laplacian*,

$$L_E(G) = \operatorname{tr}_V \left(\psi_G \psi_G^\dagger \right) = \tfrac{1}{2} \bar{M}^\dagger \bar{M}, \tag{9}$$

where the scalar of proportionality is determined by the normalization of ψ_G; this can be shown by a similar development as in Eqn. (7). While less studied than the Laplacian, a recent application of the edge Laplacian is in dynamic systems and the edge agreement problem (see [5]). The edge-Laplacian has the same positive eigenvalues as $L(G)$, but as it (usually) acts on a much larger configuration space (*i.e.* when G has more edges than vertices), it will have a larger kernel, whose dimension is the size of a cycle-basis for G.

The above discussion can be summarized as follows:

PROPOSITION 1. *Let ψ_G be an incidence vector of a graph $G = (V,E)$, as defined in Eqn. (6). Then*

$$L(G) = \operatorname{tr}_E \left(\psi_G \psi_G^\dagger \right) \quad \text{and} \quad L_E(G) = \operatorname{tr}_V \left(\psi_G \psi_G^\dagger \right). \tag{10}$$

We may interpret the vector ψ_G (or the renormalized version of this vector) as a quantum state vector on two systems of finite dimension, one of dimension at least $|V|$ and one of dimension at least $|E|$, supporting Hilbert spaces which subsume $\mathcal{H}_V$ and $\mathcal{H}_E$ respectively. Each of these systems may themselves be composed of multiple subsystems, for instance spin-1/2 particles (*i.e.* qubits), whose standard basis states are used to represent the indices $v \in V$ and $u, v \in V \times V$ for the basis vectors $\mathbf{a}_v$ and $\mathbf{d}_{u,v}$ respectively. In the standard terminology of quantum information theory, ψ_G is said to be a *purification* of $L(G)$ and $L_E(G)$. These matrices are the *reduced density matrices* with respect to $\mathcal{H}_E$ and $\mathcal{H}_V$ — albeit with the caveat that, as they are usually defined, $L(G)$ and $L_E(G)$ may have trace different from 1.

On the basis of this observation we may apply the machinery of quantum theory. The normalized Laplacian $\rho_G = \frac{1}{2|E|} L(G)$ may be then interpreted as a mixture of pure states (*i.e.* a convex combination of rank-1 operators) can be given in terms of populations and coherences, by considering how ψ_G may be interpreted as a linear combination of orthogonal vectors. In the following, we will write $|v\rangle := \mathbf{a}_v \in \mathcal{H}_V$ for standard basis states in the vertex space, and $|u,v\rangle := \mathbf{d}_{u,v} \in \mathcal{H}_E$ for standard basis states in the edge space. We may note that the expression for ψ_G in Eq. (6) represents a linear combination of states of the form

$$\psi_{uv} = \tfrac{1}{2}\Big[|u\rangle - |v\rangle\Big] \otimes \Big[|u,v\rangle - |v,u\rangle\Big] \tag{11}$$

over all edges $uv \in E(G)$; by the orthogonality of $\beta_{uv} := |u,v\rangle - |v,u\rangle$ for distinct vertex-pairs $(u,v) \in V \times V$, the density operator $\rho_G = \frac{1}{2|E|} L(G)$ is a uniformly random mixture of operators

$$\rho_G = \frac{1}{E(G)} \sum_{uv \in E(G)} \alpha_{uv} \alpha_{uv}^\dagger, \quad \text{where } \alpha_{uv} = \tfrac{1}{\sqrt{2}}|u\rangle - \tfrac{1}{\sqrt{2}}|v\rangle. \tag{12}$$

This suggests an interpretation of ρ as a uniformly random mixture of pure states in the vertex-space, where each state in the mixture corresponds to a single edge of the graph. The edge-vectors $\alpha_{uv} \in \mathcal{H}_V$ are not orthogonal vectors to one another when the edges are co-incident, and in that case would not be perfectly distinguishable from one another as quantum states.

For example, let us consider the graph $G = \{\{1,2,3\},\{\{1,2\},\{1,3\}\}\}$. We denote the standard basis vectors of $\mathcal{H}_V$ by $|1\rangle, |2\rangle$, and $|3\rangle$ corresponding to the vertex labels. To each edge $uv \in E$ we associate a unit vector

$$|\{u,v\}\rangle \propto \alpha_u^{\{u,v\}}|u\rangle + \alpha_v^{\{u,v\}}|v\rangle \in \mathcal{H}_V : \tag{13}$$

the complex argument of the scalar of proportionality does not matter. In our example,

$$|\{1,2\}\rangle = \alpha_1^{\{1,2\}}|1\rangle + \alpha_2^{\{1,2\}}|2\rangle \quad \text{and} \quad |\{1,3\}\rangle = \alpha_1^{\{1,3\}}|1\rangle + \alpha_3^{\{1,3\}}|3\rangle. \tag{14}$$

By definition,

$$|\alpha_1^{\{1,2\}}|^2 + |\alpha_2^{\{1,2\}}|^2 = |\alpha_1^{\{1,3\}}|^2 + |\alpha_2^{\{1,3\}}|^2 = 1. \tag{15}$$

A general state ρ of the system expressing a statistical mixture of $|\{1,2\}\rangle$ and $|\{1,3\}\rangle$ is described by an operator

$$\rho = \omega_{\{1,2\}}|\{1,2\}\rangle\langle\{1,2\}| + \omega_{\{1,3\}}|\{1,3\}\rangle\langle\{1,3\}|, \tag{16}$$

where $\omega_{\{1,2\}}, \omega_{\{1,3\}} \geq 0$ and $\omega_{\{1,2\}} + \omega_{\{1,3\}} = 1$. The operator ρ can the be written as

(17) $$\rho = \begin{pmatrix} \rho_{1,1} & \rho_{1,2} & \rho_{1,3} \\ \rho_{2,1} & \rho_{2,2} & \rho_{2,3} \\ \rho_{3,1} & \rho_{3,2} & \rho_{3,3} \end{pmatrix},$$

with

$$\rho_{1,1} = \omega_{\{1,2\}} \left|\alpha_1^{\{1,2\}}\right|^2 + \omega_{\{1,3\}} \left|\alpha_1^{\{1,3\}}\right|^2,$$

$$\rho_{1,2} = \rho_{2,1} = \omega_{\{1,2\}} \alpha_1^{\{1,2\}} \overline{\alpha}_2^{\{1,2\}},$$

$$\rho_{1,3} = \rho_{3,1} = \omega_{\{1,3\}} \alpha_1^{\{1,3\}} \overline{\alpha}_2^{\{1,3\}},$$

$$\rho_{2,2} = \omega_{\{1,2\}} \left|\alpha_2^{\{1,2\}}\right|^2,$$

$$\rho_{2,3} = \rho_{3,2} = 0,$$

$$\rho_{3,3} = \omega_{\{1,3\}} \left|\alpha_2^{\{1,3\}}\right|^2.$$

One conventionally interprets such an operator statistically with respect to a projective measurement process, where for some orthonormal basis $\{\mathbf{v}_j : j = 1, \ldots, n\}$ the probability of outcome j is $\mathbf{v}_j^\dagger \rho \mathbf{v}_j$, representing a realization of $\mathbf{v}_j$ as the state of the system. For instance, setting $\mathbf{v}_j = |j\rangle$ represents a measurement of ρ in the standard basis, and represents the population of the system which is in the state $|j\rangle$ (representing the vertex j in this case) for a system initialized to the state ρ. (Similarly, when measuring the state with a projective measurement with respect to standard basis, $|\alpha_1^{\{1,2\}}|^2$ represents the probability of observing the state $|1\rangle$ in the ray $|\{1,2\}\rangle$ — represented by the operator $|\{1,2\}\rangle\langle\{1,2\}|$ which is involved in the ensemble ρ.)

Therefore, $\rho_{i,i}$, with $i = 1, 2, 3$, is the probability of the state $|i\rangle$ in ρ. In other terms, if the same measurement is carried out N times (under the same initial conditions), $N\rho_{i,i}$ systems will be observed in the state $|i\rangle$. (For this reason $\rho_{i,i}$ is sometimes said to be the *population* of $|i\rangle$.) Operationally, each $\rho_{i,i}$ is the probability of getting the vertex i when "observing the graph", where the graph is itself represented by the state ρ. (It must be remarked that the observation is performed with the respect to the standard basis; projective measurement involving other bases shall give superpositions of vertices.) The cross terms of ρ indicates the subsistence of a certain amount of coherence in the system. In fact, $\rho_{i,j}$, with $i \neq j$, expresses the coherence effects between the states $|i\rangle$ and $|j\rangle$ arising from the presence of α_{uv} in the statistical mixture.

As we note above, when the mixture is equally weighted, and the states $|\{u,v\}\rangle$ are taken to be the vectors α_{uv}, i.e. uniform linear combinations up to a sign, we then obtain $\rho = \rho_G := \frac{1}{2|E|} L(G)$, the normalized Laplacian. Let

$$\omega_{\{1,2\}} = \omega_{\{1,2\}} = \tfrac{1}{2},$$

$$\alpha_1^{\{1,2\}} = \tfrac{1}{\sqrt{2}} \quad \text{and} \quad \alpha_2^{\{1,2\}} = -\tfrac{1}{\sqrt{2}},$$

$$\alpha_1^{\{1,3\}} = \tfrac{1}{\sqrt{2}} \quad \text{and} \quad \alpha_2^{\{1,3\}} = -\tfrac{1}{\sqrt{2}}.$$

Then,

$$\rho = \tfrac{1}{2}|\{1,2\}\rangle\langle\{1,2\}| + \tfrac{1}{2}|\{1,3\}\rangle\langle\{1,3\}| = \begin{pmatrix} \tfrac{1}{2} & -\tfrac{1}{4} & -\tfrac{1}{4} \\ -\tfrac{1}{4} & \tfrac{1}{4} & 0 \\ -\tfrac{1}{4} & 0 & \tfrac{1}{4} \end{pmatrix}$$

$$= \tfrac{1}{4}\left[\begin{pmatrix} 2 & 0 & 0 \\ 0 & 1 & 0 \\ 0 & 0 & 1 \end{pmatrix} - \begin{pmatrix} 0 & 1 & 1 \\ 1 & 0 & 1 \\ 1 & 1 & 0 \end{pmatrix}\right] = \frac{1}{2|E(G)|}\Big(D(G) - A(G)\Big).$$

Note that we may consider choosing $|\{u,v\}\rangle$ to be uniform superpositions with the same sign,

$$(18) \qquad |\{u,v\}\rangle = \varsigma_{uv} := \tfrac{1}{\sqrt{2}}\Big(|u\rangle + |v\rangle\Big);$$

like α_{uv}, a standard basis measurement upon the state $\varsigma_{uv} \in \mathcal{H}_V$ would yield u and v with equal probability, and all other vertex labels with probability 0. The ensemble which arises from a uniform mixture of these unsigned edge-states is then

$$(19) \qquad \varrho_G = \tfrac{1}{E(G)} \sum_{uv \in E(G)} \varsigma_{uv}\varsigma_{uv}^{\dagger} = \tfrac{1}{2|E(G)|}\,\mathrm{tr}_E\Big(\phi_G\phi_G^{\dagger}\Big),$$

where

$$(20) \qquad \phi_G = \sum_{uv \in E(G)} \varsigma_{uv} \otimes \Big(|u,v\rangle + |v,u\rangle\Big) \in \mathcal{H}_V \otimes \mathcal{H}_E.$$

By a similar analysis as that which demonstrates Proposition 1, we may then show that ϱ_G is the normalized version of the *signless Laplacian*, $L^+(G) = D(G) + A(G)$. The operators ρ_G and ϱ_G are therefore density matrices in the cases where the vertex-states have equal weighting in the pure states $|\{u,v\}\rangle$, and where each of these edge-states have equal weighting in the mixture over edge-states. Among other reasons, ρ_G is often preferred to ϱ_G because ρ_G has an all-ones eigenvector corresponding to the zero eigenvalue.

By interpreting ψ_G as a pure state, we may apply the ideas of quantum information in the graph-theoretic framework. We consider below some directions. The *Schmidt rank* of a vector $\mathbf{v} \in \mathcal{H}_A \otimes \mathcal{H}_B$ where $\mathcal{H}_A \cong \mathbb{C}^A$ and $\mathcal{H}_B \cong \mathbb{C}^B$ ($|A| = n$, $|B| = m$ and $m \geq n$) is defined as the minimum number of coefficients $\alpha_i > 0$ such that

$$(21) \qquad \mathbf{v} = \sum_{i=1}^{m} \alpha_i(\mathbf{e}_i \otimes \mathbf{f}_i),$$

where $\{\mathbf{e}_i : i = 1,\ldots,m\}$ and $\{\mathbf{f}_i : i = 1,\ldots,m\}$ are some pair of orthonormal bases. As $n \leq m$, it follows that the Schmidt rank is no larger than n. The scalars α_i are referred to as *Schmidt coefficients*. It is simple to show that the Schmidt rank of $\mathbf{v}$, denoted by $\mathrm{rank}_S(\psi_G)$, is equal to the rank of its partial traces; this follows from a direct relationship between $\mathbf{v}$ and a transformation $V : \mathcal{H}_B \to \mathcal{H}_A$ defined through its singular value decomposition,

$$(22) \qquad V = \sum_{i=1}^{m} \alpha_i\, \mathbf{e}_i \mathbf{f}_i^{\dagger}.$$

It is well-known (a consequence of the matrix-tree theorem) that the rank of the Laplacian of a graph G on n vertices is equal to $n-w(G)$, where $w(G)$ is the number of connected components of G. Directly from the definitions:

PROPOSITION 2. *Let ψ_G be an incidence vector of a graph G on n vertices. Then, the Schmidt rank of ψ_G is*

(23) $$\text{rank}_S(\psi_G) = \text{rank}(L(G)) = n - w(G).$$

We say that two vectors $\mathbf{v}, \mathbf{w} \in \mathcal{H}_A \otimes \mathcal{H}_B$ are *locally unitarily equivalent* (or LU-equivalent) if there exist unitary operators $U : \mathcal{H}_A \to \mathcal{H}_A$ and $V : \mathcal{H}_B \to \mathcal{H}_B$ such that $\mathbf{w} = (U \otimes V)\mathbf{v}$. By considering the Schmidt decompositions of two such vectors, it is clear that $\mathbf{v}$ and $\mathbf{w}$ are LU-equivalent if and only if they have the same Schmidt coefficients. Let us denote by $\text{Sp}(G)$ the *spectrum* of a graph G, which we define as the ordered sequence of eigenvalues of the Laplacian of G.

PROPOSITION 3. *Let ψ_G and ψ_H be incidence vectors of two graphs G and H, respectively, having the same number of vertices and edges. Then ψ_G and ψ_H are LU-equivalent if and only if $\text{Sp}(G) = \text{Sp}(H)$.*

Proof. If ψ_G and ψ_H are LU-equivalent, it follows that there is a unitary $U : \mathcal{H}_V \to \mathcal{H}_V$ for which

$$L(H) = \text{tr}_E(\psi_H \psi_H^\dagger) = U \, \text{tr}_E(\psi_G \psi_G^\dagger) U^\dagger = U L(G) U^\dagger$$

so that $\text{Sp}(G) = \text{Sp}(H)$. Conversely, if $\text{Sp}(G) = \text{Sp}(H)$, we have $\text{tr}_E(\psi_H \psi_H^\dagger) = L(H) = U L(G) U^\dagger = U \text{tr}_E(\psi_G \psi_G^\dagger) U^\dagger$ for some unitary U. By considering the Schmidt decompositions of ψ_G and ψ_H, it follows that there exists a unitary $V : \mathcal{H}_E \to \mathcal{H}_E$ such that $\psi_H = (U \otimes V)\psi_G$. Thus ψ_G and ψ_H are LU-equivalent. ∎

This means that graphs which are Laplacian cospectral correspond to local unitarily equivalent incidence vectors.

Recall that the edge-states α_{uv}, α_{vw} are not perfectly distinguishable from one another through any projective measurement, for any pair of edges $uv, vw \in E$ which coincide, as these vectors are not orthogonal. In particular, for measurement in the standard basis, there is a probability of $\frac{1}{2}$ that any such edge-state will give rise to the common vertex w, which is perfectly ambiguous when attempting to distinguish uv from vw. Thus, despite being a uniformly random mixture of the edge-states, the imperfect distinguishability of the edge-states implies that the *von Neumann entropy* of ρ_G,

(24) $$S(\rho_G) = -\text{tr}\Big(\rho_G \ln(\rho_G)\Big) = - \sum_{\lambda \in \text{Spec}(\rho_G)} \lambda \ln(\lambda),$$

indicates something of the structure of the graph with respect to coincidence of edges. In particular, as the Laplacian $L(G)$ and the edge-Laplacian $L_E(G)$ have the same spectrum of non-zero eigenvalues, we have

(25) $$S_V(\psi_G) := -\text{tr}(L(G) \ln L(G)) = -\text{tr}(L_E(G) \ln L_E(G)).$$

The quantity $S_V(\psi_G)$ has been recently studied in several contexts (see, *e.g.*, [3] for an application in pattern recognition and [9] for an application in loop quantum gravity; the topic has been introduced in [1] and [8]). Another term for $S_V(\psi)$, for arbitrary pure states $\psi \in \mathcal{H}_A \otimes \mathcal{H}_B$, is *entropy of entanglement*, because $S_V(\psi)$ quantifies the amount of entanglement between subsystems with Hilbert space $\mathcal{H}_A$

and $\mathcal{H}_B$. Entropy of entanglement is indeed *the* asymptotic entanglement measure for bipartite pure states. Hence the next fact, giving an interpretation to $S_V(\psi_G)$ (or, equivalently, $S(G)$):

PROPOSITION 4. *The von Neumann entropy of $\rho_G = \frac{1}{2|E|}L(G)$ is the amount of entanglement between the subsystems of ψ_G corresponding to vertices (with Hilbert space $\mathcal{H}_V$) and edges (with Hilbert space $\mathcal{H}_E$), respectively.*

Two graphs G and H are *isomorphic* if there is a bijection $f : V(G) \longrightarrow V(H)$ such that $\{i,j\} \in E(G)$ if and only if $\{f(i), f(j)\} \in E(H)$. A *permutation matrix* is a matrix with entries in the set $\{0,1\}$ and a unique 1 entry in each row and column. Then, two graphs G and H are isomorphic if and only if there is a permutation matrix P such that

$$P \rho_G P^T = \rho_H. \tag{26}$$

In this case, the matrices ρ_G and ρ_H are said to be *permutation congruent*. Two permutation congruent matrices have the same eigenvalues and so two isomorphic graphs share the same spectrum, $\mathrm{Sp}(G) = \mathrm{Sp}(H)$. It is well known that the converse is does not hold, *i.e.* there are many non-isomorphic graphs which share the same spectrum. Such pairs of graphs are called *cospectral* (see [12] for a recent review).

PROPOSITION 5. *Given two graphs with the same number of vertices. Then there exist graphs G and H with $S(G) = S(H)$ but $\mathrm{Sp}(G) \neq \mathrm{Sp}(H)$.*

We call two graphs *coentropic* if their Laplacians have the same von Neumann entropy. It is clear that the number of equivalence classes of coentropic graphs with n vertices and m edges is a lower bound to the number of (pure) bipartite entanglement classes with subsystems of corresponding dimension.

Proof. It is clear that if $\mathrm{Sp}(G) = \mathrm{Sp}(H)$ then $S(G) = S(H)$, since $S(G)$ and $S(H)$ are determined by the spectra. However, the following two graphs have spectra $[0, 3, 3, 3, 3, 6, 8, 8]/34$ and $[0, 2, 2, 4, 6, 6, 6, 8]/34$ and equal von Neumann entropy $S(G) = S(H) = \ln(34) - [18\ln(3) + 54\ln(2)]/34$:

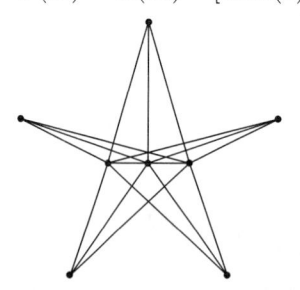

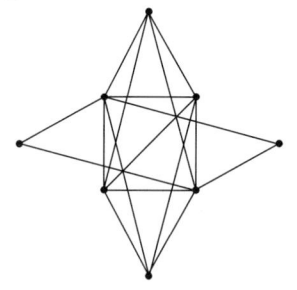

There are two pairs of non-isomorphic and non-cospectral graphs of size 8 with the same entropy. There are 8 such pairs of size 9 (enumerated in the table below) and 76 pairs of size 10. In all these cases, the pairs share the same number of edges. In this case, it sufficient that the entropy of the un-normalized Laplacian coincides:

$$\hat{S}(G) = -\sum_{\lambda \in \mathrm{Sp}(G)} \lambda \ln \lambda$$

where $\mathrm{Sp}(G)$ is here the spectrum of the un-normalized Laplacian. No examples are known for pairs with different numbers of edges.

Graph	Entropy
{{1, 8}, {1, 9}, {2, 8}, {2, 9}, {3, 8}, {3, 9}, {4, 8}, {4, 9}, {5, 8}, {5, 9}, {6, 8}, {6, 9}, {7, 8}, {7, 9}, {8, 9}} {{1, 7}, {1, 8}, {1, 9}, {2, 7}, {2, 8}, {2, 9}, {3, 7}, {3, 8}, {3, 9}, {4, 9}, {5, 9}, {6, 9}, {7, 8}, {7, 9}, {8, 9}}	$-\frac{1}{5}\ln(3)$ $+\frac{3}{5}\ln(2)$ $+\ln(5)$
{{1, 7}, {1, 8}, {1, 9}, {2, 7}, {2, 8}, {2, 9}, {3, 7}, {3, 8}, {3, 9}, {4, 7}, {4, 8}, {4, 9}, {5, 7}, {5, 8}, {5, 9}, {6, 9}, {7, 8}, {7, 9}, {8, 9}} {{1, 6}, {1, 7}, {1, 8}, {1, 9}, {2, 6}, {2, 7}, {2, 8}, {2, 9}, {3, 8}, {3, 9}, {4, 8}, {4, 9}, {5, 9}, {6, 7}, {6, 8}, {6, 9}, {7, 8}, {7, 9}, {8, 9}}	$-\frac{15}{19}\ln(3)$ $-\frac{5}{19}\ln(2)$ $+\ln(19)$
{{1, 5}, {1, 8}, {1, 9}, {2, 6}, {2, 8}, {2, 9}, {3, 7}, {3, 8}, {3, 9}, {4, 8}, {4, 9}, {5, 8}, {5, 9}, {6, 8}, {6, 9}, {7, 8}, {7, 9}, {8, 9}} {{1, 7}, {1, 8}, {1, 9}, {2, 7}, {2, 8}, {2, 9}, {3, 7}, {3, 8}, {3, 9}, {4, 7}, {4, 8}, {4, 9}, {5, 7}, {5, 8}, {5, 9}, {6, 9}, {7, 9}, {8, 9}}	$\ln(3)$ $+\frac{7}{6}\ln(2)$
{{1, 6}, {1, 7}, {1, 8}, {1, 9}, {2, 6}, {2, 7}, {2, 8}, {2, 9}, {3, 8}, {3, 9}, {4, 8}, {4, 9}, {5, 9}, {6, 7}, {6, 8}, {6, 9}, {7, 8}, {7, 9}} {{1, 7}, {1, 8}, {1, 9}, {2, 7}, {2, 8}, {2, 9}, {3, 7}, {3, 8}, {3, 9}, {4, 7}, {4, 8}, {4, 9}, {5, 7}, {5, 8}, {5, 9}, {6, 9}, {7, 8}, {7, 9}}	1.91025843
{{1, 6}, {1, 7}, {1, 8}, {1, 9}, {2, 6}, {2, 7}, {2, 8}, {2, 9}, {3, 8}, {3, 9}, {4, 8}, {4, 9}, {5, 8}, {5, 9}, {6, 7}, {6, 8}, {6, 9}, {7, 8}, {7, 9}, {8, 9}} {{1, 7}, {1, 8}, {1, 9}, {2, 7}, {2, 8}, {2, 9}, {3, 7}, {3, 8}, {3, 9}, {4, 7}, {4, 8}, {4, 9}, {5, 7}, {5, 8}, {5, 9}, {6, 8}, {6, 9}, {7, 8}, {7, 9}, {8, 9}}	$\frac{47}{20}\ln(2)$ $-\frac{6}{5}\ln(3)$ $+\ln(5)$
{{1, 6}, {1, 7}, {1, 8}, {1, 9}, {2, 6}, {2, 7}, {2, 8}, {2, 9}, {3, 8}, {3, 9}, {4, 8}, {4, 9}, {5, 8}, {5, 9}, {6, 7}, {6, 8}, {6, 9}, {7, 8}, {7, 9}} {{1, 7}, {1, 8}, {1, 9}, {2, 7}, {2, 8}, {2, 9}, {3, 7}, {3, 8}, {3, 9}, {4, 7}, {4, 8}, {4, 9}, {5, 7}, {5, 8}, {5, 9}, {6, 7}, {6, 8}, {7, 9}, {8, 9}}	$\frac{6}{19}\ln(2)$ $-\frac{7}{38}\ln(7)$ $-\frac{15}{19}\ln(3)$ $+\ln(19)$
{{1, 4}, {1, 5}, {1, 7}, {1, 8}, {1, 9}, {2, 6}, {2, 9}, {3, 6}, {3, 9}, {4, 5}, {4, 7}, {4, 8}, {4, 9}, {5, 7}, {5, 8}, {5, 9}, {6, 9}, {7, 8}, {7, 9}, {8, 9}} {{1, 5}, {1, 8}, {1, 9}, {2, 6}, {2, 7}, {2, 8}, {2, 9}, {3, 6}, {3, 7}, {3, 8}, {3, 9}, {4, 8}, {4, 9}, {5, 8}, {5, 9}, {6, 8}, {6, 9}, {7, 8}, {7, 9}, {8, 9}}	$\frac{43}{20}\ln(2)$ $-\frac{21}{20}\ln(3)$ $+\ln(5)$
{{1, 4}, {1, 6}, {1, 7}, {1, 8}, {1, 9}, {2, 5}, {2, 6}, {2, 7}, {2, 8}, {2, 9}, {3, 9}, {4, 6}, {4, 7}, {4, 8}, {4, 9}, {5, 6}, {5, 7}, {5, 8}, {5, 9}, {6, 8}, {6, 9}, {7, 8}, {7, 9}, {8, 9}} {{1, 6}, {1, 7}, {1, 8}, {1, 9}, {2, 6}, {2, 7}, {2, 8}, {2, 9}, {3, 6}, {3, 7}, {3, 8}, {3, 9}, {4, 6}, {4, 7}, {4, 8}, {4, 9}, {5, 8}, {5, 9}, {6, 7}, {6, 8}, {6, 9}, {7, 8}, {7, 9}, {8, 9}}	$\frac{59}{24}\ln(2)$ $+\frac{1}{4}\ln(3)$

References

[1] Samuel L. Braunstein, Sibasish Ghosh, and Simone Severini, *The Laplacian of a graph as a density matrix: a basic combinatorial approach to separability of mixed states*, Ann. Comb. **10** (2006), no. 3, 291–317, DOI 10.1007/s00026-006-0289-3. MR2284272 (2008f:81032)

[2] Chris Godsil and Gordon Royle, *Algebraic graph theory*, Graduate Texts in Mathematics, vol. 207, Springer-Verlag, New York, 2001. MR1829620 (2002f:05002)

[3] L. Han, F. Escolano, E. R. Hancock, R. C. Wilson, Graph characterizations from von Neumann entropy, *Pattern Recognition Letters*, **33**:15 (2012), 1958-1967.

[4] A. S. Holevo, Bounds for the quantity of information transmitted by a quantum communication channel, *Prob. Inf. Transm.* (USSR) **9**, 177–83 (1973)

[5] Mehran Mesbahi and Magnus Egerstedt, *Graph theoretic methods in multiagent networks*, Princeton Series in Applied Mathematics, Princeton University Press, Princeton, NJ, 2010. MR2675288 (2011h:93003)

[6] Michael A. Nielsen and Isaac L. Chuang, *Quantum computation and quantum information*, Cambridge University Press, Cambridge, 2000. MR1796805 (2003j:81038)

[7] James G. Oxley, *Matroid theory*, Oxford Science Publications, The Clarendon Press, Oxford University Press, New York, 1992. MR1207587 (94d:05033)

[8] F. Passerini, S. Severini, The von Neumann entropy of networks, arXiv:0812.2597 [cond-mat.dis-nn]

[9] C. Rovelli, F. Vidotto, Single particle in quantum gravity and Braunstein-Ghosh-Severini entropy of a spin network, *Phys. Rev. D* **81**, 044038 (2010).

[10] A. Uhlmann, *The "transition probability" in the state space of a * -algebra*, Rep. Mathematical Phys. **9** (1976), no. 2, 273–279. MR0423089 (54 #11072)

[11] Hassler Whitney, *2-Isomorphic Graphs*, Amer. J. Math. **55** (1933), no. 1-4, 245–254, DOI 10.2307/2371127. MR1506961

[12] Edwin R. van Dam and Willem H. Haemers, *Developments on spectral characterizations of graphs*, Discrete Math. **309** (2009), no. 3, 576–586, DOI 10.1016/j.disc.2008.08.019. MR2499010 (2010h:05178)

DEPARTMENT OF APPLIED MATHEMATICS AND THEORETICAL PHYSICS, UNIVERSITY OF CAMBRIDGE, WILBERFORCE ROAD, CAMBRIDGE, UNITED KINGDOM

NEST, SCUOLA NORMALE SUPERIORE AND ISTITUTO NANOSCIENZE-CNR, PIAZZA DEI CAVALIERI 7, I-56126 PISA, ITALY

DEPARTMENT OF COMPUTER SCIENCE, AND DEPARTMENT OF PHYSICS & ASTRONOMY, UNIVERSITY COLLEGE LONDON, WC1E 6BT LONDON, UNITED KINGDOM

DEPARTMENT OF COMPUTER SCIENCE, UNIVERSITY OF YORK, DERAMORE LANE, HESLINGTON, YORK, YO10 5GH, UNITED KINGDOM

About the relative entropy method for hyperbolic systems of conservation laws

Denis Serre and Alexis F. Vasseur

ABSTRACT. We review the relative entropy method in the context of first-order hyperbolic systems of conservation laws, in one space-dimension. We prove that contact discontinuities in full gas dynamics are uniformly stable. Generalizing this calculus, we derive an infinite-dimensional family of Lyapunov functions for the system of full gas dynamics.

1. Systems of conservation laws and entropies

We are interested in vector fields $u(x,t)$ obeying first-order PDEs. The space variable x and time t run over the physical domain $\mathbb{R}^d \times (0,T)$. The field takes values in a convex open subset $\mathcal{U}$ of $\mathbb{R}^n$.

A *conservation law* is a first-order PDE of the form
$$\partial_t a + \mathrm{div}_x \vec{b} = 0.$$
The terminology refers to the fact that weak solutions obey the identity
$$\frac{d}{dt}\int_\Omega a(x,t)\,dx + \int_{\partial\Omega} \vec{b}\cdot\nu\,ds(x) = 0,$$
for every regular open subdomain $\Omega \subset \mathbb{R}^d$. Hereabove ν is the outer normal and ds is the area element over the boundary. Actually, the PDE is often derived from the latter identity, which expresses a physical principle such as conservation of mass, momentum, species, charge, energy, ... See C. Dafermos's book [9] for a thorough description of this correspondance.

A given physical process is modelled by one or several conservation laws, which are written in a compact form as
$$\partial_t u + \mathrm{Div}_x F = 0,$$
where u is the field of conserved quantities, taking values in some open subset $\mathcal{U}$ of $\mathbb{R}^n$, and F, the flux, takes values in $(n\times d)$-matrices. Mind that the operator Div_x with a capital letter represents the row-wise divergence of a matrix-valued field. The system is closed when F is given as a function of u, in the form $F = f(u)$ where $f : \mathcal{U} \to \mathbb{R}^{n\times d}$ is a given smooth field. We then have

(1) $$\partial_t u + \mathrm{Div}_x f(u) = 0,$$

2010 *Mathematics Subject Classification.* Primary 35L65, 35L67.

which must be understood in the sense of distributions. One speaks of a *scalar equation* if $n = 1$, and of a *system* if $n \geq 2$. Mind that such a closure excludes dissipation processes such as viscous effects or thermal diffusion. Instead, it assumes a thermal equilibrium everywhere at any time.

The processes that we have in mind obey a kind of second principle of thermodynamics. Mathematically, this can be expressed in terms of one additional conservation law

$$\partial_t \eta(u) + \mathrm{div}\vec{q}(u) = 0. \tag{2}$$

We say that a function η satisfying (2) for every C^1-solution of (1) is a *mathematical entropy* of system (1), and $\vec{q}$ is its *entropy flux*. In order that (2) be compatible with (1), we have the linear differential relations

$$\frac{\partial q_\alpha}{\partial u_j} = \sum_k \frac{\partial \eta}{\partial u_k} \frac{\partial f_\alpha^k}{\partial u_j}, \qquad \alpha = 1, \ldots, d,\ j = 1, \ldots, n. \tag{3}$$

In order that (2) brings some new information, it has to follow from (1) in a non-trivial way, by multiplying the system by the differential form $\mathrm{d}\eta(u)$. In particular, η should not be affine; affine functions are called *trivial* entropies. When $n \geq 3$, or when $n, d \geq 2$, the number of constraints in (3) exceeds that of the unknown functions η, q_α, and therefore non-trivial entropies are exceptional objects : if a system (1) is chosen at random, it should not admit any non-trivial entropy. However, it is a well documented fact that systems modelling physics do admit at least one non-trivial entropy η ; somehow, they are exceptional. It turns out that most of them admit only this entropy, in the sense that the solution set of (3) is spanned by η and the trivial entropies. Mind that this is not true if either $n = 1$, or when $n = 2$ and $d = 1$, because then the number of constraints does not exceed the number of unknowns.

The additional entropy η is often a convex smooth function (and then the phase space $\mathcal{U}$ is convex), strongly convex in the sense that the Hessian matrix $D^2 \eta_a$ is positive definite at every $a \in \mathcal{U}$. It may happen that η be non convex, when (1) is an artificial first-order form of a system of second-order PDEs in several space dimensions, but then η has a related *quasi-convex* property. For the sake of simplicity, we shall assume throughout this paper that η is strongly convex. For a more general situation, one refers to the works [6, 17].

As mentionned above, if m is affine and λ is a positive constant, then $u \mapsto \lambda \eta(u) + m(u)$ is another strongly convex entropy, associated with the flux $\lambda \vec{q} + m \circ f$. This observation is at the basis of the notion of relative entropy. Given $a \in \mathcal{U}$, we define another entropy, still strongly convex, by

$$u \mapsto \eta(u|a) := \eta(u) - \eta(a) - \mathrm{d}\eta_a \cdot (u - a).$$

This is an affine correction of η. The corresponding entropy flux is

$$u \mapsto \vec{q}(u; a) := \vec{q}(u) - \vec{q}(a) - \mathrm{d}\eta(a) \cdot (f(u) - f(a)).$$

We warn the reader that the latter field is not an affine correction of $u \mapsto \vec{q}(u)$. This is why we employ a semi-colon, where we put a bar instead in the definition of $\eta(u|a)$. The notation $\vec{q}(u; a)$ is actually a bit misleading, suggesting that it depends linearly on $\vec{q}$. As a matter of fact, it may happen that $\vec{q}(u) \equiv 0$, while $\vec{q}(u; a)$ is non-trivial. We shall give later on an example of such a paradox.

The function $(u,a) \mapsto \eta(u|a)$ is the *relative entropy*. Because η is convex, $\eta(u|a)$ is positive away from the diagonal $u = a$. We point out that the partial function $a \mapsto \eta(u|a)$ is not an entropy, unless η was quadratic. The strong convexity tells us that

(4) $$\eta(u|a) \geq \omega_a |u - a|^2$$

for some $\omega_a > 0$ when u belongs to a compact neighbourhood of a. Therefore, estimating $\int \eta(u|a)\,dx$ is a way to estimate $u(\cdot, t) - a$ in L^2. Suppose for instance that $u(\cdot, t) - a$ has compact support, then the conservation law (2) tells us

$$\int_{\mathbb{R}^d} \eta(u(x,t)|a)\,dx = \int_{\mathbb{R}^d} \eta(u(x,0)|a)\,dx,$$

whence

$$\omega_a \|u(\cdot, t) - a\|^2_{L^2(\mathbb{R}^d)} \leq \int_{\mathbb{R}^d} \eta(u(x,0)|a)\,dx.$$

This expresses the L^2-stability of constant states.

The primary role of the entropy identity is therefore to provide an *a priori* estimate in L^2, or at least in some appropriate Orlicz space. It is at work in the existence theory for the Cauchy problem too, although it is far from being sufficient. Because the flux is nonlinear, the existence theory requires estimates of some derivatives. These cannot be achieved directly, but it was found by Kato, Gårding and others that H^s-estimates are valid, at least locally in time, whenever $s > 1 + \frac{d}{2}$, for systems admitting a *symmetrization* in Friedrichs' sense. A symmetrization is a change of unknowns $u \mapsto z$ by which (1) is transformed into

$$A^0(z)\partial_t z + \sum_{\alpha=1}^{d} A^\alpha(z)\partial_\alpha z = 0,$$

where the matrices $A^0(z), \ldots, A^d(z)$ are symmetric, and $A^0(z)$ is positive definite. It was observed by Godunov [13] and independently by Friedrichs & Lax [12] that systems of conservation laws admitting a strongly convex entropy can be symmetrized. One can either choose the conjugate unknowns $z = \nabla_u \eta$, or take $z = u$ and multiply the system by $A^0 = D^2 \eta(u)$. We therefore derive the fundamental local existence result, see for instance Dafermos' book [9],

THEOREM 1.1. *If (1) is endowed with a strongly convex entropy, then the Cauchy problem is locally (in time) well-posed in the space $H^s_{uloc}(\mathbb{R}^d)$ whenever $s > 1 + \frac{d}{2}$.*

The subscript "uloc" means "locally uniformly" : the norm is the supremum over all $h \in \mathbb{R}^d$ of the H^s-norm of the restriction to the unit ball centered at h. We point out that the threshold $s > d + \frac{1}{2}$ means that $H^s_{uloc}(\mathbb{R}^d) \subset \mathcal{C}^1(\mathbb{R}^d)$, by Sobolev embedding. Therefore the theorem speaks about classical solutions of (1).

2. The question of global solutions

Let $a \in \mathcal{U}$ be given. The field $\bar{a}(x,t) := a$ is a particular solution of (1). Its stability can be analyzed, in a first instance, by linearizing the system about a. One finds the constant coefficient system

$$\partial_t U + \sum_\alpha \mathrm{d}f_\alpha(a) \partial_\alpha U = 0,$$

or equivalently

$$A^0(a)\partial_t Z + \sum_{\alpha=1}^d A^\alpha(a)\partial_\alpha Z = 0.$$

This linear system propagates planar waves in direction $\xi \in S^{d-1}$ at finite velocities $\lambda_1(a;\xi) \leq \cdots \leq \lambda_n(a;\xi)$ that are the eigenvalues of the matrix

$$M(a;\xi) := A^0(a)^{-1} \sum_\alpha \xi_\alpha A^\alpha(a).$$

The fact that this matrix-valued symbol is diagonalisable with real eigenvalues is classical. The diagonalisation can be performed uniformly with respect to ξ. This is referred to as the *hyperbolicity* of the linearized system. By extension, we say also that (1) is *hyperbolic*. Therefore every closed system of first-order conservation laws endowed with a strongly convex entropy is hyperbolic.

By linear superposition, one may design linear waves that are non-planar. They still propagate at finite velocity in the direction normal to the front.

Because of the non-linearity, the velocities depend on the state $a \in \mathcal{U}$. This usually causes a loss of regularity in finite time for the solutions of (1). This can be seen even in the simplest nonlinear equation, named after Burgers,

$$\partial_t u + \partial_x \frac{u^2}{2} = 0, \qquad (n = d = 1).$$

Then the space derivative $y := \partial_x u$ satisfies

$$(\partial_t + u\partial_x)y + y^2 = 0.$$

The former equation, written as $(\partial_t + u\partial_x)u = 0$, is a transport equation, telling us that $u(X(t),t) \equiv u(X(0),0)$ along the integral curves of $\frac{dX}{dt} = u(X,t)$ (the wave velocity is therefore u itself). The latter is a Riccati's equation along the same curves, whose solution blows up in finite time whenever $y(X(0),0)$ is negative. Therefore the solution of the forward Cauchy problem with initial data u_0 blows up in finite time in the C^1-norm, unless u_0 is monotone non-decreasing.

We infer that for most systems of the form (1), and for most initial data, the classical solution exists only on some finite time interval. For practical applications it is however necessary to consider large-time, say global, solutions. This requires considering non-differentiable solution, for which (1) is interpreted in the distributional sense. To this end, it suffices that u be a locally measurable bounded field. Within this framework, the blow-up described above is usually resolved by the development of discontinuities, which propagate along hypersurfaces. Let Σ be such a discontinuity front in $\mathbb{R}^d \times (0,T)$, with unit normal $\mathbf{n} = (n_1, \ldots, n_d, n_0)$, and let $u_\pm$ be the limits of $u(x,t)$ at some point $(\bar{x},\bar{t}) \in \Sigma$ from either side. Then it is well-known that $(u_-, u_+; \mathbf{n})$ satisfies the *Rankine–Hugoniot condition*

(5) $$n_0(u_+ - u_-) + \sum_\alpha n_\alpha(f(u_+) - f(u_-)) = 0.$$

Notice that for a genuine discontinuity, one has $(n_1, \ldots, n_d) \neq 0$. Denoting $[h] := h(u_+) - h(u_-)$, Rankine–Hugoniot can therefore be written in the compact form

$$[f(u)] \cdot \nu = \sigma[u],$$

where $\nu \in S^{d-1}$ is the normal of the spatial trace of the front and σ is the normal velocity.

The possibility for solutions to (1) to admit discontinuities is not the end of the story. Although this extension makes up for the lack of classical solutions, it has the flaw that rough solutions are way too many. This is where the second principle comes into play, with the role of selecting the physically admissible solutions. Mathematically speaking, one first observes that the additional conservation law (2) is not any more a consequence of (1), because multiplying (1) by $d\eta(u)$ does not make sense, and the chain rule does not apply for non-Lipschitz fields. Instead, one argues that first-order conservation laws are actually the limit (as some parameter $\epsilon > 0$ tends to zero) of some second-order system that includes a natural diffusion process. For instance, the Euler equation of a perfect fluid is the limit of the Navier-Stokes-Fourier system. The latter system has a parabolic type which makes the solution smooth for $t > 0$; it is therefore liable to multiply it by $d\eta(u)$, and apply the chain rule. One finds then that its solutions satisfy some differential inequality

$$\partial_t \eta(u^\epsilon) + \text{div}_x \vec{q}(u^\epsilon, \epsilon \nabla_x u^\epsilon) \leq 0.$$

Passing to the limit, we obtain, at least formally, the distributional inequality

(6) $$\partial_t \eta(u) + \text{div}_x \vec{q}(u) \leq 0,$$

called the *entropy inequality*. We therefore declare that an admissible solution (1) must satisfy (6). We speak of an *entropy solution*. Remark that (6) does not yield any additional information in zones where u is a Lipschitz function. However, it tells us that across a discontinuity Σ, $(u_-, u_+; \mathbf{n})$ satisfies

(7) $$[\vec{q}(u)] \cdot \nu \leq \sigma[\eta(u)],$$

where

$$u_\pm = \lim_{h \to 0^+} u(x \pm h\nu, t).$$

The role of the entropy condition is therefore to restore the uniqueness in the Cauchy problem, while leaving the door open to the existence. We don't know so far if this role is successful, except in a few instances. We have a complete and satisfactory theory for scalar equations in any dimension, due to Kruzkhov. A. Bressan and collaborators showed that the Cauchy problem for one-dimensional systems is well-posed as long as the initial data has small total variation, see [1]. When $d = 1$ and $n = 2$, R. DiPerna succeeded to prove the existence of an entropy solution for large initial data, if the system is sufficiently non-linear, by using the compensated compactness method initiated by L. Tartar. When $n \geq 3$, the mere existence of an entropy solution for large data is still unknown. The uniqueness of entropy solution is doubtful in absence of BV regularity ; C. de Lellis and L. Székelyhidi provide actually counter-examples [10].

3. The method of relative entropy

The method of relative entropy is a technique devised by R. DiPerna [11] and C. Dafermos [7, 8] in order to estimate the L^2-distance between two solutions, one of both being smooth or at least not too weak. It adapts to the context of first-order conservation laws the *weak-strong* estimates that is well-known for 3-D Navier-Stokes equation, for instance.

If u and v are smooth solutions of (1), say Lipschitz continuous, then the relative entropy, evaluated at (u, v), satisfies the identity

$$\partial_t \eta(u|v) + \mathrm{div}_x \vec{q}(u;v) = -\mathcal{R}$$

where

$$\mathcal{R} := \sum_\alpha \partial_\alpha (d\eta(v)) \cdot f_\alpha(u|v),$$

with the obvious definition for $f(u|v)$ (affine correction of $f(u)$). If u and v agree outside of some relatively compact domain $\Omega \subset \mathbb{R}^d$, we deduce

$$\frac{d}{dt} \int_\Omega \eta(u|v) \, dx = -\int_\Omega \mathcal{R} \, dx.$$

The source term $\mathcal{R}$ can be bounded by

$$C \|\nabla_x v\|_\infty |u - v|^2.$$

Using (4), we infer that

$$\frac{d}{dt} \int_\Omega \eta(u|v) \, dx \leq C' \int_\Omega \eta(u|v) \, dx, \qquad C' := \frac{C \|\nabla_x v\|_\infty}{\omega}.$$

The Gronwall inequality thus gives an estimate

$$\int_\Omega \eta(u|v) \, dx \leq e^{C't} \int_\Omega \eta(u_0|v_0) \, dx,$$

where u_0, v_0 are the initial data. This is the way uniqueness is proved for Lipschitz solutions, but also stability : if u_0 and v_0 are close to each other, then so are $u(t)$ and $v(t)$. Mind however that the distance between $u(t)$ and $v(t)$ may increase unboundedly as $t \to +\infty$. We speak of *finite-time stability*. In addition, because of the propagation of the information at finite velocity, the correct estimate is instead (again, see Dafermos' book)

$$\int_\Omega \eta(u|v) \, dx \leq e^{C't} \int_{\Omega + B(0;V)} \eta(u_0|v_0) \, dx,$$

where V is a suitable constant, larger than the wave velocities.

When v is Lipschitz but u is only a bounded entropy solution, one still has

$$\partial_t \eta(u|v) = \partial_t \eta(u) - \partial_t \eta(v) - d\eta(v) \cdot (\partial_t u - \partial_t v) - D^2 \eta_v (\partial_t v, u - v),$$

with a analogous formula for $\mathrm{div}_x \vec{q}(u;v)$. Using then the system for u and v, the entropy inequality for u and the entropy *equality* for v, we obtain the distributional inequality

(8) $$\partial_t \eta(u|v) + \mathrm{div}_x \vec{q}(u;v) \leq -\mathcal{R}.$$

From (8), we can reach the same conclusion as when u and v were both Lipschitz. We therefore have uniqueness and finite-time stability as long as the Cauchy problem admits a Lipschitz continuous solution v. This leaves open the uniqueness question beyond the blow-up time (except in the cases covered by Bressan's work).

Such a result is very similar to the weak-strong stability/uniqueness statement for the incompressible Navier-Stokes in dimension three.

When v is globally defined, that is for all $t > 0$, and is Lipschitz, one may wonder whether the source term $\mathcal{R}$ remains non-negative, in which case the total relative entropy

$$\int_{\mathbb{R}^d} \eta(u|v) \, dx$$

is a non-increasing function of time. We then say that v is *uniformly stable*. This happens in gas dynamics for solutions of the 1-d Riemann problem when only rarefaction waves occur (G.-Q. Chen & coll. [2–5]), and also for multi-dimensional expanding flows (D. Serre [19]).

3.1. Stability of discontinuities. From now on, we restrict to the one-dimensional situation, but we consider the stability of a one-dimensional simple discontinuity

$$v(x,t) = \begin{cases} u_\ell, & \text{if } x < \sigma t, \\ u_r, & \text{if } x > \sigma t. \end{cases}$$

Without loss of generality, we may assume $\sigma = 0$, up to replacing x by $x - \sigma t$ and $f(u)$ by $f(u) - \sigma u$. We thus restrict to a steady discontinuity, meaning that $f(u_\ell) = f(u_r)$ and $q(u_r) \leq q(u_\ell)$.

Away from $x = 0$, the inequality (8) is still valid, while $\mathcal{R} \equiv 0$, because v is constant on either sides of $x = 0$. We therefore obtain

$$\frac{d}{dt} \int_{\mathbb{R}} \eta(u|v) \, dx \leq q(u_+; u_r) - q(u_-; u_\ell),$$

where $u_\pm$ stand for the right and left limits of $u(x,t)$ as $x \to 0$. The discontinuity will therefore be stable if the right-hand side, a numerical function of (u_-, u_+), is non-positive. Unfortunately, this is usually not the case; the interested reader might compute a Taylor expansion and find that its sign is non constant, but let us present a more illuminating argument. Suppose (1) is a scalar equation with $f'' > 0$. Take $u^0 = v^0 + \phi$, where ϕ is bounded, compactly supported. It is classical that the initial disturbance is absorbed by the shock: after some finite time $T > 0$, u coincides with the shifted shock $v(\cdot - h)$. The conservation of mass determines $h = \frac{1}{[u]} \int_{\mathbb{R}} \phi \, dx$. With $\eta(s) = \frac{1}{2} s^2$, that is $\eta(a|b) = \frac{1}{2}(a-b)^2$, we deduce that

$$\int_{\mathbb{R}} \eta(u|v) \, dx = \frac{1}{2} \left| [u] \int_{\mathbb{R}} \phi \, dx \right|.$$

On the other hand, we have $\eta(u^0|v^0) = \frac{1}{2}\phi^2$. Because the inequality

$$\left| [u] \int_{\mathbb{R}} \phi \, dx \right| \leq \int_{\mathbb{R}} \phi^2 \, dx$$

is violated for some (many) disturbances ϕ, we see that $t \mapsto \int_{\mathbb{R}} \eta(u|v) \, dx$ is not a decreasing function in general.

N. Leger & A. Vasseur [14, 15] introduced the following refinement of the method. The example above suggests to estimate the distance between $u(t)$ and v

by the relative entropy, *up to a shift*. We thus define a functional

$$E(t) := \inf_{h \in \mathbb{R}} \int_{\mathbb{R}} \eta(u(x,t)|v(x+h))\,dx = \inf_{h \in \mathbb{R}} \left(\int_{-\infty}^{h} \eta(u|u_\ell)\,dx + \int_{h}^{+\infty} \eta(u|u_r)\,dx \right).$$

The quantity in parenthesis is a continuous function of h, which tends to $+\infty$ as $h \to \pm\infty$. Therefore the infimum is attained at some $\bar{h}$. If u has left and right limits at $\bar{h}$, the left and right h-derivatives at $\bar{h}$ must be negative and positive respectively:

$$\eta(u_-|u_\ell) \leq \eta(u_-|u_r), \qquad \eta(u_+|u_\ell) \geq \eta(u_+|u_r).$$

If u is continuous at $\bar{h}$, this tells us $\eta(u|u_\ell) = \eta(u|u_r)$; in other words, $\bar{u} := u(\bar{h})$ belongs to the hyperplane defined by $[d\eta] \cdot \bar{u} = [d\eta(u) \cdot u - \eta(u)]$, which separates u_ℓ from u_r. If $u_+ \neq u_-$, Rankine–Hugoniot gives instead $f(u_+) - f(u_-) = X'(u_+ - u_-)$, where we may suppose that $X \equiv h$ locally (isolated times are not meaningful for the derivative), and therefore $X' = h'$. In conclusion, (u_-, u_+) obey to severe constraints in both situations. Because

$$E'(t) \leq q(u_+; u_r) - q(u_-; u_\ell) - h'(\eta(u_+|u_r) - \eta(u_-|u_\ell)) =: D(u_\pm; u_{\ell,r}),$$

we see that the discontinuity v is stable up to a shift if $D \leq 0$ for every pair (u_-, u_+, h') satisfying the constraints above. We point out that the definition of D is not ambiguous: h' is well-defined (by R.-H.) in the discontinuous case, while $D = q(u; u_r) - q(u; u_\ell)$ in the continuous case. With this approach, Leger and Vasseur proved

THEOREM 3.1. *For a scalar equation with a convex flux, shocks are uniformly stable up to a shift.*

In terms of the entropy $\frac{1}{2}u^2$, this can be viewed as an L^2-contraction property: $t \mapsto \inf_h \|u - v(\cdot - h)\|_{L^2}$ is non-increasing. We warn the reader that this is true only if v is a pure shock. When u, v are arbitrary entropy solutions, we only know the contraction in the L^1-norm, which is a part of Kruzkhov's theory.

4. Using two entropies

In [20], the authors applied the idea described above to systems ($n \geq 2$). Unfortunately, most systems resist to the method. A new tool was therefore needed, which has been elaborated recently by Vasseur. Instead of one entropy η, one uses two entropies $\eta_\pm$ to measure the distance from u to v. Let us consider the quantity

$$E(t) := \int_{-\infty}^{0} \eta_-(u|u_\ell)\,dx + \int_{0}^{+\infty} \eta_+(u|u_r)\,dx.$$

Of course, we may incorporate a shift:

$$E_{\min}(t) := \inf_h \left(\int_{-\infty}^{h} \eta_-(u|u_\ell)\,dx + \int_{h}^{+\infty} \eta_+(u|u_r)\,dx \right).$$

In practice, η_+ and η_- are proportional, because most systems admit only one independent non-trivial entropy. Say $\eta_\pm = a_\pm \eta$ with $a_\pm$ positive constants. Remark that the uniform stability (in terms of the decay of E) implies the stability up to a shift, due to the space-shift invariance of the Cauchy problem.

A calculation analogous to that in the scalar case can be made. If v^0 is a steady Lax shock associated with a genuinely nonlinear field, and if $u^0 = v^0 + \phi$ is a compactly supported perturbation where ϕ is small, then the asymptotic behaviour has been described by T.-P. Liu [16]. It consists of a superposition of so-called N-waves (which decay in L^2-norm like $t^{-1/4}$), of linear waves that just propagate at constant velocity, and of a shift of the shock from $x = 0$ to $x = h$. To determine h, we split the mass $m = \int_{\mathbb{R}} \phi \, dx$ into three parts $X_- + h[u] + X_+$, where X_- (resp. X_+) belongs to the stable (resp. unstable) subspace of $df(u_\ell)$ (resp. $df(u_r)$). All these waves asymptotically separate from each other and therefore the limit of $E(t)$ consists of the sum of contributions of the shift and of the linear waves. In particular, a decay of $E(t)$ for $t > 0$ would imply that $|h|\eta(u_r|u_\ell)$, or $|h|\eta(u_\ell|u_r)$ is not greater than

$$\int_{-\infty}^{0} \eta_-(u_\ell + \phi|u_\ell) \, dx + \int_{0}^{+\infty} \eta_+(u_r + \phi|u_r) \, dx \sim \int_{\mathbb{R}} \phi^2 \, dx.$$

This is obviously false, and therefore the uniform stability of shock waves cannot be obtained without involving a shift. As a matter of fact, Vasseur showed recently that the stability up to a shift may hold true for "extreme shocks", that is the slowest and the fastest shocks.

The argument presented above does not settle the case of contact discontinuities, because then the asymptotics does not involve a shift. We have at least a positive result in the case of full gas dynamics. The system consists of conservation of mass, momentum and energy. In Lagrangian variables (where x is mass), it writes

(9) $\quad \partial_t \tau = \partial_x w, \quad \partial_t w + \partial_x p(\tau, e) = 0, \quad \partial_t(e + \frac{1}{2}w^2) + \partial_x(wp) = 0,$

where τ is the specific volume, w the flow velocity and e the internal energy. The pressure p is given by an equation of state $p(e, \tau)$. Thermodynamics tells us that there exists a positive function $\theta(e, \tau)$ (the temperature) and a concave function $s(e, \tau)$ (the entropy) satisfying the Gibbs relation $\theta ds = de + p d\tau$. The second principle is then that $\partial_t s \geq 0$ for observable flows. In other words, $\eta = -s$ is an entropy in the mathematical sense, and admissible flows should satisfy the entropy inequality. Let us point out that $q \equiv 0$ in this example.

THEOREM 4.1. *Consider the 1-D system of full gas dynamics in Lagrangian variables. Then the contact discontinuity (u_ℓ, u_r) is uniformly stable in the sense that*

$$t \mapsto E(t) := -\int_{-\infty}^{0} \theta_\ell s(u|u_\ell) \, dx - \int_{0}^{+\infty} \theta_r s(u|u_r) \, dx$$

is non-increasing, for every solution u equal to u_ℓ/u_r as x is sufficiently large and negative/positive.

Of course, we may just ask that $u(\cdot, t) - u_\ell \in L^2(\mathbb{R}^-)$ and $u(\cdot, t) - u_r \in L^2(\mathbb{R}^+)$. Our statement has a counterpart when one uses the Eulerian variables, where x is then a space coordinate.

Proof
For $x > 0$, we have $\partial_t s(u|u_r) + \partial_x q(u; u_r) \leq 0$, where $q(a; b) = ds_b \cdot (f(a) - f(b))$. We deduce

$$E'(t) \leq \theta_\ell ds_\ell \cdot (f_- - f(u_\ell)) - \theta_r ds_r \cdot (f_+ - f(u_r))$$

where $f_\pm$ denote the right/left traces of $f(u)$ along $x = 0$. The conservation laws (1) tells us, not only that these traces are well defined as bounded measurable functions, but also that they coincide. Because $f(u_r) = f(u_\ell)$, we have $w_r = w_\ell$ and $p_r = p_\ell$, and therefore $\theta_\ell \mathrm{d}s_\ell = \theta_r \mathrm{d}s_r$. We thus have

$$E'(t) \leq (\theta \mathrm{d}s)_{\ell,r} \cdot (f_- - f_\ell - f_+ + f_r) = 0.$$

■

Remarks. – Full gas dynamics admits infinitely many independent entropies $g \circ s$, where g is any numerical function. Such entropies are strongly convex when $g' < 0$ and $g'' \geq 0$. We could therefore have chosen $\eta_\pm = g_\pm \circ s$ in the estimate above. However the derivation of the entropy inequality from the Navier-Stokes-Fourier system works only for one entropy, namely $-s$ (see [18]). Therefore our estimates with $\eta_\pm = -\theta_{\ell,r} s$ are the only one with physically relevance. – The calculation above can be adapted to more general systems for which $\lambda \equiv 0$ is a simple eigenvalue of $\mathrm{d}f$. The image of $\mathcal{U}$ under $u \mapsto (f(u), q(u))$ is a manifold of codimension 2. We ask that it be contained in a hyperplane, that is $\alpha q + L \circ f \equiv \mathrm{cst}$, with $\alpha \neq 0$, say $\alpha > 0$. Up to replacing η by $\alpha \eta + L(u)$, we may assume that $q \equiv 0$. One shows that there exists an analogue of the temperature $\theta(u)$ such that if (u_ℓ, u_r) is a steady contact discontinuity, then $[\theta \mathrm{d}\eta] = 0$. Therefore the analogue of Theorem 4.1 holds true:

$$t \mapsto E(t) := \int_{-\infty}^{0} \theta_\ell \eta(u|u_\ell) \, dx + \int_{0}^{+\infty} \theta_r \eta(u|u_r) \, dx$$

is non-increasing, for every solution u equal to u_ℓ/u_r as x is sufficiently large and negative/positive.

5. Lyapunov function for full gas dynamics

The calculus above suggests a fruitful generalization. On the one hand, we consider special solutions of (9) of the form $v \equiv v(x)$ where the velocity $\bar{w}$ and the pressure $\bar{p}$ are constant. For instance, v could be a pure contact discontinuity as above. Such fields are parametrized by the choice of a positive measurable $\tau(x)$, bounded by below and above, and by the constants $(\bar{w}, \bar{p})$. The internal energy is obtained by solving the equation $p(\tau(x), e(x)) = \bar{p}$. These solutions are precisely what we called *linear waves* in Section 4.

Given a linear wave v, we may define a functional

$$L_v[u] := -\int_{\mathbb{R}} \theta(v) s(u|v) \, dx,$$

provided that $u - v \in L^2(\mathbb{R})^3$. In the periodic case, where the domain is $\mathbb{R}/M\mathbb{Z}$, we define instead

$$L_v[u] := -\int_{0}^{M} \theta(v) s(u|v) \, dx.$$

We have a seemingly new result:

THEOREM 5.1. *The functionals L_v are Lyapunov functions: for every entropy solution of (9) such that $u(t) - v \in L^2$, the function $t \mapsto L_v[u(t)]$ is non-increasing.*

Proof

We first prove the result when τ is a smooth function. We have
$$\begin{aligned}\partial_t(\theta(v)s(u|v)) &= \theta(v)(\partial_t s(u) - \mathrm{d}s_v \partial_t u)\\ &\geq \theta(v)\mathrm{d}s_v \partial_x f(u)\\ &= (\mathrm{d}(e + \frac{1}{2}w^2) - \bar{w}\mathrm{d}w + \bar{p}\mathrm{d}\tau)\partial_x f(u) =: \omega(\partial_x f(u)).\end{aligned}$$

The differential form has constant coefficients and therefore we have $\partial_t(\theta(v)s(u|v)) \geq \partial_x(\omega \circ f(u))$. Integrating in space, it comes $\frac{d}{dt}L_v[u(t)] \leq 0$.

If τ is not smooth, we may approach every function τ by a sequence $\tau_\epsilon = \tau * \rho(\cdot/\epsilon)$, which converges boundedly almost everywhere. Then $L_{v_\epsilon}[u(t)] \to L_v[u(t)]$. The monotony passes to the limit.

∎

References

[1] Alberto Bressan, Graziano Crasta, and Benedetto Piccoli, *Well-posedness of the Cauchy problem for $n \times n$ systems of conservation laws*, Mem. Amer. Math. Soc. **146** (2000), no. 694, viii+134, DOI 10.1090/memo/0694. MR1686652 (2000m:35122)

[2] Gui-Qiang Chen, *Vacuum states and global stability of rarefaction waves for compressible flow*, Methods Appl. Anal. **7** (2000), no. 2, 337–361. Cathleen Morawetz: a great mathematician. MR1869289 (2002k:76083)

[3] Gui-Qiang Chen and Jun Chen, *Stability of rarefaction waves and vacuum states for the multidimensional Euler equations*, J. Hyperbolic Differ. Equ. **4** (2007), no. 1, 105–122, DOI 10.1142/S0219891607001070. MR2303477 (2008f:35300)

[4] Gui-Qiang Chen and Hermano Frid, *Uniqueness and asymptotic stability of Riemann solutions for the compressible Euler equations*, Trans. Amer. Math. Soc. **353** (2001), no. 3, 1103–1117 (electronic), DOI 10.1090/S0002-9947-00-02660-X. MR1804414 (2001m:35211)

[5] Gui-Qiang Chen and Yachun Li, *Stability of Riemann solutions with large oscillation for the relativistic Euler equations*, J. Differential Equations **202** (2004), no. 2, 332–353, DOI 10.1016/j.jde.2004.02.009. MR2068444 (2005e:35155)

[6] Yann Brenier, *Hydrodynamic structure of the augmented Born-Infeld equations*, Arch. Ration. Mech. Anal. **172** (2004), no. 1, 65–91, DOI 10.1007/s00205-003-0291-4. MR2048567 (2005a:35264)

[7] C. M. Dafermos, *The second law of thermodynamics and stability*, Arch. Rational Mech. Anal. **70** (1979), no. 2, 167–179, DOI 10.1007/BF00250353. MR546634 (80j:73004)

[8] C. Dafermos. Stability of motions of thermoelastic fluids. J. Thermal Stresses, **2** (1979), pp. 127–134.

[9] Constantine M. Dafermos, *Hyperbolic conservation laws in continuum physics*, 3rd ed., Grundlehren der Mathematischen Wissenschaften [Fundamental Principles of Mathematical Sciences], vol. 325, Springer-Verlag, Berlin, 2010. MR2574377 (2011i:35150)

[10] Camillo De Lellis and László Székelyhidi Jr., *On admissibility criteria for weak solutions of the Euler equations*, Arch. Ration. Mech. Anal. **195** (2010), no. 1, 225–260, DOI 10.1007/s00205-008-0201-x. MR2564474 (2011d:35386)

[11] Ronald J. DiPerna, *Uniqueness of solutions to hyperbolic conservation laws*, Indiana Univ. Math. J. **28** (1979), no. 1, 137–188, DOI 10.1512/iumj.1979.28.28011. MR523630 (80i:35119)

[12] K. O. Friedrichs and P. D. Lax, *Systems of conservation equations with a convex extension*, Proc. Nat. Acad. Sci. U.S.A. **68** (1971), 1686–1688. MR0285799 (44 #3016)

[13] S. K. Godunov, *An interesting class of quasi-linear systems* (Russian), Dokl. Akad. Nauk SSSR **139** (1961), 521–523. MR0131653 (24 #A1501)

[14] Nicholas Leger, *L^2 stability estimates for shock solutions of scalar conservation laws using the relative entropy method*, Arch. Ration. Mech. Anal. **199** (2011), no. 3, 761–778, DOI 10.1007/s00205-010-0341-7. MR2771666 (2012a:35196)

[15] Nicholas Leger and Alexis Vasseur, *Relative entropy and the stability of shocks and contact discontinuities for systems of conservation laws with non-BV perturbations*, Arch. Ration. Mech. Anal. **201** (2011), no. 1, 271–302, DOI 10.1007/s00205-011-0431-1. MR2807139 (2012k:35334)

[16] Tai-Ping Liu, *Pointwise convergence to N-waves for solutions of hyperbolic conservation laws*, Bull. Inst. Math. Acad. Sinica **15** (1987), no. 1, 1–17. MR947772 (89h:35202)

[17] Denis Serre, *Hyperbolicity of the nonlinear models of Maxwell's equations*, Arch. Ration. Mech. Anal. **172** (2004), no. 3, 309–331, DOI 10.1007/s00205-003-0303-4. MR2062427 (2005g:35282)

[18] Denis Serre, *The structure of dissipative viscous system of conservation laws*, Phys. D **239** (2010), no. 15, 1381–1386, DOI 10.1016/j.physd.2009.03.014. MR2658332 (2012f:35345)

[19] D. Serre. Long-time stability in systems of conservation laws, using relative entropy/energy. Accepted by *Arch. Rational Mech. Anal.*

[20] D. Serre, A. Vasseur. L^2-type contraction for systems of conservation laws. *Journal de l'École polytechnique; Mathématiques*, **1** (2014), pp 1–28.

UMPA, ENS-LYON 46, ALLÉE D'ITALIE 69364 LYON CEDEX 07, FRANCE

UNIVERSITY OF TEXAS AT AUSTIN, 1 UNIVERSITY STATION C1200, AUSTIN, TEXAS 78712-0257

Smoothing techniques for exact penalty methods

Christian Grossmann

ABSTRACT. In this paper numerical methods based on exact penalties for the treatment of nonlinear programming problems in finite dimensions as well as in Hilbert spaces are under considered. The advantage of exact penalties is that under suitable conditions already for sufficiently large, but finite penalty parameters optimal solutions of the original constrained problem are obtained. However, as a rule, exact penalty functions are not differentiable. This restricts the applicability or at least the efficiency of the methods, like Newton's method, for solving the generated unconstrained problems. To overcome this drawback various smoothing techniques are proposed in the literature and intensively studied till now.

Smoothing techniques replace the original penalty functions by differentiable smooth ones. First, an overview of different types of smoothing of exact penalties known from the literature are presented and its approximation properties investigated. Among all the various types the method based upon the smoothing $\sqrt{t^2 + s^2}$ for $|t|$ with the smoothing parameter $s \to 0+$ appears preferable because of its arbitrarily often differentiability and its low growth behavior. This convergence behavior of this method is analyzed in the case its application to finite dimensional optimization problem. In particular, by means of implicit function technique first order convergence could be obtained. Under the second order sufficiency condition these improved convergence results with respect to the smoothing parameter are derived and its extension to optimal control problems discussed. Numerical examples in finite dimensional as well as discretized optimal control cases show the proved convergence order.

1. Introduction

Classical penalty and barrier functions are widely used in nonlinear programming. In the monograph of Fiacco/McCornmick[1968] a wide class of these techniques has been described and analyzed. While smooth penalties like the quadratic loss function allow to a large extend the application of standard optimization methods to the generated auxiliary problems. This however is not possible if non-smooth penalties are used. Their advantage is that the wanted optimal solution of the original problem can be solved already for finite penalty parameters or even for any positive parameter. On the other hand the missing differentiability requires adapted solution techniques, e.g. like bundle methods, or is overcome by appropriate smoothing techniques. In the last years such smoothing techniques gained renewed attention and different smooth approximations of exact penalties have been proposed.

2010 *Mathematics Subject Classification*. Primary 90C25, 90C51, 65K05.

©2016 American Mathematical Society

The aim of the first part of the present paper is to summarize some classical results on exact penalties and its smoothing. For more recent comprehensive studies of nonlinear programming algorithms we refer e.g. to the Monographs of Bonnans et al. [2006], Kelly [1999].

For the sake of simplicity in the first two sections we make rather strong assumptions upon the considered optimization problem and refer to the original papers for further relaxations. We concentrate our attention upon certain generic properties. For the proofs we refer to the literature. As main result of our paper in Section 4 we develop a path-following Newton method for smoothed exact penalty methods and study its convergence. While this method is new, the related convergence results extend directly the approach developed first for the linearly constrained case and later to quite general nonlinear constraints. Here we provide convergence investigations for smoothing functions of exact penalties. The related analysis proving the existence of a differentiable path for this rather different class of algorithms rests widely on the proofs of the papers Grossmann/Zadlo [2005] for classical penalty/barrier methods. However, since for the new case of smoothing for exact penalty methods some parts had to be appropriately adapted we repeat the essential steps of the proofs along this line to make the paper self-contained. In Section 4 we apply the obtained results to develop a path-following large step Newton method. Finally in Section 5 some numerical examples illustrate given theoretical results.

2. Exact Penalties in Nonlinear Programming - Revisited

2.1. Problem Setting, Basic Properties.
Let be given some C^2 functions $f : \mathbb{R}^n \to \mathbb{R}$, $g_j : \mathbb{R}^n \to \mathbb{R}^m$, $j \in J$ with $J := \{1, 2, \ldots, m\}$ with Lipschitz-continuous second order derivatives. For simplicity of the presentation the functions f is assumed to be uniformly convex, i.e.

$$f(x) + \nabla f(x)^T(z - x) + c\|z - x\|^2 \leq f(z) \qquad \forall x, z \in \mathbb{R}^n$$

with some constant $c > 0$. Further, we assume that the functions g_j, $j \in J$ are convex. Let denote

$$G := \{x \in \mathbb{R}^n : g_j(x) \leq 0, j \in J\}, \qquad G^0 := \{x \in \mathbb{R}^n : g_j(x) < 0, j \in J\}$$

and suppose $G^0 \neq \emptyset$.

In the first part of our paper we consider the classical finite dimensional convex programming problem

(1) $$f(x) \to \min ! \qquad \text{subject to } x \in G.$$

Our strong assumptions guarantee that problem (1) possesses a unique solution $\bar{x}$, i.e.

$$\bar{x} \in G, \qquad f(\bar{x}) \leq f(x) \quad \forall x \in G,$$

and that at least one related optimal Lagrange multiplier $\bar{y} \in \mathbb{R}_+^m$ exists such that that pair $(\bar{x}, \bar{y}) \in G \times \mathbb{R}_+^m$ forms a saddle point of the Lagrangian related to (1), i.e.

(2) $$L(\bar{x}, y) \leq L(\bar{x}, \bar{y}) \leq L(x, \bar{y}) \qquad \forall x \in \mathbb{R}^n, y \in \mathbb{R}_+^m,$$

with

$$L(x, y) := f(x) + \sum_{j \in J} y_j \, g_j(x), \qquad x \in \mathbb{R}^n, y \in \mathbb{R}_+^m.$$

Later the requested properties will be relaxed by assuming that they hold only in some neighborhood of a strict local minimizer.

Next we deal with the well known exact penalty which incorporates the constraints $g_j(x)$, $j \in J$ into an auxiliary objective function of the penalty problem

$$(3) \qquad F(x,q) := f(x) + \sum_{j \in j} q_j \, \phi(g_j(x)) \to \min ! \qquad \text{s.t.} \quad x \in \mathbb{R}^n$$

with some parameter vector $q \in \mathbb{R}_{++}^m$ and the exact penalty function $\phi(t) = \max\{0, t\}$, $t \in \mathbb{R}$. Here $\mathbb{R}_{++}$ denotes the positive reals. Under the made convexity (strong convexity respectively) assumptions we immediately obtain that for any $r \in \mathbb{R}_{++}^m$ the auxiliary problem (3) possesses a unique solution $\bar{x}(p)$, i.e.

$$\bar{x}(q) \in \mathbb{R}^n, \qquad F(\bar{x}(q), q) \leq F(x, q) \qquad \forall x \in \mathbb{R}^n.$$

THEOREM 1. *Let the penalty parameter vector satisfy $q > \bar{y}$ for some optimal Lagrange vector $\bar{y}$. Then the solution $\bar{x}(q)$ of the auxiliary problem (3) is also the solution of the considered original problem (1), i.e. $\bar{x}(q) = \bar{x}$.*

2.2. Smoothing of Exact Penalties. The lack of differentiability of the exact penalty function requires adapted methods for the treatment of the related auxiliary problems (3). A popular approach to overcome this difficulty is the application of smoothing techniques which approximate the exact penalty $\phi(t)$ by some appropriate smooth function $\Phi(t,s)$, where $s > 0$ denotes a smoothing parameter. This allows to apply standard methods for the treatment of differentiable optimization problems to the generated approximate smooth problems. The smoothing of exact penalties has recently gained a renewed attention and various techniques have been proposed in the literature. Let us refer e.g. to [16], .[20], [22], [28]. Later in Subsection 2.3 some of these smoothing functions will be described and their properties studied.

A function $\Phi : \mathbb{R} \times \mathbb{R}_{++} \to \mathbb{R}_+$ is called a smoothing function for ϕ if the following properties hold:
- For any $s > 0$ the function $\Phi(\cdot, s)$ is continuously differentiable, convex and nondecreasing;
- There exists some function $\delta : \mathbb{R}_{++} \to \mathbb{R}_{++}$ with

$$(4) \qquad \lim_{s \to 0+} \delta(s) = 0 \quad \text{and} \quad |\Phi(t,s) - \phi(t)| \leq \delta(s) \quad \forall t \in \mathbb{R}.$$

Replacing ϕ by $\Phi(\cdot, s)$ from (3) we obtain the parametric smoothed auxiliary problems

$$(5) \qquad F(x,q,s) := f(x) + \sum_{j \in J} q_j \, \Phi(g_j(x), s) \to \min ! \qquad \text{s.t.} \quad x \in \mathbb{R}^n$$

with the smoothing parameter $s > 0$. The made assumptions for the smoothing function Φ guarantee that $F(\cdot, q, s)$ is strongly convex for any $q \in \mathbb{R}_{++}^m$ and $s > 0$. Thus (5) possesses a unique solution $\bar{x}(q,s)$. Please notice that F as well as $\bar{x}$ describe different functions and solutions, respectively, depending upon the values of the occurring parameters.

THEOREM 2. *Let the penalty parameter vector satisfy $q > \bar{y}$ for some optimal Lagrange vector $\bar{y}$. Then the solution $\bar{x}(q,s)$ of the smoothed auxiliary problem (5) converges to the solution of the considered original problem (1), i.e. $\lim\limits_{s \to 0+} \bar{x}(q,s) = \bar{x}$. Further, define the penalty multiplier by*

$$(6) \qquad \bar{y}_j(q,s) := q_j \frac{\partial}{\partial t}\Phi(g_j(\bar{x}(s)), s), \qquad j \in J.$$

Then the family $\{\bar{y}(q,s)\}_{s>0}$ is bounded and any accumulation point $\bar{y}$ of $\{\bar{y}(q,s)\}_{s \to 0+}$ forms an optimal Lagrange multiplier at $\bar{x}$, i.e. it satisfies

$$(7) \qquad \bar{y} \in \mathbb{R}^m_+, \quad g(\bar{x}) \leq 0, \quad \bar{y}^T g(\bar{x}) = 0, \quad \nabla f(\bar{x}) + \sum_{j=1}^{m} \bar{y}_j \nabla g_j(\bar{x}) = 0.$$

We notice that the MFCQ constraint qualification ensures the boundedness of the set $\bar{Y}$ of all optimal Lagrange multipliers. Further, the proof of Theorem 2 can be obtained by the concept of Fiacco/McCormick[10] for the convergence of primal and dual variables of parametric penalty techniques, see also [13].

THEOREM 3. *Let the smoothed exact penalty Φ satisfy*

$$(8) \qquad \frac{\partial}{\partial t}\Phi(0,s) \geq c > 0 \qquad \forall s > 0$$

with some $c > 0$ and let the parameter $q \in \mathbb{R}^m_{++}$ satisfy the additional condition

$$(9) \qquad q > \frac{1}{c}\bar{y} \qquad \forall \bar{y} \in \bar{Y},$$

where $\bar{Y}$ denotes the set of all optimal multipliers. Then some $\tilde{s} > 0$ exists such that

$$(10) \qquad \bar{x}(q,s) \in G^0 \qquad \forall s \in (0, \tilde{s}].$$

2.3. Selected Smoothing Functions. A well known smooth approximation of $|t|$ is given by

$$(11) \qquad |t| \approx \sqrt{t^2 + s^2}, \quad s > 0$$

for $s \to 0$. It should be noticed that the use of s^2 instead of simply s as parameter provides a natural scaling which will be discussed later.

Kaplan[18] applied (11) to the identity $\phi(t) = \frac{1}{2}(t + |t|)$ and obtained the smoothing function

$$(12) \qquad \Phi(t,s) := \frac{1}{2}\left(t + \sqrt{t^2 + s^2}\right), \quad t \in \mathbb{R}, \ s > 0.$$

The function $\Phi(\cdot, s)$ is infinitely often differentiable and satisfies

$$(13) \qquad 0 \leq \Phi(t,s) - \phi(t) = \frac{1}{2}\left(\sqrt{t^2 + s^2} - \sqrt{t^2}\right) = \frac{1}{2}\frac{s^2}{\sqrt{t^2 + s^2} + \sqrt{t^2}} \leq \frac{s}{2}.$$

Thus, the function Φ defined by (12) is a smoothing function with $\delta(s) = \frac{1}{2}s$.

In [24] the following smoothing function is investigated:

$$(14) \qquad \Phi(t,s) = \begin{cases} 0 & \text{if } t < 0, \\ t^2/(2s) & \text{if } 0 \leq t \leq s, \\ t - s/2 & \text{if } t > s. \end{cases}$$

The function $\Phi(\cdot)$ for any parameter $s > 0$ is continuously differentiable, but only one time. Its second derivative possesses jumps at $t = 0$ as well as at $t = s$.

In [**20**] the following smoothing function

$$\Phi(t,s) = \begin{cases} \frac{3}{2} s\, e^{t/s} & \text{if } t \leq 0, \\ t - \frac{1}{2} s\, e^{-t/s} & \text{if } t > 0 \end{cases} \tag{15}$$

has been proposed. As the function (14) also (15) is continuously differentiable, but its second order derivative has a jump at the point $t = 0$.

In connection with semi-infinite programming in [**22**] further smoothing functions are mentioned

$$\Phi(t,s) = s \log(1 + e^{t/s}); \tag{16}$$

$$\Phi(t,s) = \begin{cases} 2s\, e^{t/s} & \text{if } t < 0, \\ t + s\left(\log(1 + t/s) + 2\right) & \text{if } t \geq 0; \end{cases} \tag{17}$$

$$\Phi(t,s) = \begin{cases} s\, e^{t/s} & \text{if } t < 0 \\ t + s & \text{if } t \geq 0; \end{cases} \tag{18}$$

In [**21**] as smoothing function

$$\Phi(t,s) = \begin{cases} \frac{1}{2} s\, e^{t/s} & \text{if } t \leq 0, \\ t + s\, e^{-t/s} & \text{if } t > 0 \end{cases} \tag{19}$$

has been applied. It should be mentioned that (19) is at least twice continuously differentiable. The smoothing function (16) was proposed by Gugat/Herty [**15**] (see also [**17**]) to treat state constraints in optimal control problems for partial differential equations.

In [**26**] smoothed exact penalties have been applied to global optimization.

In [**28**] and further publications lower order exact penalties like $\phi(t) = t^\nu$ with $\nu \in (0,1)$ and related smoothing function are considered. However such low order exact penalties are not in the focus of the present study.

All the function given above possess an important natural scaling by the following identity

$$\frac{\partial}{\partial t}\Phi(t,s) = \Psi(\frac{t}{s}) \qquad \forall t \in \mathbb{R},\ s \in R_{++} \tag{20}$$

with some function $\Psi : \mathbb{R} \to \mathbb{R}$. This important relation was introduced in [**14**] to develop a general local convergence theory of penalty/barrier methods which enables to establish large step Newton path-following methods for the related iterations. The functions Ψ are called the generating functions for Φ.

In the sequel we concentrate our attention upon the smoothing function (12) which was introduced be Kaplan [**18**]. In this paper the convergence theory of the smoothing of exact penalty functions has been established similar to the one given above. However, the further analysis based on the relation (20) is new and extends the investigations given in [**14**] to the slightly different situation of smoothing exact penalties. The smoothing (12) is related to the generating function

$$\Psi(r) = \frac{1}{2}\left(1 + \frac{r}{\sqrt{r^2+1}}\right) \qquad r \in \mathbb{R}. \tag{21}$$

It is important for the further investigation, in particular for the convergence proof of a path-following scheme based on (12), that its first order derivative

$$\Psi'(r) = \frac{1}{2}\frac{1}{(r^2+1)^{3/2}} \qquad r \in \mathbb{R} \tag{22}$$

is globally Lipschitz continuous. To simplify the notation in the sequel we abbreviate

$$\Phi_j(t,s) := q_j\, \Phi(t,s) \quad \text{and} \quad \Psi_j(t,s) := q_j\, \Psi(t,s) \quad j \in J, \tag{23}$$

i.e. we include the weights into the notion of the functions Φ and Ψ, respectively. Further, consistently the weight parameter $p \in \mathbb{R}^m_{++}$ will be omitted in the notations.

3. Existence and Differentiability of a Local Path

Let us consider a more general problem and study the behavior of smoothed exact penalty methods applied to a stable local minimizer $\bar{x} \in G$ of (1). We assume that the objective function f as well as all constraint functions g_j, $j \in J$ are sufficiently smooth. Further, the stronger constraint qualification LICQ, i.e. the linear independence of the gradients of the active constraints at $\bar{x}$ is assumed. In addition, the strict complementarity

$$\bar{y}_j > 0 \iff g_j(\bar{x}) = 0 \tag{24}$$

and

$$\left.\begin{array}{l} w \in \mathbb{R}^n, \quad w \neq 0, \\ \nabla g_j(\bar{x})^T w = 0,\ j \in J_0 \end{array}\right\} \Rightarrow w^T \nabla^2_{xx} L(\bar{x}, \bar{y}) w > 0 \tag{25}$$

should be satisfied. The conditions above provide just the well known second order sufficiency optimality conditions (cf. [**10**]) which imply that $\bar{x}$ is an isolated local minimizer of (1) and that $\bar{x}$ is stable under perturbations.

THEOREM 4. *Let the assumptions made at the beginning of Section 3 be fulfilled and let the generating function Ψ continuously differentiable. Then there exist some $\bar{s} \in (0, \hat{s}]$, $\delta > 0$ such that for each $s \in (0, \bar{s}]$ the parametric system*

$$\nabla f(x(s)) + \sum_{j \in J} \Psi_j(g_j(x(s))/s)\, \nabla g_j(x(s)) = 0. \tag{26}$$

of nonlinear equations possesses a unique solution $x(s) \in U_\delta(\bar{x})$, where $U_\delta(\bar{x})$ denotes an open ball with the radius δ and the center $\bar{x}$. Let $y(s)$ be related to $x(s)$ by

$$y_j(s) := \Psi_i(g_j(x(s))/s), \quad j \in J. \tag{27}$$

then we have

$$\lim_{s \to 0+}(x(s), y(s)) = (\bar{x}, \bar{y})$$

and the functions $x(\cdot), y(\cdot)$ are continuously differentiable in $(0, \bar{s}]$. In addition, the derivatives $\dot{x}(\cdot), \dot{y}(\cdot)$ are bounded as $s \to 0+$. With

(28) $$x(0) := \bar{x} \quad \text{and} \quad y(0) := \bar{y}$$

$x(\cdot), y(\cdot)$ possess a right side derivative at $s = 0$. The derivative of $x(s), y(s)$ is given by

(29) $$\begin{pmatrix} \dot{x}(s) \\ \dot{y}(s) \end{pmatrix} = H(s)^{-1} \begin{pmatrix} 0 \\ q(s) \end{pmatrix},$$

where

$$H(s) := \begin{pmatrix} \nabla_{xx} L(x(s), y(s)) & \nabla g_1(x(s)) & \nabla g_2(x(s)) & \cdots & \nabla g_m(x(s)) \\ \Psi'_1\left(\frac{g_1(x(s))}{s}\right) \nabla g_1(x(s))^T & -s & 0 & \cdots & 0 \\ \Psi'_2\left(\frac{g_2(x(s))}{s}\right) \nabla g_2(x(s))^T & 0 & -s & \cdots & 0 \\ \vdots & \vdots & \vdots & \ddots & \vdots \\ \Psi'_m\left(\frac{g_m(x(s))}{s}\right) \nabla g_m(x(s))^T & 0 & 0 & \cdots & -s \end{pmatrix}$$

and $q(s) := (q_1(s), \ldots, q_m(s))^T$ with $q_j(s) := s^{-1} \Psi'_i\left(\frac{g_j(x(s))}{s}\right) g_j(x(s))$.

This theorem essentially rests on the implicit function theorem applied to the following perturbed KKT-system:

$$\nabla f(x(s,r)) + \sum_{j \in J_0} y_j(s,r) \nabla g_j(x(s,r)) = r,$$
$$s \Psi_j^{-1}(y_j(s,r)) = g_j(x(s,r)), \quad j \in J_0.$$

Here we omit the proof and refer to [11] for technical details. For equality constraint problems a local convergence analysis can be found in [2].

COROLLARY 5. *Under the given assumptions, there exist some constants $s_0 \in (0, \bar{s}]$ and $c_L > 0$ such that*

(30) $$\left. \begin{array}{l} \|x(s) - x(t)\| \leq c_L |s - t| \\ \|y(s) - y(t)\| \leq c_L |s - t| \end{array} \right\} \quad \forall s, t \in [0, s_0].$$

REMARK 6. *The implicit function theorem and the regularity of the matrix $H(s)$ also imply that the mappings $x(s), y(s)$ possess higher order derivatives provided the functions f and $g_j, j \in J$ are sufficiently smooth.*

4. Path-Following Newton Technique

Unlike the classical penalty/barrier-methods as discussed e.g. in [10], [12], in path-following each of the auxiliary problems (26) is not solved up to a preselected accuracy, but at each parameter level $s > 0$ only a limited number of iteration steps of the solver is performed. As in long step path-following studies, here we consider particularly just one Newton step applied to

$$\nabla_x F(x, s_k) = 0 \quad \text{where} \quad F(x,s) := \nabla f(x) + \sum_{j \in J} \Psi_j(g_j(x)/s)\, \nabla g_j(x).$$

Path-Following Algorithm.

Step 1: Select parameters $\varepsilon > 0$ and c, $s_0 > 0$, sufficiently small and $\nu \in (0,1)$.

Find $x^0 \in U_\delta$ such that

(31) $$\|x^0 - x(s_0)\| \leq c\, s_0.$$

Set $k := 0$.

Step 2: Determine $x^{k+1} \in \mathbb{R}^n$ via the linear system

(32) $$\begin{aligned}\nabla^2_{xx} F(x^k, s_k)\, d^k &= -\nabla_x F(x^k, s_k) \\ \text{and} \quad x^{k+1} &:= x^k + d^k\end{aligned}$$

Step 3: If $s_k \leq \varepsilon$ then stop. Otherwise set $s_{k+1} := \nu s_k$ and go to **Step 2** with $k := k+1$.

Newton's method applied to the auxiliary problem (26) converges quadratically for any fixed parameter $s = s_k$ provided the initial guess x^k is sufficiently close to $x(s_k)$ and the parameter $s_k > 0$ is sufficiently small. However, the radius of convergence of Newton's method depends upon s_k due to the asymptotically singular embedding for $s \to 0+$.

Let the smoothing parameter $s > 0$ be fixed and let $x \in U_\delta$ denote an approximation of the solution $x(s)$ of (26). Related to x we define vectors

$$u = (u_1, \ldots, u_m)^T \in \mathbb{R}^m \quad \text{and} \quad v = (v_1, \ldots, v_m)^T \in \mathbb{R}^m$$

by

(33) $$\left.\begin{aligned} u_j &:= u_j(x,s) = \Psi_i(g_i(x)/s) \\ v_j &:= v_j(x,s) = s^{-1}\Psi'_i(g_i(x)/s) \end{aligned}\right\} \quad \forall\, s \in (0, \hat{s}],\ j \in J.$$

In the particular case $x = x(s)$, definition (27) yields $y(s) = u(x(s), s)$.

Starting from $x \in U_\delta(\bar{x})$, one Newton-step defines a new approximate $\tilde{x}$ of $x(s)$ as the solution of the linear system

(34) $$\nabla_x F(x,s) + \nabla^2_{xx} F(x,s)\, (\tilde{x} - x) = 0.$$

Following the second order sufficiency criterion (25) the system matrix

$$\nabla^2_{xx} F(x,s) = \nabla^2_{xx} L(x,u) + \sum_{j \in J} v_j\, \nabla g_j(x)\, \nabla g_j^T(x)$$

is regular for sufficiently small smoothing parameters $s > 0$ along the local path $x(s)$, $s \in (0, s_0]$ and for all neighboring points which satisfy $\|x - x(s)\| \leq \omega s$. Moreover, this condition also guarantees $\tilde{x} \in U_s(\bar{x})$ for the solution $\tilde{x}$ of the Newton equation (34).

An important fact are the following lemmas (see [**14**]) which form an essential tool to establish improved convergence results for $s \to 0+$ similar to [**23**], [**8**], but with a different analysis.

LEMMA 7. *There exist constants $s_0 \in (0, \bar{s}]$ and $\omega, \sigma > 0$ such that*
$$\|x - x(s)\| \leq \omega s, \ s \in (0, s_0] \quad \Rightarrow \quad \begin{cases} x \in B_s, \\ v^T \nabla^2_{xx} F(x,s) v \geq \sigma \|v\|^2, \ \forall v \in \mathbb{R}^n. \end{cases}$$

LEMMA 8. *Let $\bar{\gamma} \geq \gamma > 0$ denote given constants and let $\{a_i\}_{i=1}^m \subset \mathbb{R}^n$ be some fixed set of linearly independent vectors. Then for any symmetric (n,n)-matrix C satisfying*

(35) $$\gamma \|z\|^2 \leq z^T C z, \quad \forall z \in \mathbb{R}^n \quad \text{and} \quad \|C\| \leq \bar{\gamma},$$

and for any $\alpha_i, \beta_i \in \mathbb{R}$ with $\alpha_i > 0$, $\alpha_i \geq |\beta_i|$, $i = 1, \ldots m$, the linear system

(36) $$\left(C + \sum_{i=1}^m \alpha_i \, a_i \, a_i^T \right) x = \left(\sum_{i=1}^m \beta_i \, a_i \, a_i^T \right) u$$

defines a unique mapping $\mathbb{R}^n \ni u \mapsto x \in \mathbb{R}^n$. Moreover, some constant $\omega > 0$ exists such that

(37) $$\|x\| \leq \omega \left(1 + \frac{\bar{\gamma}}{\gamma}\right) \frac{\alpha_{max}}{\alpha_{min}} \|u\|,$$

where $\alpha_{max} := \max\limits_{1 \leq i \leq m} \alpha_i$, $\alpha_{min} := \min\limits_{1 \leq i \leq m} \alpha_i$.

The proof can be found in [**1**].

THEOREM 9. *There exist constants $s_* \in (0, \bar{s}]$, $c_\delta, c_\rho > 0$ such that*
$$\left. \begin{array}{l} x \in \mathbb{R}^n, \quad s \in (0, s_*], \\ \|x - x(s)\| \leq c_\rho s \end{array} \right\} \Longrightarrow \begin{cases} \text{System (34) possesses a unique solution } \tilde{x} \text{ and} \\ \|\tilde{x} - x(s)\| \leq c_\delta \, s^{-1} \|x - x(s)\|^2. \end{cases}$$

Proof: As mentioned above, for certain penalty/barrier methods, but not directly for smoothed exact penalty methods the proof can be found in [**1**], [**14**]. Thus, here we sketch only the essential steps of the proof and concentrate our attention on the new parts which are connected with nonlinear constraints.

Let us recall that holds
$$\nabla_x F(x,s) = \nabla_x L(x, u(x,s)) \quad \text{and}$$
$$\nabla^2_{xx} F(x,s) = \nabla^2_{xx} L(x, u(x,s)) + \sum_{j \in J} v_j(x,s) \, \nabla g_j(x) \, \nabla g_j^T(x).$$

Newton's method yields

(38) $$\nabla^2_{xx} F(x,s)(\tilde{x} - x(s)) = \nabla_x F(x(s), s) - \nabla_x F(x,s) - \nabla^2_{xx} F(x,s)(x(s) - x).$$

We notice that by Lemma 7 the matrix $\nabla^2_{xx} F(x,s)$ is positive definite for sufficiently small $s_0, \omega > 0$ and
$$\|x - x(s)\| \leq \omega s, \quad s \in (0, s_0].$$
In the sequel we suppose this choice of s, ω and x. Hence, Newton's method is well defined.

By Taylor's formula we have

$$\nabla_x F(x(s), s) = \nabla_x F(x, s) + \int_0^1 \nabla_{xx}^2 F(x + \tau(x(s) - x)) \, d\tau \, (x(s) - x)$$

and, with (38), it follows that

(39) $$\nabla_{xx}^2 F(x, s)(\tilde{x} - x(s)) = \sum_{j=1}^4 q_j,$$

where the components $q_j \in \mathbb{R}^n$, $j = 1, \ldots, 4$ are defined by

$$q_1 := \int_0^1 (\nabla^2 f(x + \tau(x(s) - x)) - \nabla^2 f(x)) \, d\tau \, (x(s) - x),$$

$$q_2 := \sum_{i \notin I_0} \int_0^1 [v_j(x + \tau(x(s) - x), s) \nabla g_j(x + \tau(x(s) - x)) \nabla g_j^T(x + \tau(x(s) - x))$$
$$- v_j(x, s) \nabla g_j(x) \nabla g_j^T(x)] \, d\tau \, (x(s) - x),$$

$$q_3 := \sum_{j \in J_0} \int_0^1 [v_j(x + \tau(x(s) - x), s) \nabla g_j(x + \tau(x(s) - x)) \nabla g_j^T(x + \tau(x(s) - x))$$
$$- v_j(x, s) \nabla g_j(x) \nabla g_j^T(x)] \, d\tau \, (x(s) - x),$$

$$q_4 := \sum_{j \in J} \int_0^1 [u_j(x + \tau(x(s) - x), s) \nabla^2 g_j(x + \tau(x(s) - x))$$
$$- u_j(x, s) \nabla^2 g_j(x)] \, d\tau \, (x(s) - x).$$

To estimate $\|\tilde{x} - x(s)\|$ we study the linear systems

(40) $$\nabla_{xx}^2 F(x, s) \, p_j = q_j, \quad j = 1, \ldots, 4$$

separately. Before we continue let us remark that a similar split was applied in [25] to analyze the convergence of a path-following algorithm, but of logarithmic barrier-type only. On the other hand, general analysis was carried out in [1], [11] for linearly constrained problems. This resulted in simplified terms q_2, q_3 and q_4 missing.

Now, let us estimate the solutions p_j of (40) by studying the right hand sides q_j and applying Lemma 8 in some cases. Since f is assumed to be twice Lipschitz continuously differentiable, we obtain immediately that

(41) $$\|p_1\| \leq c_1 \|x - x(s)\|^2$$

for some constant $c_1 > 0$.

Next, we turn to q_2. For $i \notin I_0$ and with $\hat{x} := x + \tau(x(s) - x)$ we have

$$\|v_j(\hat{x}, s) \nabla g_j(\hat{x}) \nabla g_j^T(\hat{x}) - v_j(x, s) \nabla g_j(x) \nabla g_j^T(x)\|$$

$$\leq \|(v_j(x, s) - v_j(\hat{x}, s)) \nabla g_j(\hat{x}) \nabla g_j^T(\hat{x})\|$$

$$+ \|v_j(x, s)(\nabla g_j(\hat{x}) \nabla g_j^T(\hat{x}) - \nabla g_j(x) \nabla g_j^T(x))\|$$

The assumed local Lipschitz continuity of Ψ' and g_j yield

$$|v_j(x,s) - v_j(\hat{x},s)| \leq L \left| \frac{g_j(\hat{x}) - g_j(x)}{g_j(x)g_j(\hat{x})} \right|$$

$$\leq c\delta^{-2} L |x - x(s)|, \quad i \notin I_0,$$

with some constants c, $\delta > 0$ and the Lipschitz constant L. Further, we have

$$\|\nabla g_j(\hat{x}) \nabla g_j^T(\hat{x}) - \nabla g_j(x) \nabla g_j^T(x)\| \leq \tilde{c} \|x - x(s)\| \quad \text{with } \tilde{c} > 0.$$

Finally,

$$\|p_2\| \leq c_2 \|x - x(s)\|^2, \tag{42}$$

with some $c_2 > 0$.

With the Lipschitz continuity of Ψ' for $j \in J_0$ we have

$$|v_j(x,s) - v_j(\hat{x},s)| = s^{-1} \left| \Psi'_i\left(\frac{g_j(x)}{s}\right) - \Psi'_i\left(\frac{g_j(\hat{x})}{s}\right) \right|$$

$$\leq cs^{-2} L \|x(s) - x\|, \quad j \in J_0.$$

Hence

$$q_3 = \sum_{j \in J_0} \int_0^1 \nu_j(s) \nabla g_j(\hat{x}) \nabla g_j^T(\hat{x}) + v_j(x,s)(\nabla g_j(\hat{x}) \nabla g_j(\hat{x})^T - \nabla g_j(x) \nabla g_j(x)^T) \, d\tau \, (x(s) - x),$$

with

$$|\nu_j(s)| \leq c s^{-2} \|x - x(s)\| \tag{43}$$

and some $c > 0$.

Next, we split q_3 further by $q_3 = q_{31} + q_{32} + q_{33}$ with

$$q_{31} := \sum_{j \in J_0} \nu_j(s) \nabla g_j(x) \nabla g_j^T(x)(x - x(s)),$$

$$q_{32} := \sum_{j \in J_0} \int_0^1 \nu_j(s)(\nabla g_j(\hat{x}) \nabla g_j(\hat{x})^T - \nabla g_j(x) \nabla g_j(x)^T) \, d\tau (x - x(s)),$$

$$q_{33} := \sum_{j \in J_0} \int_0^1 v_j(x,s)(\nabla g_j(\hat{x}) \nabla g_j(\hat{x})^T - \nabla g_j(x) \nabla g_j(x)^T) \, d\tau (x - x(s)).$$

After a short transformation we can apply Lemma 8 to

$$q_{31} = \frac{\max_{i \in I_0} \nu_j(s)}{\min_{i \in I_0} \alpha_j(s)} \sum_{j \in J_0} \frac{\nu_j(s)}{\max_{j \in J_0} \nu_j(s)} \min_{i \in I_0} \alpha_j(s) \nabla g_j(x) \nabla g_j^T(x)(x - x(s)).$$

To match the notation of Lemma 8 for $j \in J_0$ we define:

$$\alpha_j(s) := \alpha_j = v_j(x,s) - \tilde{c} = s^{-1}\Psi'_j(\Psi_j^{-1}(u_j(x,s))) - \tilde{c},$$

$$\beta_j(s) := \frac{\nu_j(s)}{\max_{j \in J_0} \nu_j(s)} \min_{j \in J_0} \alpha_j(s)$$

and we set $A_j := \nabla g_j(x) \nabla g_j(x)^T$, $C := \nabla^2_{xx} F(x,s) + \tilde{c} \sum_{j \in J_0} A_j$, $u := x - x(s)$,

where $\tilde{c} > 0$ denotes some sufficiently large constant (see proof of Lemma 7).

Now, there exist some $c_3 > 0$ and $s_1 \in (0, s_0]$ such that
$$\alpha_j(s) \geq c_3 \, s^{-1}, \qquad s \in (0, s_1], \quad j \in J_0.$$
By definition we have $v_j(x,s) = s^{-1} \Psi'_j\left(\dfrac{g_j(x)}{s}\right)$, $j \in J$. and the properties of Ψ_j and Ψ'_j lead to
$$(44) \qquad 0 < s^{-1} \Psi'_j(\Psi_j^{-1}(\underline{\kappa})) \leq v_j(x,s) \leq s^{-1} \Psi'_j(\Psi_j^{-1}(\overline{\kappa})), \qquad j \in J_0,$$
with $\overline{\kappa} > \underline{\kappa} > 0$. With the definition α_j this leads to the estimate
$$\frac{\max\limits_{j \in J_0} \alpha_j}{\min\limits_{j \in J_0} \alpha_j} \leq \frac{\max\limits_{j \in J_0} \Psi'_j(\Psi_j^{-1}(\overline{\kappa})) - \tilde{c}\, s}{\min\limits_{j \in J_0} \Psi'_j(\Psi_j^{-1}(\underline{\kappa})) - \tilde{c}\, s}.$$
Hence, the quotient $\max\limits_{j \in J_0} \alpha_j / \min\limits_{j \in J_0} \alpha_j$ remains bounded for $s \to 0+$. Similarly, (43) and $\|x - x(s)\| \leq \omega s$ yield the boundedness of the quotient $\max\limits_{j \in J_0} \nu_j / \min\limits_{j \in J_0} \alpha_j$ for $s \to 0+$. Now Lemma 8 can be applied and we obtain
$$\|p_{31}\| = c_4 \, s^{-1} \|x - x(s)\|^2$$
with some $c_4 > 0$.

Next we estimate q_{32}. The Lipschitz continuity of ∇g_j. guarantees
$$\|q_{32}\| \leq c_5 \, s^{-2} \|x - x(s)\|^3$$
and we obtain
$$\|p_{32}\| = c_6 \, s^{-1} \|x - x(s)\|^2$$
with constants $c_5, c_6 > 0$.

For q_{33} we have
$$\|q_{33}\| \leq \sum_{j \in J_0} \int_0^1 \|v_j(x,s)(\nabla g_j(\hat{x}) \nabla g_j(\hat{x})^T - \nabla g_j(x) \nabla g_j(x)^T)\| \, d\tau \, \|x - x(s)\|$$
$$\leq c \|x - x(s)\|^2.$$
Now, using the triangle inequality this yields
$$(45) \qquad \|p_3\| \leq c_7 \, s^{-1} \|x - x(s)\|^2$$
with some $c_7 > 0$.

Furthermore
$$\|u_j(\hat{x}, s) \nabla^2 g_j(\hat{x}) - u_j(x, s) \nabla^2 g_j(x)\|$$
$$\leq \|(u_j(\hat{x}, s) - u_j(x, s)) \nabla^2 g_j(\hat{x})\| + \|u_j(x, s)(\nabla^2 g_j(\hat{x}) - \nabla^2 g_j(x))\|,$$
and with the Lipschitz continuity of $\nabla^2 g_j$ and Ψ_i and by (44) it follows that
$$(46) \qquad \|p_4\| \leq c_8 \|x - x(s)\|^2,$$
for some $c_8 > 0$.

In view of (39), we have
$$\tilde{x} - x(s) = \sum_{j=1}^4 p_j,$$
and using (41), (42), (45), (46) we conclude that some $c_\delta, c_\rho > 0$ exist with
$$(47) \quad \|\tilde{x} - x(s)\| \leq c_\delta \, s^{-1} \|x - x(s)\|^2 \quad \forall x \in U_s, \, \|x - x(s)\| \leq c_\rho \, s, \quad s \in (0, s_0].$$
In addition, due to Lemma 7, all the constants are selected such that $\tilde{x} \in U_s$. ∎

In addition, in case we ensure that
$$\|x^{k+1} - x(s_{k+1})\| \leq \omega\, s_{k+1} \tag{48}$$
holds for the new iterate x^{k+1}, then Lemma 7 also guarantees $x^{k+1} \in B_{s_{k+1}}$. Hence, to be able to apply these results the constant $c > 0$ of the algorithm has to be selected according to
$$c \in (0, \min\{c_\rho, \omega\}]. \tag{49}$$

THEOREM 10. *Let $s_0 > 0$ be sufficiently small, $c > 0$ satisfy (49) and let the parameter $\nu \in (0,1)$ be selected such that*
$$\nu^{-1}\left(c_\delta\, c^2 + c_L(1-\nu)\right) \leq c, \tag{50}$$
where c_δ, c_L denote the constants defined in Theorem 9 and Corollary 5, respectively. Then the given path-following algorithm is well defined and generates iterates $x^k \in \mathbb{R}^n$ that satisfy
$$\|x^k - x(s_k)\| \leq c\, s_k, \qquad k = 0, 1, \ldots. \tag{51}$$
Furthermore, the algorithm terminates after at most $k^* := \lceil \ln(\varepsilon/s_0)/\ln(\nu) \rceil$ steps and the estimate
$$\|x^{k^*} - \bar{x}\| \leq (c_L + c)\,\varepsilon \tag{52}$$
holds.

Proof: We apply induction to show (51). If this inequality holds for some k then Lemma 7 yields $x^k \in B_{s_k}$ and the positive definiteness of the matrix $\nabla^2_{xx} F(x^k, s_k)$ occurring in the Newton step (32) of the algorithm.

For $k = 0$, inequality (51) holds because of the selection of the initial iterate x^0 in step 1.

Now we consider the induction step. Let (51) be true for some k. Since $c \in (0, \omega]$ and by Lemma 7 the new iterate $x^{k+1} \in \mathbb{R}^n$, as mentioned already, is uniquely defined. Now, the triangle inequality, (49), Theorem 9 and Corollary 5 imply
$$\begin{aligned}
\|x^{k+1} - x(s_{k+1})\| &\leq \|x^{k+1} - x(s_k)\| + \|x(s_k) - x(s_{k+1})\| \\
&\leq c_\delta\, c^2\, s_k + c_L(s_k - s_{k+1}) \\
&= \frac{s_k}{s_{k+1}}\left(c_\delta\, c^2 + c_L\right) s_{k+1} - c_L\, s_{k+1} \\
&= \nu^{-1}\left(c_\delta\, c^2 + c_L(1-\nu)\right) s_{k+1} \leq c\, s_{k+1},
\end{aligned}$$
with the constants c_δ, c_L from Theorem 9 and Corollary 5. This completes the induction.

The second part of the assertion is an immediate consequence of the parameter reduction procedure in step 3 of the method, the termination criterion $s_{k^*} \leq \varepsilon$ and the estimate (52), which follows from Corollary 5 and (51). ∎

REMARK 11. *The smaller the reduction factor $\nu \in (0,1)$ is the faster the algorithm converges. Depending upon the value of $c > 0$, we obtain the related minimal ν that satisfies condition (50) by*
$$\nu_{min} = \nu_{min}(c) = \frac{c_\delta\, c^2 + c_L}{c + c_L}. \tag{53}$$

Since $\nu_{min}(0) = 0$ and $\nu'_{min}(0) < 0$, it is always possible to select c, ν according to the conditions of the theorem above.

5. Numerical Examples

In this section we present numerical examples that illustrate the convergence of the algorithm as the smoothing parameter s tends to zero.

Example 1

The well known Rosen-Suzuki problem

$$f(x) = x_1^2 + x_2^2 + 2\,x_3^2 + x_4^2 - 5\,x_1 - 5\,x_2 - 21\,x_3 + 7\,x_4 \rightarrow \min !$$
$$\text{s.t.} \quad x_1^2 + x_2^2 + x_3^2 + x_4^2 + x_1 - x_2 + x_3 - x_4 \leq 8,$$
$$x_1^2 + 2\,x_2^2 + x_3^2 + 2\,x_4^2 - x_1 + x_4 \leq 10,$$
$$2\,x_1^2 + x_2^2 + x_3^2 + 2\,x_1 - x_2 - x_4 \leq 5$$

has the optimal solution $\bar{x} = (0, 1, 2, -1)^T \in \mathbb{R}^4$ with $f(\bar{x}) = 44$.

Numerical experiments reflect perfectly the theoretical result $\|\bar{x}(s) - \bar{x}\| = O(s)$ for $s \rightarrow 0+$ as shown on the following Fig 1: The red curve shows the numerically observe behavior of the constant in the error estimate. The blue curve shows the absolute error in dependence of s while the green curve represents (for comparison reasons only) the graph of the function $z = s$.

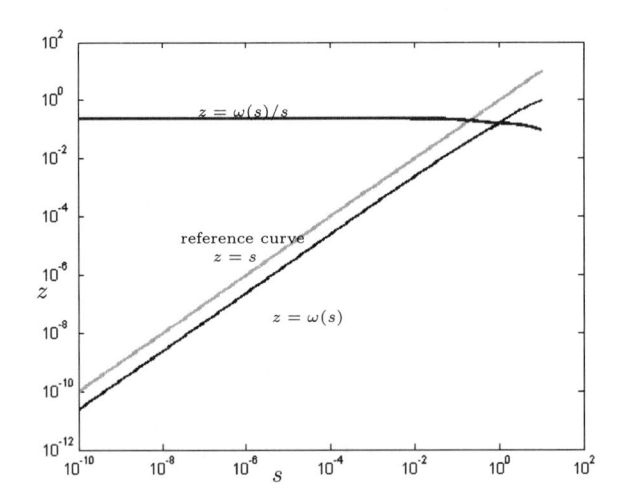

Fig. 1 Asymptotic behavior of $\omega(s) := \|\bar{x}(s) - \bar{x}\|$

Example 2

We consider the variational formulation

(54)
$$f(x) = \frac{1}{2}\int_0^1 (x'(t))^2\,dt - \int_0^1 p(t)\,x(t)\,dt \rightarrow \min !$$
$$\text{s.t.} \quad x \in G := \{\,x \in X := H_0^1(0,1) : x \geq q \text{ a.e. in } (0,1)\,\}$$

as an abstract optimization problem in the appropriate function space. Here $p, q \in L^2(0,1)$ are given functions and $H_0^1(0,1) \subset L^2(0,1)$ denotes the usual Sobolev space of functions having square integrable weak derivatives. In the considered example we choose

$$p(x) \equiv -10, \quad q(x) \equiv -1 + 0.7\,x, \quad x \in [0,1].$$

Problem (54) forms a one-dimensional obstacle problem. Its analytical solution is

$$(55) \quad \bar{x}(t) = \begin{cases} 5t^2 - (\sqrt{20} - 0.7)\,t, & \text{if } t \in \left(0, \frac{\sqrt{20}}{10}\right], \\ -1 + 0.7\,t, & \text{if } t \in \left(\frac{\sqrt{20}}{10}, 1 - \sqrt{\frac{0.3}{5}}\right), \\ 5(1-t)^2 - (10\sqrt{\frac{0.3}{5}} + 0.7)(1-t), & \text{if } t \in \left(1 - \sqrt{\frac{0.3}{5}}, 1\right]. \end{cases}$$

First, we discretize (54) by conforming piecewise linear elements $\Phi_j \in C[0,1]$, $j = 1, \ldots, N-1$ over a uniform grid $t_j := j\,h$, $j = 0, \ldots, N$ with $h := 1/N$. The underlying space X is discretized by $X_h := \text{span}\{\Phi_j\}_{j=1}^{N-1}$ and the feasible domain G by

$$G_h := \{\,x_h \in X_h : x_h \geq q_h\,\}$$

with a piecewise linear approximation $q_h \in C[0,1]$ of q.

So, we have $q_h = q$, and obtain the finite dimensional approximation

$$(56) \quad f(x_h) = \frac{1}{2} \int_0^1 (x_h'(t))^2\,dt - \int_0^1 p(t)\,x_h(t)\,dt \to \min ! \quad \text{s.t.} \quad x_h \in G_h.$$

of problem (54). With the standard continuous hat-functions Φ_j that form a Lagrange basis of X_h, i.e. $\Phi_j(t_i) = \delta_{ij}$ with Kronecker's delta, we may identify x_h with the vector $(x_j)_{j=1}^{N-1} \in \mathbb{R}^{N-1}$ of nodal values. In this way, the discrete optimization problem (56) can be equivalently represented by

$$(57) \quad f(x_h) = \frac{1}{2} x_h^T A_h x_h - p_h^T x_h \to \min ! \quad \text{s.t.} \quad x_j \geq q_h(t_j), \quad j = 1, \ldots, N-1,$$

with

$$A_h = (a_{ij})_{i,j=1}^{N-1} \in \mathcal{L}(\mathbb{R}^{N-1}), \qquad a_{ij} := \int_0^1 \Phi_i'(t)\,\Phi_j'(t)\,dt,$$

$$p_h = (f_j)_{j=1}^{N-1} \in \mathbb{R}^{N-1}, \qquad p_j := \int_0^1 p(t)\,\Phi_j(t)\,dt.$$

With our data this yields

$$a_{ji} = \begin{cases} 2/h, & i = j, \\ -1/h, & |i-j| = 1, \\ 0, & \text{else}, \end{cases} \qquad p_j = -10\,h, \; j = 1, \ldots, N-1.$$

Numerical experiments confirm the theoretically established result $\|\bar{x}(s) - \bar{x}\|_h \leq c_N\,s$ for $s \to 0+$. Moreover the constant c_N shows a mesh independence as the following table shows.

s	N=50	N=500	N=5000	N=50000
8.19e-02	9.048e-01	8.967e-01	8.959e-01	8.958e-01
2.68e-02	9.576e-01	9.502e-01	9.493e-01	9.492e-01
8.80e-03	9.898e-01	9.865e-01	9.858e-01	9.857e-01
2.88e-03	9.985e-01	1.010e+00	1.009e+00	1.009e+00
9.44e-04	9.676e-01	1.025e+00	1.024e+00	1.024e+00
3.09e-04	8.349e-01	1.032e+00	1.034e+00	1.034e+00

Table 1 Numerically obtained results $\|\bar{x}_h(s) - \Pi_h \bar{x}\|_h / s$

We applied the L_2 compatible norm $\|x_h\|_h^2 := \frac{1}{N} \sum x_i^2$ and scaled q_j to adjust them with the dual variables which change with refining the grid. The obtained numerical results indicate a mesh-independent behavior. However this cannot be concluded from the arguments in Section 3 and Section 4 because there is no positive gap to inactive constraints. So, to establish such a principle further investigations are required.

Acknowledgment

Let me thank my friend and colleague A. Kaplan for the long standing cooperation in particular for all the fruitful discussions on penalty as well as smoothing methods. Further I thank the unknown referees for their constructive remarks.

References

[1] D. Al-Mutairi, C. Grossmann, and Kim Tuan Vu, *Path-following barrier and penalty methods for linearly constrained problems*, Optimization **48** (2000), no. 3, 353–374, DOI 10.1080/02331930008844510. MR1811752 (2001k:90046)

[2] A. Auslender, R. Cominetti, and M. Haddou, *Asymptotic analysis for penalty and barrier methods in convex and linear programming*, Math. Oper. Res. **22** (1997), no. 1, 43–62, DOI 10.1287/moor.22.1.43. MR1436573 (98e:90112)

[3] A. Ben-Tal and M. Teboulle, *A smoothing technique for nondifferentiable optimization problems*, Optimization (Varetz, 1988), Lecture Notes in Math., vol. 1405, Springer, Berlin, 1989, pp. 1–11, DOI 10.1007/BFb0083582. MR1036540 (91g:49006)

[4] Bertsekas, D.P.: Non-differentiable optimization via approximation, Math. Programming Study, 3 (1973), 1-25.

[5] Dimitri P. Bertsekas, *Necessary and sufficient condition for a penalty method to be exact*, Math. Programming **9** (1975), no. 1, 87–99. MR0384144 (52 #5021)

[6] J. Frédéric Bonnans, J. Charles Gilbert, Claude Lemaréchal, and Claudia A. Sagastizábal, *Numerical optimization*, 2nd ed., Universitext, Springer-Verlag, Berlin, 2006. Theoretical and practical aspects. MR2265882 (2007e:90001)

[7] Christakis Charalambous, *A lower bound for the controlling parameters of the exact penalty functions*, Math. Programming **15** (1978), no. 3, 278–290, DOI 10.1007/BF01609033. MR514613 (80a:90119)

[8] Peter Deuflhard, *Newton methods for nonlinear problems*, Springer Series in Computational Mathematics, vol. 35, Springer-Verlag, Berlin, 2004. Affine invariance and adaptive algorithms. MR2063044 (2005h:65002)

[9] G. Di Pillo and L. Grippo, *An exact penalty function method with global convergence properties for nonlinear programming problems*, Math. Programming **36** (1986), no. 1, 1–18, DOI 10.1007/BF02591986. MR862065 (88h:90187)

[10] Anthony V. Fiacco and Garth P. McCormick, *Nonlinear programming: Sequential unconstrained minimization techniques*, John Wiley and Sons, Inc., New York-London-Sydney, 1968. MR0243831 (39 #5152)

[11] Christian Grossmann, *Penalty/barrier path-following in linearly constrained optimization*, Discuss. Math. Differ. Incl. Control Optim. **20** (2000), no. 1, 7–26, DOI 10.7151/dmdico.1001.

German-Polish Conference on Optimization—Methods and Applications (Żagań, 1999). MR1752567 (2001h:90080)

[12] Christian Großmann and Aleksander A. Kaplan, *Strafmethoden und modifizierte Lagrangefunktionen in der nichtlinearen Optimierung* (German), BSB B. G. Teubner Verlagsgesellschaft, Leipzig, 1979. Teubner-Texte zur Mathematik. [Teubner Texts in Mathematics]; With English, French and Russian summaries. MR581367 (82k:90099)

[13] Christian Großmann and Johannes Terno, *Numerik der Optimierung* (German), Teubner Studienbücher Mathematik. [Teubner Mathematical Textbooks], B. G. Teubner, Stuttgart, 1993. MR1297959 (96h:90001)

[14] C. Grossmann and M. Zadlo, *A general class of penalty/barrier path-following Newton methods for nonlinear programming*, Optimization **54** (2005), no. 2, 161–190, DOI 10.1080/02331930412331326310. MR2132735 (2005m:90162)

[15] Martin Gugat and Michael Herty, *The smoothed-penalty algorithm for state constrained optimal control problems for partial differential equations*, Optim. Methods Softw. **25** (2010), no. 4-6, 573–599, DOI 10.1080/10556780903002750. MR2724157 (2011j:49068)

[16] Martin Gugat and Michael Herty, *A smoothed penalty iteration for state constrained optimal control problems for partial differential equations*, Optimization **62** (2013), no. 3, 379–395, DOI 10.1080/02331934.2011.588230. MR3028186

[17] M. Herty, A. Klar, A. K. Singh, and P. Spellucci, *Smoothed penalty algorithms for optimization of nonlinear models*, Comput. Optim. Appl. **37** (2007), no. 2, 157–176, DOI 10.1007/s10589-007-9011-6. MR2325655 (2008c:90072)

[18] A. A. Kaplan, *Algorithms for convex programming using smoothing of exact penalty functions* (Russian), Sibirsk. Mat. Zh. **23** (1982), no. 4, 53–64, 219. MR668335 (84b:90083)

[19] C. T. Kelley, *Iterative methods for optimization*, Frontiers in Applied Mathematics, vol. 18, Society for Industrial and Applied Mathematics (SIAM), Philadelphia, PA, 1999. MR1678201 (2000h:90003)

[20] Shu-jun Lian, *Smoothing approximation to l_1 exact penalty function for inequality constrained optimization*, Appl. Math. Comput. **219** (2012), no. 6, 3113–3121, DOI 10.1016/j.amc.2012.09.042. MR2992010

[21] Bingzhuang Liu, *On smoothing exact penalty functions for nonlinear constrained optimization problems*, J. Appl. Math. Comput. **30** (2009), no. 1-2, 259–270, DOI 10.1007/s12190-008-0171-z. MR2496616 (2010b:90139)

[22] Qian Liu, Changyu Wang, and Xinmin Yang, *On the convergence of a smoothed penalty algorithm for semi-infinite programming*, Math. Methods Oper. Res. **78** (2013), no. 2, 203–220, DOI 10.1007/s00186-013-0440-y. MR3121011

[23] Yurii Nesterov and Arkadii Nemirovskii, *Interior-point polynomial algorithms in convex programming*, SIAM Studies in Applied Mathematics, vol. 13, Society for Industrial and Applied Mathematics (SIAM), Philadelphia, PA, 1994. MR1258086 (94m:90005)

[24] Mustafa Ç. Pinar and Stavros A. Zenios, *On smoothing exact penalty functions for convex constrained optimization*, SIAM J. Optim. **4** (1994), no. 3, 486–511, DOI 10.1137/0804027. MR1287812 (95d:90051)

[25] Stephen J. Wright, *On the convergence of the Newton/log-barrier method*, Math. Program. **90** (2001), no. 1, Ser. A, 71–100, DOI 10.1007/PL00011421. MR1819787 (2001k:90082)

[26] Z. Y. Wu, H. W. J. Lee, F. S. Bai, and L. S. Zhang, *Quadratic smoothing approximation to l_1 exact penalty function in global optimization*, J. Ind. Manag. Optim. **1** (2005), no. 4, 533–547, DOI 10.3934/jimo.2005.1.533. MR2186624 (2006i:90055)

[27] Willard I. Zangwill, *Non-linear programming via penalty functions*, Management Sci. **13** (1967), 344–358. MR0252040 (40 #5265)

[28] Wenling Zhao and Ranran Li, *A second-order differentiable smoothing approximation lower order exact penalty function*, Appl. Math. Inf. Sci. **6** (2012), no. 2, 275–279. MR2914089

Technische Universität Dresden, Institut für Numerische Mathematik, D-01062 Dresden, Germany

E-mail address: christian.grossmann@tu-dresden.de

Numerical aspects of sonic-boom minimization

Navid Allahverdi, Alejandro Pozo, and Enrique Zuazua

ABSTRACT. The propagation of the sonic-boom produced by supersonic aircrafts can be modeled by means of a nonlocal version of the viscous Burgers equation. In this paper, motivated by the sonic-boom minimization problem, we analyze the numerical aspects of the optimization process, with a focus on the large-time dynamics of the underlying model. We develop an adjoint methodology at the numerical level allowing to recover accurate approximations of the minimizers. We observe however that some of the minima are hard to achieve due to the intrinsic nonlinear and viscous effects.

1. Motivation

1.1. Sonic-boom minimization. When flying above the speed of sound, supersonic airplanes create pressure disturbances in the atmosphere resulting from air compression due to the moving volume of the plane and the aerodynamic lift, which is required to maintain the flight altitude. Some portion of the pressure disturbances, which imply a significant amount of acoustic energy, propagates in the atmosphere and reaches the ground level, resulting in the so-called sonic-boom [2]. Sonic-boom, perceived on the ground as two subsequent loud bangs with a short time lapse in between, is annoying for people living in the sonic-boom carpet under the flight track and may potentially damage building facade and glazing in extreme cases. For all these reasons sonic-boom has been one of the main obstacles when it comes to commercializing supersonic air travels.

Throughout the last decades of the 20th century, it became clear that sonic-boom reduction technologies had to be advanced, in order to make flights of supersonic airplanes a reality. In fact, the DARPA/NASA/Northrop-Grumman Shaped Sonic Boom Demonstrator (SSBD) project [22] showed that the sonic-boom could be partially mitigated by tailoring the shape of the aircraft, confirming it experimentally for the first time.

In this paper, we confine ourselves to the simulation of the propagation of the sonic-boom from the near-field of the plane down to the ground level. We refer the

---

2010 *Mathematics Subject Classification.* Primary 49M25, 35Q35, 35B40.

This research is supported by the Advanced Grant NUMERIWAVES/FP7-246775 of the European Research Council Executive Agency, FA9550-14-1-0214 of the EOARD-AFOSR, FA9550-15-1-0027 of AFOSR, the BERC 2014-2017 program of the Basque Government, the MTM2011-29306 and SEV-2013-0323 Grants of the MINECO and a Humboldt Award at the University of Erlangen-Nürnberg. A. Pozo is also supported by the Basque Government (PREDOC Program 2012, 2013 and 2014).

reader to [2] for a detailed review on the state of the art regarding the complete sonic-boom minimization problem, including shape optimization of the aircraft as well.

It is well known that the pressure signature evolves into an N-wave (see Figure 1) if the time of propagation is long enough. The N-wave refers to the shape that results from the collapse of the multiple shocks into a leading and a trailing shock separated by a nearly linear pressure expansion, perceived by humans as the aforementioned sonic-boom. At the beginning, it was thought that the N-wave was unavoidable. But McLean proved theoretically that it was possible to tailor the near-field signature of the aircraft so that the N-wave was not developed by the time the signature had reached the ground level [19], thus, mitigating the sonic-boom. Nevertheless, as mentioned above, this was not empirically verified until the SSBD tests was done in 2003.

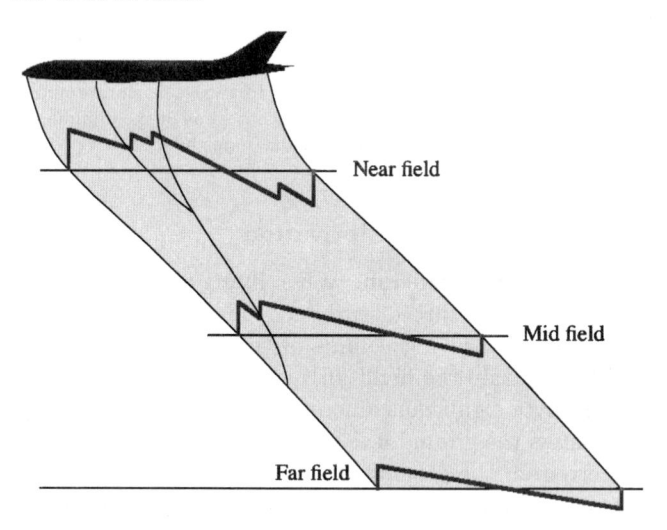

FIGURE 1. Diagram of the propagation of the sonic-boom. Shocks created in the near-field collapse into two single shocks, forming an N-wave in the far-field.

The high cost of realistic experiments like the SSBD project promoted the development of theoretical and computational tools that could handle the simulation of sonic-boom minimization problem more efficiently. Historically, linear theory was used to model the sonic-boom propagation, following the seminal works done by Hayes [9] and Whitham [30] in the late 1940s and early 1950s and, until recently, most of the analytical and numerical research done followed the pioneering classical work of Jones-Seebass-George-Darden theory for sonic-boom minimization [6, 15, 26–28].

More recently, new nonlinear physical models have been utilized to improve the sonic-boom propagation from the near-field to the ground. Two such methods are worth to mention. One approach, proposed by Ozcer [21], advocates the use of the full potential equation in the region from the near-field to the ground. The other approach, adopted initially by Cleveland [5] and then by Rallabhandi [24], uses an augmented Burgers equation within the context of ray-tracing/geometrical acoustics to propagate the source signatures to the ground, including viscous effects that lead to shock discontinuities with non-zero thickness. The models based on the latter approach arise from the theory of ray propagation in a moving medium [16, 23, 29].

1.2. The augmented Burgers equation (ABE). In this paper we focus on the latter approach. The augmented Burgers equation –or extended Burgers equation [5]– takes into account nonlinear effects, such as waveform steepening and variable-speed wave propagation, as well as molecular relaxation phenomena, ray tube spreading and atmospheric stratification.

The equation is given by:

$$(1.1) \quad \frac{\partial P}{\partial \sigma} = P \frac{\partial P}{\partial \tau} + \frac{1}{\Gamma} \frac{\partial^2 P}{\partial \tau^2} + \sum_\nu C_\nu \frac{1}{1+\theta_\nu \frac{\partial}{\partial \tau}} \frac{\partial^2 P}{\partial \tau^2} - \frac{1}{2G} \frac{\partial G}{\partial \sigma} P + \frac{1}{2\rho_0 c_0} \frac{\partial (\rho_0 c_0)}{\partial \sigma} P,$$

were $P = P(\sigma, \tau)$ is the dimensionless perturbation of the pressure distribution. The distance σ and time τ of the perturbation are also dimensionless. The operator appearing in the summation corresponds to the molecular relaxation, and it is equivalent to the following integral transform:

$$\frac{1}{1+\theta_\nu \frac{\partial}{\partial \tau}} f(\tau) = \frac{1}{\theta_\nu} \int_{-\infty}^{\tau} e^{(\xi-\tau)/\theta_\nu} f(\xi) d\xi,$$

with $f = \frac{\partial^2 P}{\partial \tau^2}$, in this case. Function $G(\sigma)$ denotes the ray-tube area, which can be evaluated assuming cylindrical or spherical wave propagation. The atmosphere conditions are given by density ρ_0, and speed of sound c_0, both being a function of the altitude. Other parameters include a dimensionless thermo-viscous parameter Γ, a dimensionless relaxation time θ_ν and dispersion parameter C_ν for each relaxation mode. Typically, there are two relaxation modes, one corresponding to the relaxation occurred in Oxygen molecules and the other one in Nitrogen ones. We refer to Section 4 for more details on the parameters used in the model.

The sonic-boom minimization problem we address here consists of, roughly, given a desired ground signature P^* and the distance of the propagation Σ –which is closely related to the altitude of the flight–, recovering the near-field signature that reproduces P^* as best as possible with regard to a norm. This is commonly formulated as an optimal control problem, through a least square approach [25], as follows:

$$(1.2) \quad \min_{P_0} \mathcal{S}(P_0) = \min_{P_0} \frac{1}{2} \int_{\mathbb{R}} \left(P(\Sigma, \tau) - P^*(\tau) \right)^2 d\tau.$$

Here P_0 lies in a set of admissible near-field signatures –which is usually constrained by the design variables associated to the geometry of the aircraft [20, 25]– and P is the solution of (1.1) with $P(\sigma_0, \tau) = P_0(\tau)$ for all $\tau \in \mathbb{R}$, where σ_0 would be the distance of the measure point in the near-field to the plane. The perceived loudness (PLdB) or the shock over-pressures are other functionals that are usually considered for minimization too [20].

1.3. Contents of the paper. The sonic-boom minimization problem involves various different time scales: the perturbation of the pressure takes place for less than half a second, while the propagation can last up to a minute, depending on the flight conditions [24]. As we shall see, this makes the computational treatment a hard task and motivates taking into account the large-time asymptotic behavior of the employed numerical schemes because of its possible impact on the minimization process.

The connection between large-time behavior and the efficiency of the numerical minimization tools has been the object of earlier investigation. For instance, in [1], in the context of Burgers equation, it is shown that numerical schemes that do not behave correctly as time tends to infinity, may produce significant errors in the computation of minimizers for long-time horizon control problems.

Several analytical and numerical important issues arise in the context of the sonic-boom minimization problem. The large-time dynamics of solutions and its numerical counterparts is one of them. As we shall see, the asymptotic profile of solutions as $\sigma \to \infty$ is a gaussian, which is a manifestation of an asymptotic simplification towards a linear dynamics. But this is compatible with the intermediate asymptotics, given by the N-wave. In [17], the authors show that, for the viscous Burgers equation, the N-wave endures for a time period of order $O(1/\varepsilon)$, where ε is the viscosity coefficient. They also analyze the transition from the N-wave metastable state to the asymptotic one, which does not change sign. Both empirical and numerical simulations show that the propagation of the sonic boom, and in particular the augmented Burgers equation, exhibits a similar behavior, without contradicting our asymptotic result –the fact that N-wave arises and remains, but it would also disappear eventually if extending the atmosphere and propagation distance *large enough*.

The rest of the paper is structured as follows. In Section 2 we describe the numerical methods that we use to solve (1.1)-(1.2). Then, in Section 3 we elaborate on some of the numerical difficulties one faces when dealing with optimal control problems for Burgers-like equations. Finally, in Section 4, we present a numerical simulation of the sonic-boom minimization problem (1.1)-(1.2) using realistic parameters.

2. Discretization of the augmented Burgers equation

For the numerical approximation of (1.1), we follow [5, 25], using a splitting method for each of the operators appearing in the equation. However, in our case we select Engquist-Osher for the nonlinear term, which has been shown to be appropriate for large-time evolution problems [13]. With respect to the optimization method, we opt for a conjugate gradient descent method. In particular, we calculate the gradient based on the adjoint methodology via optimize-then-discretize approach. Let us remark that we pay special attention to the discretization and settings of the algorithm, as detailed in [1].

2.1. Optimization. In order to solve the optimal control problem (1.1)-(1.2), we resort to gradient-based techniques; in particular, to a conjugate gradient descent method (CG). We evaluate the gradient of the objective functional in (1.2) by means of the adjoint variable $Q(\sigma, \tau)$, which is governed by:

$$(2.1) \quad -\frac{\partial Q}{\partial \sigma} = -P\frac{\partial Q}{\partial \tau} + \frac{1}{\Gamma}\frac{\partial^2 Q}{\partial \tau^2} + \sum_\nu \frac{C_\nu}{1-\theta_\nu \frac{\partial}{\partial \tau}}\frac{\partial^2 Q}{\partial \tau^2} - \frac{1}{2G}\frac{\partial G}{\partial \sigma}Q + \frac{1}{2\rho_0 c_0}\frac{\partial(\rho_0 c_0)}{\partial \sigma}Q,$$

The adjoint equation is solved backward in σ, starting from the final propagation distance Σ to the initial distance σ_0. The initial condition provided to (2.1) is

$$Q(\Sigma, \tau) = P(\Sigma, \tau) - P^*(\tau).$$

In that case, $Q_0(\tau) = Q(\sigma_0, \tau)$ provides the gradient of the objective function in (1.2) with respect to $P_0(\tau)$.

2.2. Operator splitting. In solving forward augmented Burgers equation (1.1), or its corresponding adjoint equation (2.1), we employ operator splitting techniques. In this type of methods, the equation is decomposed into simpler subproblems corresponding to different terms generating the overall dynamics; each of them being handled independently by an appropriate discretization and numerical integration scheme. We refer readers to [10, 14] for the basic theory of operator splitting techniques.

For solving the diffusion subproblem, a central difference discretization is used, together with Crank-Nicolson integration:

$$\frac{P_j^{n+\frac{1}{5}} - P_j^n}{\Delta\sigma} = \frac{1}{\Gamma}\left(\frac{\alpha}{\Delta\tau^2}(P_{j-1}^n - 2P_j^n + P_{j+1}^n) + \frac{1-\alpha}{\Delta\tau^2}(P_{j-1}^{n+\frac{1}{5}} - 2P_j^{n+\frac{1}{5}} + P_{j+1}^{n+\frac{1}{5}})\right),$$

where $\alpha \in [0,1]$. In our case, we take $\alpha = 1/2$.

Next, we adopt Engquist-Osher's scheme for discretizing the nonlinear term in (1.1), in combination with explicit Euler integration:

$$\frac{P_j^{n+\frac{2}{5}} - P_j^{n+\frac{1}{5}}}{\Delta\sigma} = \frac{1}{4\Delta\tau}\left(P_j^{n+\frac{1}{5}}(P_j^{n+\frac{1}{5}} - |P_j^{n+\frac{1}{5}}|) + P_{j+1}^{n+\frac{1}{5}}(P_{j+1}^{n+\frac{1}{5}} + |P_{j+1}^{n+\frac{1}{5}}|) \right.$$
$$\left. - P_{j-1}^{n+\frac{1}{5}}(P_{j-1}^{n+\frac{1}{5}} - |P_{j-1}^{n+\frac{1}{5}}|) - P_j^{n+\frac{1}{5}}(P_j^{n+\frac{1}{5}} + |P_j^{n+\frac{1}{5}}|)\right).$$

Each of the relaxation terms is discretized independently using central differences, along with Crank-Nicolson integration:

$$\frac{P_j^{n+\frac{3}{5}} - P_j^{n+\frac{2}{5}}}{\Delta\sigma} + \frac{\theta_{\nu_O}}{\Delta\tau}\left(\frac{P_{j+1}^{n+\frac{3}{5}} - P_{j+1}^{n+\frac{2}{5}}}{\Delta\sigma} - \frac{P_{j-1}^{n+\frac{3}{5}} - P_{j-1}^{n+\frac{2}{5}}}{\Delta\sigma}\right)$$
$$= C_{\nu_O}\left(\frac{\alpha}{\Delta\tau^2}(P_{j-1}^{n+\frac{2}{5}} - 2P_j^{n+\frac{2}{5}} + P_{j+1}^{n+\frac{2}{5}}) + \frac{1-\alpha}{\Delta\tau^2}(P_{j-1}^{n+\frac{3}{5}} - 2P_j^{n+\frac{3}{5}} + P_{j+1}^{n+\frac{3}{5}})\right),$$

and

$$\frac{P_j^{n+\frac{4}{5}} - P_j^{n+\frac{3}{5}}}{\Delta\sigma} + \frac{\theta_{\nu_N}}{\Delta\tau}\left(\frac{P_{j+1}^{n+\frac{4}{5}} - P_{j+1}^{n+\frac{3}{5}}}{\Delta\sigma} - \frac{P_{j-1}^{n+\frac{4}{5}} - P_{j-1}^{n+\frac{3}{5}}}{\Delta\sigma}\right)$$
$$= C_{\nu_N}\left(\frac{\alpha}{\Delta\tau^2}(P_{j-1}^{n+\frac{3}{5}} - 2P_j^{n+\frac{3}{5}} + P_{j+1}^{n+\frac{3}{5}}) + \frac{1-\alpha}{\Delta\tau^2}(P_{j-1}^{n+\frac{4}{5}} - 2P_j^{n+\frac{4}{5}} + P_{j+1}^{n+\frac{4}{5}})\right),$$

where subindexes O and N stand for Oxygen and Nitrogen respectively. In both cases we choose $\alpha = 1/2$.

For the last two terms in (1.1), leading to ordinary differential equations, we simply perform analytical integration:

$$(2.2) \qquad P_j^{n+1} = k^n P_j^{n+\frac{4}{5}},$$

where k^n is the scaling factor due to the ray-tube spreading and atmospheric stratifications [5], given by

$$k^n = \sqrt{\frac{G(\sigma_n)}{G(\sigma_{n+1})}}\sqrt{\frac{\rho(\sigma_{n+1})c_0(\sigma_{n+1})}{\rho(\sigma_n)c_0(\sigma_n)}}.$$

For solving the adjoint equation in (2.1), we adopt the same splitting process. Discretizing the adjoint equation is done the same way as the forward equation except for the linear convective term. Proper flux differencing was used to follow the propagation direction in the convective term.

3. Critical issues for the discrete methods

Optimal control problems like (1.1)-(1.2) are well-known to be extremely challenging from a numerical point of view. In fact, the nonlinearity is the dominant term and (1.1) exhibits a hyperbolic-like dynamics for large times, as we shall see.

3.1. Large-time dynamics.
Motivated by the different time scales appearing in the dynamics of the sonic-boom phenomenon, in this section we analyze the large-time behavior of the augmented Burgers equation (1.1).

Note that equation (1.1) fits in the following type of equations:

$$(3.1) \quad \begin{cases} u_t = uu_x + \varepsilon u_{xx} + \sum_j c_j(K_{\eta_j} * u_{xx}) + (H(t))_t \, u, & (t,x) \in (0,\infty) \times \mathbb{R}, \\ u(0,x) = u_0(x), & x \in \mathbb{R}, \end{cases}$$

where

$$K_\eta(z) = \begin{cases} \dfrac{1}{\eta} e^{-z/\eta}, & z > 0, \\ 0, & \text{elsewhere.} \end{cases}$$

In fact, it is is enough to take $t = \Gamma^2(\sigma - \sigma_0)$, $x = \Gamma\tau$ and

$$P(\sigma,\tau) = \Gamma\, u(\Gamma^2(\sigma - \sigma_0), \Gamma\tau)$$

in (3.1), as well as $\varepsilon = 1/\Gamma$, $c_j = C_\nu$ and $\eta_j = \Gamma\theta_\nu$. Moreover, in our case, function H would be given by

$$H(t) = \ln \sqrt{\frac{\rho_0(\frac{t}{\Gamma^2} + \sigma_0)\, c_0(\frac{t}{\Gamma^2} + \sigma_0)}{G(\frac{t}{\Gamma^2} + \sigma_0)}}.$$

To analyze the large-time dynamics of these models, we remark that the change of variables $v = ue^{-H}$ transforms (3.1) into

$$(3.2) \quad \begin{cases} v_t = e^H vv_x + \varepsilon v_{xx} + \sum_j c_j(K_{\eta_j} * v_{xx}), & (t,x) \in (0,\infty) \times \mathbb{R}, \\ v(0,x) = v_0(x) = e^{-H(0)}u_0(x), & x \in \mathbb{R}. \end{cases}$$

A similar equation, with a constant coefficient as multiplicative factor of the nonlinearity, was already addressed in [11] where the authors proved the well-posedness of the Cauchy problem for L^1 initial data and obtained its large-time behavior.

The same arguments can be applied to (3.2) and, hence, also to (3.1). However, to do that, further assumptions are required regarding functions ρ_0, c_0 and G. If we assume that ρ_0 and c_0 are bounded from below and from above and that G corresponds to cylindrical or spherical waves (thus, $G_t/G = r/t$ with $r = 1/2$ for cylindrical waves and $r = 1$ for spherical ones [5]), we can prove the following result.

THEOREM 3.1. *For any initial data $u_0 \in L^1(\mathbb{R})$, there exists a unique solution $u \in C([0,\infty), L^1(\mathbb{R}))$ of (3.1). Moreover, for any $p \in [1,\infty]$, u satisfies*

$$e^{-H(t)} t^{\frac{1}{2}(1-\frac{1}{p})} \|u(t) - e^{H(t)} v_M(t)\|_p \longrightarrow 0, \quad \text{as } t \to \infty,$$

where $v_M(t,x)$ is the solution of the following heat equation:

$$(3.3) \quad \begin{cases} v_t = (\varepsilon + \sum_j c_j) v_{xx}, & x \in \mathbb{R}, t > 0, \\ v(0) = M\delta_0. \end{cases}$$

Here δ_0 denotes the Dirac delta at the origin and M is the mass of the initial data, $M = \int_{\mathbb{R}} v_0(x) dx$.

SKETCH OF THE PROOF. The key point is that H is bounded from above for all $t \geq 0$ and that it goes to $-\infty$ as $t \to \infty$. Therefore, all the estimates obtained in [11] are applicable to this case. For the same reason, we can also apply the scaling procedure to obtain the asymptotic profile of (3.2). The only difference now is that the nonlinear term vanishes, since $e^H \to 0$ as $t \to \infty$. Thus, the equation for the first term in the asymptotic expansion is the heat equation. Moreover, reversing the change of variable $v = ue^{-H}$, we can obtain the large-time behavior of the solution u to (3.1). □

It is important to emphasize that, due to the range of parameters entering in (1.1) (e.g. [5, 24, 25]), the diffusive profile v_M arising from (3.3) will not be achieved, the relevant time-horizon not being large enough, in realistic situations, in which N-waves are usually obtained. The coexistence of these two phenomena, N-waves for intermediate times and viscous waves for large times, was observed and analyzed in [17] in the context of the viscous Burgers equation

$$w_t = ww_x + \varepsilon w_{xx},$$

where it was observed that N-waves are intermediate metastable states that endure until $t = O(1/\varepsilon)$. Numerical experiments for (1.1) show that the behavior of its solutions is similar, being dominated by the nonlinear effects. As a matter of fact, Theorem 3.1 states that the molecular relaxation terms act as additional viscous terms as $t \to \infty$.

As we mentioned above, the range of parameters for (1.1) in the optimization problem under consideration, leads to a hyperbolic-like dynamics, while the other terms are still present. Thus, there are two key points that need to be taken into account at the numerical level:

- The nonlinear flux has to be discretized appropriately. Numerical fluxes that introduce too much numerical viscosity will affect, not only the accuracy of the forward problem, but the efficiency of optimization algorithms too (e.g. [1, 11, 13]).
- The molecular relaxation term behaves, as time tends to infinity, as an additional viscous term. Moreover, being a nonlocal operator, it has to be treated carefully to avoid undesired large-time effects [11, 12].

3.2. Drawbacks of the adjoint methodology and the least squares functional (L^2-norm).
In the context of the inviscid Burgers equation it is well known that (1.2) has multiple minimizers (see [3]). For instance, all three different initial conditions shown in Figure 2 give rise to the same solution at final time T, where a shock is present. In [1], the authors show that, among the various possible minimizers, the classical adjoint methodology tends to recover a compression wave in which the shock will not appear until the final time (Figure 2, top). The alternating descent method proposed in [3] leads to initial data that create the shock at the initial time (Figure 2, middle). Other strategies can also be defined so to have a shock arising at some intermediate instant (Figure 2, bottom).

In the viscous case (and also for (1.1)), the equation under consideration enjoys the property of backward uniqueness [8] and, therefore, the lack of uniqueness for the optimization problem is no longer valid. Nevertheless, when the viscosity

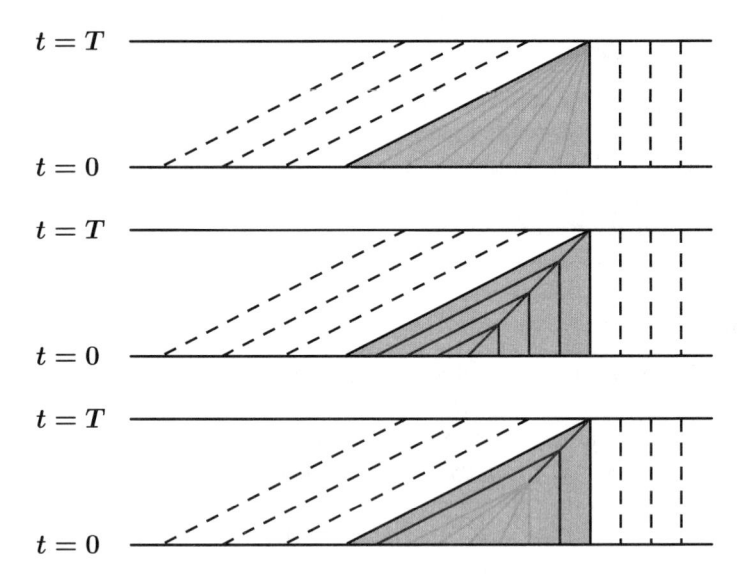

FIGURE 2. Different initial conditions which create the same target function with a shock at final time, T, for the inviscid Burgers equation: from a compression wave (top), from a shock at the initial data (middle) and from a shock that arises sometime in between (bottom).

is small, at the numerical level, non-uniqueness issues still arise. Furthermore, the asymptotic simplification properties of Burgers-type equations for large times reduce the sensitivity with respect to perturbations of the initial data. In Figure 3 we show two different initial data that lead to very similar profiles (up to an order of 10^{-6} in the L^2-norm of their distance). The smooth one has been obtained by means of an adjoint-based conjugate gradient method, following a discretize-then-optimize approach. Thus, the adjoint methodology is biased in preferring the smooth initial datum over some other possible minimizers.

On the other hand, the strong time-irreversibility of viscous systems and the difficulty of recovering high-frequencies produces a second phenomenon that needs to be handled. In the previous example we showed that two different initial data –the smooth one and the piecewise linear one– can produce very similar profiles. When using the adjoint method to compute the gradient of the functional, the viscous effects are enhanced, since numerical viscosity is also present when solving the equation backwards. Thus, it is computationally expensive to recover initial data with peaks.

This pathology is very clear for the heat equation, where the influence of the nonlinearity is put aside. In Figure 4 we can observe that a target function that appears from a piecewise linear function is well reproduced by a smoother initial data. The latter has been computed using an adjoint-based conjugate gradient method, within a discretize-then-optimize approach.

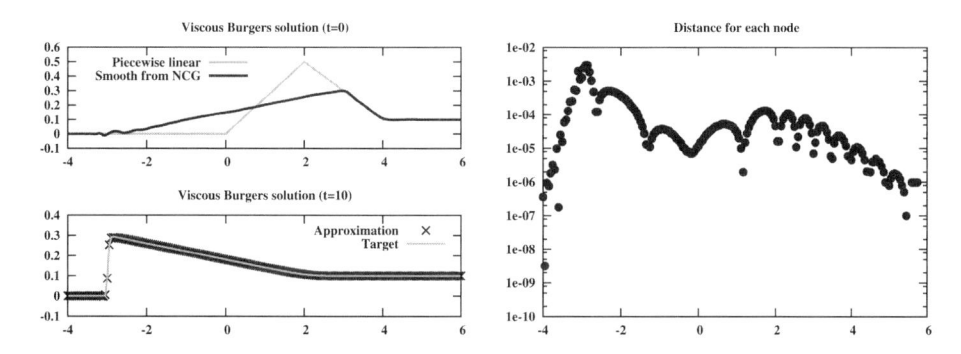

FIGURE 3. On the left, solutions of the viscous Burgers equation $u_t = uu_x + \nu u_{xx}$ with $\nu = 10^{-4}$ (bottom) for two different initial data (top). On the right, the L^2-norm of the difference of both solutions.

In short, the presence of viscosity in the model desensitizes the least square objective functional to the presence of high frequency components in the initial data. This insensitivity or flatness of the objective functional with respect to high frequency components makes the optimization algorithm to be inefficient when trying to achieve singular minimizers.

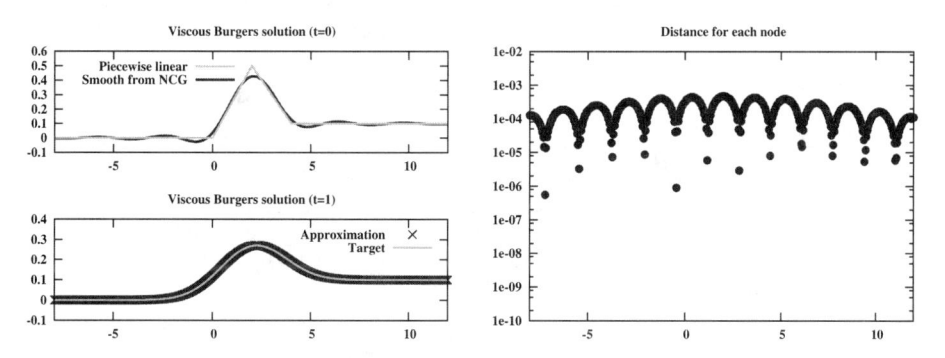

FIGURE 4. On the left, solutions of the heat equation $u_t = u_{xx}$ (bottom) for two different initial data (top). On the right, the L^2-norm of the difference of both solutions.

As we shall see in the next section, the problem of minimizing the sonic-boom within the class of F-function initial data exhibits similar symptoms.

4. Numerical results

In this section we present the numerical simulation of the sonic-boom minimization problem (1.1)-(1.2). The optimization problem consists of searching for a near-field pressure that gives rise to a prescribed far-field target. In doing so, we pick a target function which has not developed into a "*perfect*" N-wave at the front (nose) shock – see Figure 6-right for the target function. This target function was used by Darden [7] to optimize the nose of supersonic planes for nose bluntness via using modified linear theory based on the parameterized F-function. We attempt to solve the same minimization problem using augmented Burger's equation (ABE) for the sonic boom propagation. In addition, we do not make any assumption on the form of the near-field pressure such as a priori known parameterized F-function or whatsoever. We give a brief summary on F-function in next paragraphs for completeness. We only use F-function as parameterized by Jones-Seebass-George-Darden theory to construct the near-field pressure and initial condition for ABE in obtaining the target function. Our optimization procedure does not use F-function, or its parameterized form by any means.

From practical point of view, the near-filed pressure depends on the geometry of the airplane. One of the common ways of representing the near-field pressure is using the F-function theory developed by Whitham [30] . This theory stipulates that the near-field pressure can be approximated using F-functions, the Abel transform of the cross-sectional area of the plane, $A(x)$:

$$F(x) = \frac{1}{2\pi} \int_0^x \frac{A''(\xi)}{\sqrt{x-\xi}}.$$

In F-function theory, the airplane's geometry is considered as a body of revolution along x-axis. Despite all inherent simplifications, the F-function theory provides a straightforward approach to identify the near-field when the geometry is known or, inversely, to approximate the optimal geometry when optimal near-field is given.

We construct the F-function based on reference test case provided by Darden. We refer readers to [7, 20] for detailed information on near-field calculation. In Table 1, we summarized the main parameters used to obtain the initial near-field pressure signature.

Mach number, M	2.7
Body length, L	91.5 m
Flight altitude, z	18288 m
Near-field distance measured from airplane, $r = 2L$	183 m

TABLE 1. Flight parameters to compute the near-field pressure from F-function (taken from [7, 20]).

Regarding the physical parameters present in the ABE equation, we take $\Gamma = 8 \times 10^6$. The relaxation parameters are shown in Table 2. Density and speed of sound vary as a function of altitude. In Figure 5, we show the variation of density and speed of sound in the atmosphere. Moreover, we consider cylindrical ray-tube spreading [5], which implies that $G_\sigma/G = 1/(2\sigma)$. Note that the initial signature is measured at some distance from the plane and, thus, $\sigma_0 > 0$.

	Oxygen	Nitrogen
C_ν	1.7×10^{-5}	1.2×10^{-4}
θ_ν	4.6×10^{-8}	1.6×10^{-6}

TABLE 2. Parameters used for molecular relaxation (taken from [5, 24, 25]).

The minimizer we obtain is shown in Figure 6, as well as its corresponding signature at the ground level, in comparison with the exact solution. Even though,

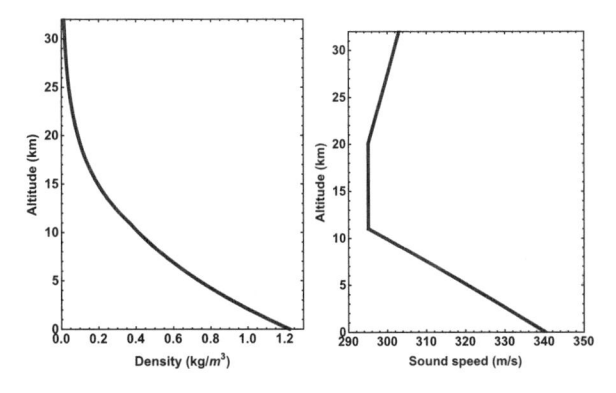

FIGURE 5. Density of the air (left) and sound speed (right) of the atmosphere [5].

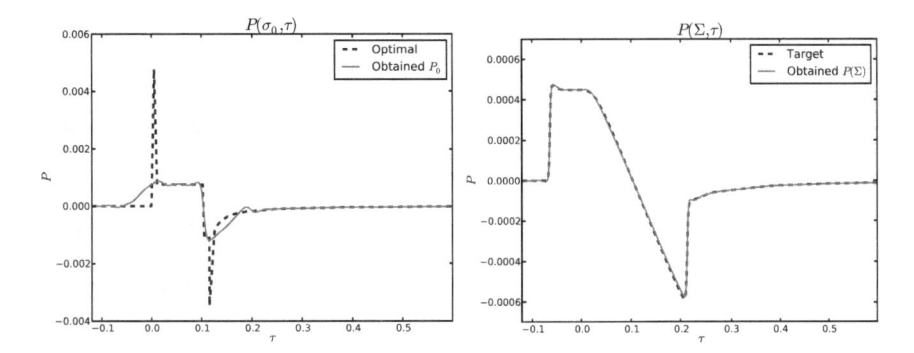

FIGURE 6. The solid green lines are the obtained minimizer (left) and its corresponding solution at the ground level (right). We compare them with the a-priori known optimal initial data and the corresponding target function it produces (blue dashed lines).

the final profile is closely matching the target function, the initial data are different. The obtained initial data lacks the peaks present in the original F-function. In other words, the L^2-norm used in the objective functional is not sensitive to high frequency modes in the near-field. This is due to the presence of viscosity in the model and the large-time evolution. We also observe the selective nature of the adjoint methodology when dealing with multiple solutions for inverse problems and its bias towards selecting smooth initial solutions.

5. Conclusions and future research

In this paper we implemented a least square approach to recover an admissible ground signature mitigating the sonic-boom. From a mathematical point of view, the accuracy of the obtained results is sufficient, as we were able to reproduce the target function with a very high precision level. Moreover, we were able to handle efficiently the long time horizon that the propagation of the sonic-boom involves, which requires preserving the large-time asymptotic dynamics.

Nevertheless, further developments are required to obtain more satisfactory results in solving the sonic-boom minimization problem. In short, we identify two major drawbacks in our approach. Namely, the objective functional is highly insensitive to high frequency modes, which is, indeed, exacerbated by the biased nature of the adjoint methodology. The preference for smooth initial solutions accentuates the difficulty of recovering F-functions as initial data both due to the diffusive term and the nonlinearity in the model.

One possible way of bypassing these difficulties would be to consider classes of parameterized initial data of F-function type, in contrast with the present work where we did not impose any restriction on the near-field signature. Another added remedy would be to consider functionals capable of reproducing not only the ground signature, but also intermediate states.

References

[1] N. Allahverdi, A. Pozo, and E. Zuazua, Numerical aspects of large-time optimal control of Burgers equation. *ESAIM: Mathematical Modelling and Numerical Analysis* (accepted).

[2] J. J. Alonso and M. R. Colonno, *Multidisciplinary optimization with applications to sonic-boom minimization*, Annual review of fluid mechanics. Volume 44, 2012, Annu. Rev. Fluid Mech., vol. 44, Annual Reviews, Palo Alto, CA, 2012, pp. 505–526, DOI 10.1146/annurev-fluid-120710-101133. MR2882607 (2012m:76076)

[3] C. Castro, F. Palacios, and E. Zuazua, *An alternating descent method for the optimal control of the inviscid Burgers equation in the presence of shocks*, Math. Models Methods Appl. Sci. **18** (2008), no. 3, 369–416, DOI 10.1142/S0218202508002723. MR2397976 (2009c:35392)

[4] C. Castro, F. Palacios, and E. Zuazua, *Optimal control and vanishing viscosity for the Burgers equation*, Integral methods in science and engineering. Vol. 2, Birkhäuser Boston, Inc., Boston, MA, 2010, pp. 65–90, DOI 10.1007/978-0-8176-4897-8_7. MR2663150

[5] R. O. Cleveland, Propagation of sonic booms through a real, stratified atmosphere. *PhD thesis*, University of Texas at Austin (1995).

[6] C. Darden, Sonic boom theory: its status in prediction and minimization. *Journal of Aircraft* **14** (1977), 569–576.

[7] C. Darden, Sonic-Boom Minimization With Nose-Bluntness Relaxation. *NASA TP-1348* (1979).

[8] C. Fabre, J.-P. Puel, and E. Zuazua, *On the density of the range of the semigroup for semilinear heat equations*, Control and optimal design of distributed parameter systems (Minneapolis, MN, 1992), IMA Vol. Math. Appl., vol. 70, Springer, New York, 1995, pp. 73–91, DOI 10.1007/978-1-4613-8460-1_4. MR1345629 (97a:35097)

[9] W. D. Hayes, Linearized Supersonic Flow. *PhD thesis*, California Institute of Technology, Pasadena, California (1947).

[10] H. Holden, K. H. Karlsen, K.-A. Lie, and N. H. Risebro, *Splitting methods for partial differential equations with rough solutions*, EMS Series of Lectures in Mathematics, European Mathematical Society (EMS), Zürich, 2010. Analysis and MATLAB programs. MR2662342 (2011j:65002)

[11] L. I. Ignat and A. Pozo, A semi-discrete large-time behavior preserving scheme for the augmented Burgers equation. *Submitted* (2014).

[12] L. I. Ignat and A. Pozo, A splitting method for the augmented Burgers equation. *Submitted* (2015).

[13] L. I. Ignat and E. Zuazua, *Asymptotic expansions for anisotropic heat kernels*, J. Evol. Equ. **13** (2013), no. 1, 1–20, DOI 10.1007/s00028-012-0166-y. MR3020134

[14] A. Iserles, *A first course in the numerical analysis of differential equations*, Cambridge Texts in Applied Mathematics, Cambridge University Press, Cambridge, 1996. MR1384977 (97m:65003)

[15] L. B. Jones, Lower bounds for sonic bangs. *Journal of the Royal Aeronautical Society* **65**, 606 (1961), 433–436.

[16] A. Loubeau and F. Coulouvrat, Effects of Meteorological Variability on Sonic Boom Propagation from Hypersonic Aircraft. *AIAA Journal 47*, 11 (2009), 2632–2641.

[17] Y. J. Kim and A. E. Tzavaras, *Diffusive N-waves and metastability in the Burgers equation*, SIAM J. Math. Anal. **33** (2001), no. 3, 607–633 (electronic), DOI 10.1137/S0036141000380516. MR1871412 (2002i:35121)

[18] J.-L. Lions and B. Malgrange, *Sur l'unicité rétrograde dans les problèmes mixtes paraboliques* (French), Math. Scand. **8** (1960), 277–286. MR0140855 (25 #4269)

[19] F. E. McLean, Some nonasymptotic effects on the sonic boom of large aircraft. *Tech. Rep. D-2877*, NASA (June 1965).

[20] A. Minelli, I. S. el Din, and G. Carrier, Advanced optimization approach for supersonic low-boom design. *18th AIAA/CEAS Aeroacoustics Conference (33rd AIAA Aeroacoustics Conference)* (Colorado Springs, 2012).

[21] I. Ozcer, Sonic boom prediction using Euler/full potential methodology. *45th AIAA Aerospace Sciences Meeting and Exhibit*, Aerospace Sciences Meetings, American Institute of Aeronautics and Astronautics (2007).

[22] J. W. Pawlowski, D. H. Graham, C. H. Boccadoro, P. G. Coen, and D. J. Maglieri, Origins and overview of the shaped sonic boom demonstration program. *Paper 2005-5*, American Institute of Aeronautics and Astronautics (2005).

[23] A. D. Pierce, Acoustics: an introduction to its physical principles and applications. Acoustical Society of America (1989).

[24] S. K. Rallabhandi, Advanced sonic boom prediction using augmented Burger's equation. *Journal of Aircraft* **48**, 4 (2011), 1245–1253.

[25] S. K. Rallabhandi, Sonic boom adjoint methodology and its applications. In *29th AIAA Applied Aerodynamics Conference*, American Institute of Aeronautics and Astronautics (June 2011).

[26] R. Seebass, Minimum sonic boom shock strengths and overpressures. *Nature* **221** (1969), 651–653.

[27] R. Seebass, Sonic boom theory. *Journal of Aircraft* **6** (1969), 177–184.

[28] R. Seebass and A. R. George, Sonic-boom minimization. *Journal of the Acoustical Society of America* **51** (1972), 686–694.

[29] P. A. Thompson, Compressible-fluid dynamics. *Advanced Engineering Series* (February 1972).

[30] G. B. Whitham, *The flow pattern of a supersonic projectile*, Comm. Pure Appl. Math. **5** (1952), 301–348. MR0050442 (14,330d)

NEW YORK CITY COLLEGE OF TECHNOLOGY
E-mail address: nh8@njit.edu, nhajiallahverdipur@citytech.cuny.edu

CBT - INNOVALIA, CARRETERA DE ASÚA, 6, 48930 LAS ARENAS, GETXO, SPAIN
E-mail address: alejandropozo@gmail.com

DEPARTAMENTO DE MATEMÁTICAS, UNIVERSIDAD AUTÓNOMA DE MADRID, 28049 MADRID, SPAIN
E-mail address: enrique.zuazua@unam.es

Selected Published Titles in This Series

658 Carlos M. da Fonseca, Dinh Van Huynh, Steve Kirkland, and Vu Kim Tuan, Editors, A Panorama of Mathematics: Pure and Applied, 2016

655 A. C. Cojocaru, C. David, and F. Pappalardi, Editors, SCHOLAR—a Scientific Celebration Highlighting Open Lines of Arithmetic Research, 2015

654 Carlo Gasbarri, Steven Lu, Mike Roth, and Yuri Tschinkel, Editors, Rational Points, Rational Curves, and Entire Holomorphic Curves on Projective Varieties, 2015

653 Mark L. Agranovsky, Matania Ben-Artzi, Greg Galloway, Lavi Karp, Dmitry Khavinson, Simeon Reich, Gilbert Weinstein, and Lawrence Zalcman, Editors, Complex Analysis and Dynamical Systems VI: Part 1: PDE, Differential Geometry, Radon Transform, 2015

652 Marina Avitabile, Jörg Feldvoss, and Thomas Weigel, Editors, Lie Algebras and Related Topics, 2015

651 Anton Dzhamay, Kenichi Maruno, and Christopher M. Ormerod, Editors, Algebraic and Analytic Aspects of Integrable Systems and Painlevé Equations, 2015

650 Jens G. Christensen, Susanna Dann, Azita Mayeli, and Gestur Ólafsson, Editors, Trends in Harmonic Analysis and Its Applications, 2015

649 Fernando Chamizo, Jordi Guàrdia, Antonio Rojas-León, and José María Tornero, Editors, Trends in Number Theory, 2015

648 Luis Álvarez-Cónsul, José Ignacio Burgos-Gil, and Kurusch Ebrahimi-Fard, Editors, Feynman Amplitudes, Periods and Motives, 2015

647 Gary Kennedy, Mirel Caibăr, Ana-Maria Castravet, and Emanuele Macrì, Editors, Hodge Theory and Classical Algebraic Geometry, 2015

646 Weiping Li and Shihshu Walter Wei, Editors, Geometry and Topology of Submanifolds and Currents, 2015

645 Krzysztof Jarosz, Editor, Function Spaces in Analysis, 2015

644 Paul M. N. Feehan, Jian Song, Ben Weinkove, and Richard A. Wentworth, Editors, Analysis, Complex Geometry, and Mathematical Physics, 2015

643 Tony Pantev, Carlos Simpson, Bertrand Toën, Michel Vaquié, and Gabriele Vezzosi, Editors, Stacks and Categories in Geometry, Topology, and Algebra, 2015

642 Mustapha Lahyane and Edgar Martínez-Moro, Editors, Algebra for Secure and Reliable Communication Modeling, 2015

641 Maria Basterra, Kristine Bauer, Kathryn Hess, and Brenda Johnson, Editors, Women in Topology, 2015

640 Gregory Eskin, Leonid Friedlander, and John Garnett, Editors, Spectral Theory and Partial Differential Equations, 2015

639 C. S. Aravinda, William M. Goldman, Krishnendu Gongopadhyay, Alexander Lubotzky, Mahan Mj, and Anthony Weaver, Editors, Geometry, Groups and Dynamics, 2015

638 Javad Mashreghi, Emmanuel Fricain, and William Ross, Editors, Invariant Subspaces of the Shift Operator, 2015

637 Stéphane Ballet, Marc Perret, and Alexey Zaytsev, Editors, Algorithmic Arithmetic, Geometry, and Coding Theory, 2015

636 Simeon Reich and Alexander J. Zaslavski, Editors, Infinite Products of Operators and Their Applications, 2015

635 Christopher W. Curtis, Anton Dzhamay, Willy A. Hereman, and Barbara Prinari, Editors, Nonlinear Wave Equations, 2015

634 Steven Dougherty, Alberto Facchini, André Leroy, Edmund Puczyłowski, and Patrick Solé, Editors, Noncommutative Rings and Their Applications, 2015

For a complete list of titles in this series, visit the
AMS Bookstore at www.ams.org/bookstore/conmseries/.